Mathematical Engineering

Today, the development of high-tech systems is unthinkable without mathematical modeling and analysis of system behavior. As such, many fields in the modern engineering sciences (e.g. control engineering, communications engineering, mechanical engineering, and robotics) call for sophisticated mathematical methods in order to solve the tasks at hand.

The series Mathematical Engineering presents new or heretofore little-known methods to support engineers in finding suitable answers to their questions, presenting those methods in such manner as to make them ideally comprehensible and applicable in practice.

Therefore, the primary focus is—without neglecting mathematical accuracy—on comprehensibility and real-world applicability.

To submit a proposal or request further information, please use the PDF Proposal Form or contact directly: Dr. Thomas Ditzinger (thomas.ditzinger@springer.com)

Indexed by SCOPUS, zbMATH, SCImago.

Mangey Ram · Shristi Kharola · Akshay Kumar
Editors

Advances in Mathematics for Engineering Sciences

Springer

Editors
Mangey Ram
Department of Mathematics
Graphic Era Deemed to be University
Dehradun, Uttarakhand, India

Shristi Kharola
Department of Mathematics
Graphic Era Deemed to be University
Dehradun, Uttarakhand, India

Akshay Kumar
Department of Mathematics
Graphic Era Hill University
Dehradun, Uttarakhand, India

ISSN 2192-4732 ISSN 2192-4740 (electronic)
Mathematical Engineering
ISBN 978-3-031-95692-8 ISBN 978-3-031-95693-5 (eBook)
https://doi.org/10.1007/978-3-031-95693-5

This Springer imprint is published by the registered company Springer Nature Switzerland AG
The registered company address is: Gewerbestrasse 11, 6330 Cham, Switzerland

Preface

This book is an inspiration to future mathematicians and researchers who are looking for a blend of mathematical solutions to the problems that are currently viewed as one of the most challenging in the world. From history to health, demand to digital, transportation to traveling salesman problem, this book covers a dynamic range of mathematical models. The authors have contributed their original work from different parts of world, and their efforts have shaped this book to be a success. To briefly explain the context, this book explains some new mathematical assignments that talk about powerful tools that help in navigating through mathematical problems. All the chapters are well structured and have been compiled using simple English language for a better understanding of the concepts. The proposed book is edited in good faith and for the betterment of science and mankind.

Chapter "Advanced Integration of Artificial Intelligence and IoT for Enhancing the Reliability in Renewable Energy Systems" aims to enhance energy demand by studying the role of integration Artificial Intelligence (AI) and the Internet of Things (IoT) in enhancing the performance and the reliability of renewable energy systems. Chapter "Three New Considerations Applied to a Transportation Model: Accelerated Convergence, Ordered Optimal Solutions, and Analysis of a Protean Transportation System" recognizes the transportation model as an assignment model and is solved by the Hungarian method of assignment. Chapter "Specified Number of Nodes Travelling Salesman Problem" introduces the specified number of nodes traveling salesman problem (SNN-TSP), a variant of the traveling salesman problem to determine a minimum cost tour that visits specified number of nodes exactly once. Chapter "Mathematical Analysis of Real Gas Effects on the Accuracy of Microfluidic Gas Pressure Profiles" models' microfluidic behavior with a new mathematical nonlinear model that incorporates a coupled pressure-density-viscosity equation of state. A numerical method is developed for solving a nonlinear Reynolds integral equation that incorporates real gas effects. Chapter "A Novel Multi-objective Programming Model for Sustainable Agriculture Planning Based on the Water-Energy-Food Nexus Framework" aims to present a multi-objective mathematical programming model using the WEF nexus approach to achieve sustainable agricultural development. Chapter "Proposed a Food Insecurity Index: An Economic and Mathematical Modeling Approach" studies food insecurity and presents a mathematical programming model to optimize the FGT index by increasing income through the implementation of an optimal cropping pattern. Chapter "Development of Quantum Information Using Matrix Representation of Single and Multi-Qubit Quantum Gates" studies Quantum computation to explore individual quantum gates (single and multi-qubit) and their series-parallel combinations through their matrix representation, to analyze their role in computing state evolution, commutation relation, probability measurement, and expectation values. Chapter "Computational Approach to Solve Ricci Solitons on Kenmotsu Manifolds" aims to study the Ricci-soliton on the Kenmotsu manifold post studying the Ricci Flow, Kenmotsu manifolds, M-projective curvature tensor, pseudo-projective curvature tensor, and semi-symmetric properties of Ricci tensor. Chapter "A Study on the Performance Analysis of Handoff Requirement in a Wireless Vehicular Communication Network for Urban Areas: Case of Artificial Neural Network (ANN)" proposes an innovative service handoff scheme for efficient wireless communication systems that ensures seamless transitions between different service providers for Safe-as-a-Service provisioning in road transportation. The study is based on a real-life scenario in the city of Dehradun, Uttarakhand. Chapters "Mathematical Insights on Influenza Virus: Unveiling Dynamics and Informing Control Strategies" provides an overview of mathematical insights into influenza virus dynamics, including compartmental models, evolutionary models, and agent-based models. Chapter "Using Mathematical Approaches for Predictive Modelling of Herbal-Drug Interactions" identifies the potential use of mathematical models to assess the neuroprotective and metabolic effects of herbal treatments in Diabetes Mellitus (DM) and Alzheimer's Disease (AD) which are two main burden chronic diseases with a remarkably high prevalence rate worldwide. Chapter "Enhanced Nodal Variation Approach for Fracture Analysis Using Meshless Methods" discusses modeling and simulation of crack problems under mechanical loads with enhanced element-free Galerkin method. Chapter "Securing and Authenticating Vedic Documents: A Blockchain-Based Approach" proposes a blockchain technology-based framework to verify and validate Vedic texts, which can enhance the digitized Vedic manuscripts in further deformation from originality. Chapter "DeepFM-Driven Personalized Recommendations for Children: Tackling Data Sparsity with Negative Sampling" develops an anime recommendation platform for children featuring a hybrid Deep Factorization Machines (DeepFM) and negative sampling structure which resolves data sparsity problems. Chapter "Behavioural Analytics with Digitized Sustainability" discusses the importance of digital platforms as powerful tools for promoting behavior change towards sustainability and a collaborative space for knowledge exchange, where users can learn from experts, participate in discussions, and contribute to collective problem-solving.

Dehradun, India

Mangey Ram
Shristi Kharola
Akshay Kumar

Contents

About the Editors

Prof. Dr. Mangey Ram received the Ph.D. degree major in Mathematics and minor in Computer Science from G. B. Pant University of Agriculture and Technology, Pantnagar, India, in 2008. He has been a Faculty Member for around 17 years and has taught several core courses in pure and applied mathematics at undergraduate, postgraduate, and doctorate levels. He is currently the *Research Professor* at Graphic Era (Deemed to be University), Dehradun, India. Before joining the Graphic Era, he was a Deputy Manager (Probationary Officer) with Syndicate Bank for a short period. He is Editor-in-Chief of the *International Journal of Mathematical, Engineering and Management Sciences*; *Journal of Reliability and Statistical Studies*; *Journal of Graphic Era University*; Series Editor of six Book Series with *Elsevier, CRC Press-A Taylor & Frances Group, Walter De Gruyter Publisher Germany, River Publisher* and the Guest Editor and Associate Editor with various journals. He has published 400 plus publications (journal articles/books/book chapters/conference articles) in *IEEE, Taylor & Francis, Springer Nature, Elsevier, Emerald, World Scientific* and many other national and international journals and conferences. Also, he has published more than 60 books (authored/edited) with international publishers like *Elsevier, Springer Nature, CRC Press-A Taylor and Frances Group, Walter De Gruyter Publisher Germany, River Publisher*. His fields of research are reliability theory and applied mathematics. Dr. Ram is a Senior Member of the IEEE, the Senior Life Member of the Operational Research Society of India, the Society for Reliability Engineering, Quality and Operations Management in India, Indian Society of Industrial and Applied Mathematics, He has been a member of the organizing committee of a number of international and national conferences, seminars, and workshops. He has been conferred with the "*Young Scientist Award*" by the Uttarakhand State Council for Science and Technology, Dehradun, in 2009. He has been awarded the "*Best Faculty Award*" in 2011; the "Research Excellence Award" in 2015; "*Outstanding Researcher Award*" in 2018 for his significant contribution to academics and research at Graphic Era Deemed to be University, Dehradun, India.

Also, he has received the "*Excellence in Research of the Year-2021 Award*" by the Honorable Chief Minister of Uttarakhand State, India, and the "Emerging Mathematician of Uttarakhand" state award by the Director, Uttarakhand Higher Education. Recently, he received the "*Distinguished Service Award-2023*" for the subject and nation development by Vijñāna Parishad of India.

Dr. Shristi Kharola is a committed and industrious researcher and educator in the field of Applied Mathematics, with specialized expertise in Reliability Engineering, Operations Research, and Decision-Making. Currently serving as an Assistant Professor (Research) at Graphic Era (Deemed to be University), Dehradun, she has consistently demonstrated academic excellence and research rigor in her early career. Dr. Kharola earned her Ph.D. in Mathematics with a focus on sustainable systems and reliability modeling and has contributed to high-impact publications and international collaborations, including a project funded by USERC involving partners from India, Turkey, and the UK on reducing food waste through Industry 4.0 and circular economy approaches. Her research portfolio includes multiple publications in SCIE- and Scopus-indexed journals, authored and edited book chapters with international publishers like Elsevier, Springer, IGI Global, and River publishers, and active participation in prestigious conferences. She has also played key organizational roles in various academic events and serves as an early-career editor for a reputed international journal. She also plays an active role in the academic community as a regular reviewer for reputed journals and IEEE conferences. Notably, she served as one of the editors for *Advances in Soft Computing Applications*, an internationally indexed book contributing to the field of intelligent systems. In addition to her research, Dr. Kharola is deeply committed to teaching and mentoring. She is currently guiding students from undergraduate, postgraduate, and two Ph.D. students, and is known for her systematic and student-centered approach in the classroom.

Dr. Akshay Kumar received the B.Sc. and M.Sc. degrees in science from Chaudhary Charan Singh University, Meerut, India, in 2010 and 2012, respectively, and the Ph.D. degree major in Mathematics and minor in Computer science from G. B. Pant University of Agriculture and Technology, Pantnagar, India, in 2017. He is currently working as an Assistant Professor in the Department of Mathematics, Graphic Era Hill University, Dehradun in India. He is currently teaching LLP and Probability and Statistics in B.Sc. (H) and have taught Group Theory, Numerical Methods, Business Statistics, Discrete Mathematics, and Engineering Mathematics-III. In P.G. level he has taught OR, Linear Algebra and Discrete Structure. He is a regular Reviewer for more than 50 international journals including Elsevier, Springer, Emerald, MDPI, Cambridge University Press, World Scientific, River Publisher, IGI Global Publisher and many other publishers. He has published 100 plus research papers in Communications in Statistics—Simulation and Computation, Emerald, Inderscience, International Journal of Mathematical, Engineering and Management Sciences, IGI Global Publisher, MDPI, Cambridge Scientific Publishers, Springer nature, Elsevier Multimedia Tools and Applications and many other national and international journals of repute and presented his works at national and international conferences. Also, he has published six books (edited) with international publishers like River Publisher, De Gruyter publisher, Elsevier, CRC Press Taylor & Francis and IGI Global Publisher. His fields of research are reliability theory, fuzzy reliability, signature reliability, and applied Mathematics. He has been awarded recently "Outstanding Researcher Award" in 2019–2020 at Graphic Era Hill University Dehradun, India, and in 15th and 16th Uttarakhand State Science and Technology Congress 2020–2022 for his significant contribution in academics and research at Graphic Era Deemed to be University, Dehradun, India.

Advanced Integration of Artificial Intelligence and IoT for Enhancing the Reliability in Renewable Energy Systems

Ghalia Nassreddine, Mahmoud El Samad, Mohamad Nassereddine, and Obada Al-Khatib

Abstract The energy demand has recently increased due to technological advancement and the expanded number of smart cities. This growth may create new challenges for environmental and infrastructure sustainability. Thus, the demand for green and reliable energy sources increases. However, these sources, such as solar and wind, are unpredictable. This chapter examines the role of integration Artificial Intelligence (AI) and the Internet of Things (IoT) in enhancing the performance and the reliability of renewable energy systems. IoT and AI play an important role in improving grid stability by detecting faults, predicting demand, and optimizing energy generation. Indeed, AI-based systems can provide an accurate and robust prediction of energy demand and power produced. Thus, they can assist in holding the balance between supply and demand. However, IoT devices have the capability of real-time data gathering and monitoring. Nevertheless, the integrating of AI- and IoT-based systems produces new challenges like privacy considerations, data quality, high costs, scalability, power consumption, and regulatory compliance. This renders the process of integrating IoT and AI into renewable energy systems complex. Despite these challenges, the incorporation between AI and IoT can develop the role of renewable energy systems and produce a new path for a more sustainable and strong energy future.

G. Nassreddine (✉)
Rafik Hariri University, Meshref, Lebanon
e-mail: nassreddinega@rhu.edu.lb

M. El Samad
Lebanese International University, Beirut, Lebanon
e-mail: Mahmoud.samad@liu.edu.lb

M. Nassereddine · O. Al-Khatib
University of Wollongong in Dubai, Dubai, UAE
e-mail: mohamadnassereddine@uowdubai.ac.ae

O. Al-Khatib
e-mail: ObadaAlKhatib@uowdubai.ac.ae

M. Ram et al. (eds.), *Advances in Mathematics for Engineering Sciences*, Mathematical Engineering, https://doi.org/10.1007/978-3-031-95693-5_1

Keywords Artificial intelligence · Internet of Things · Data collection · Predictive analysis · Renewable energy system · Power demand prediction · Weather parameters · Sensors

1 Introduction

In the last decades, energy demand has witnessed a huge increase due to the expansion of smart cities, technological advancement, and industrialization. This increase in electricity demand produces many challenges, especially for the electricity grid, existing infrastructure, and environmental sustainability. Therefore, finding a reliable, efficient, sustainable, and scalable energy solution that covers all the demands without any negative impact on the environment becomes an essential need for smart cities and modern societies. Renewable energy systems (RESs) can handle these challenges by producing clean and reliable power (Nassereddine and Khang 2024) that is generated from natural and non-depletable sources, such as solar and wind plants. Hence, the use of such energy sources may lead to minimal or zero carbon emissions. Recently, the use of renewable energy sources has become critical for resolving the energy crisis and facing the climate change challenge (Nasserddine et al. 2023).

However, the adoption of renewable sources faces many challenges due to their unpredictability. Indeed, green sources such as wind and solar depend on weather conditions and geographical location. Thus, accurately predicting the amount of energy generated by RES could be a complex task as it depends on many meteorological and chronological parameters such as humidity, sunlight, and wind speed (Nassreddine et al. 2024). In addition, the amount of electricity generated from alternative energy is unstable and may have huge fluctuations. This necessitates further plans, such as using storage devices or backup systems to ensure a consistent supply. Therefore, controlling and integrating these RESs into existing grids can be complicated and challenging (Mazzeo et al. 2021).

Recently, the vast advancement in technology led to new science and tools, such as artificial intelligence (AI) and the Internet of Things (IoT). AI is a branch of computer science that develops systems able to mimic human intelligence. AI-based systems can perform advanced tasks such as predicting values, comprehending text, and decision-making. Many sectors, including healthcare and power fields, widely utilize AI systems (Zhang and Lu 2021). In RESs, AI-based systems can analyze the data collected from renewable sources, such as weather parameters, to predict the accuracy of the energy produced. This prediction can play an important role in optimizing energy production and enhancing grid stability.

Furthermore, thanks to machine learning techniques and the advanced analytics capability of AI systems, fault detection in RES becomes easier. AI systems analyze the data collected from the different parts of the RES to monitor its performance and detect any need for maintenance before failure. This real-time monitoring allows an easy integration of renewable energy into existing grids and produces more reliable and cost-effective energy solutions. Serban et al. (2020) address the integration of AI-based systems in the European Union's renewable energy sector. They examine the efficacy of converting renewable energy into the energy chain from Gross Inland Consumption to Final Energy Usage. They also examine the consequences of the structure of renewable energy sources, including solar, wind, and biomass, as well as labor productivity in the renewable energy industry relative to the economy and its relationship with investment levels. Furthermore, they study the consequences of implementing AI for renewable energy in the building of smart future cities. Many researchers have developed new AI systems to forecast accurately the energy produced by RES (Elsaraiti and Merabet 2022; Zazoum 2022). Other studies developed an AI-based system for detecting faults in these systems (Amiri et al. 2024; Sohail et al. 2023).

Due to the size and potential noise generated by renewable energy plants, most of them are in rural areas. Thus, the monitoring and control of these plants become challenging. Therefore, the integration of IoT platforms becomes essential for controlling and maintaining the performance of RESs in remote locations. IoT sensors and devices collect real-time data related to performance, weather conditions, and equipment status. These data can be examined by an AI system to enable efficient supervision and optimization. Furthermore, IoT technology facilitates effective energy management and grid integration, ensuring that energy from these remote plants is utilized and distributed efficiently (Ullah et al. 2023).

The chapter is divided into six main sections. First, we will describe RES in Sect. 2. Next, we will present IoT systems in Sect. 3, followed by an illustration of AI technology in Sect. 4. Section 5 will explore the role of the integration of AI and IoT in enhancing the reliability of RES. Finally, the chapter will conclude in Sect. 6.

2 Renewable Energy Systems

Energy is commonly defined as a system's ability to exert external forces, such as applying a force across a distance. The energy available to a body is limited by the input or production of work. Different types of energy include mechanical, potential, kinetic, thermal, magnetic, electrical, radiation, nuclear, and chemical energy. An energy system is the process chain (or subset of it) that connects the extraction of primary energy to the use of final energy to deliver goods and services. Renewable power is defined as energy derived from raw materials that replenish faster than they are depleted. Examples of such sources include sunlight and wind. RESs are comprehensive energy supply and demand systems that rely on renewable energy sources rather than fossil fuels and nuclear power (Twidell 2021).

The use of RES in daily activities such as transportation, heating, and electricity generation will help prevent the average global temperature from rising by more than 1.5 °C. Renewable energy sources enable nearly complete decarburization of power generation to achieve net-zero emissions by 2050 (Anika et al. 2022). Meanwhile, substantial emissions reductions in industry, buildings, and transportation are facilitated by renewable heat and transportation fuels. Over the next few years, the globe is expected to add more renewable capacity than has been installed since the construction of the first commercial renewable energy power plant. Based on legislation that promotes this in over 130 countries, nearly 3,700 GW of new renewable power is projected to be installed and operational between 2023 and 2028. Several significant milestones in renewable energy are anticipated within the next five years (Bojek 2023):

- Wind and solar PV will generate more electricity in 2024 than hydropower.
- In 2025, renewable energy sources will surpass coal as the primary source of electricity production.
- By 2026, wind and solar photovoltaics will exceed nuclear power generation.
- By 2028, renewable energy sources will contribute more than 42% of the world's electricity, with solar and wind PV accounting for a 25% share.

In this section, the four main types of RES will be presented.

2.1 *Photovoltaic (PV) System*

A photovoltaic (PV) cell is a non-mechanical device that transforms sunlight into electricity. Certain PV cells may also generate power using artificial light. PV systems use silicon solar panels to turn solar radiation into electrical energy. These panels are made up of arrays of PV cells combined in parallel and serial. A photovoltaic cell is an energy-gathering device that uses the photovoltaic effect to convert solar energy into electricity (Fig. 1). The PV effect involves the process of converting sunlight into electric current in a photovoltaic cell. The p-type and n-type semiconductors combined into these cells are merged to make a p–n junction. At this junction, electrons transfer from the p-side to the n-side, generating an electric field. This field causes positively and negatively charged particles to be pushed in opposite directions. Sunlight is composed of small energy bundles known as photons. When these cells are exposed to light of a specific wavelength, the energy from the photons is transmitted to electrons in the semiconducting material, elevating them to a higher energy state known as the conduction band. As a result, these electrons can move through the material, creating an electric current within the cell (Nassreddine et al. 2023).

There are two main types of installation of PV systems: Off-Grid and On-grid. On-grid systems add to the electricity you get from your utility company by connecting to the electric grid. Off-grid solutions, on the other hand, depend on battery storage and are completely autonomous.

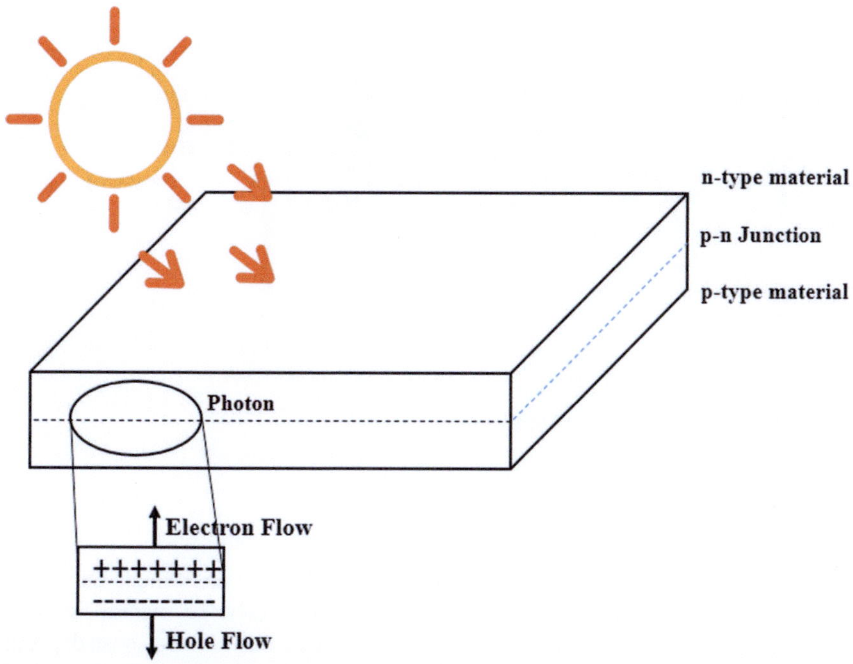

Fig. 1 Photovoltaic effect in PV panel

2.2 *Wind Power System*

Wind is additional green energy source. A wind power system includes a wind turbine mounted on a tower to improve access to wind currents. Figure 2 shows the key components of the wind electric system. The wind turbine turns wind energy into mechanical energy, which is ultimately converted into electricity by a generator. A transformer and transmission wires are then utilized to either store the electrical energy in a battery or transmit it to the grid. Wind turbines are typically divided into two types: horizontal and vertical (He et al. 2021). The horizontal variant is recommended because it is more adaptable for aerodynamic efficiency and provides efficient mechanical balance support. Wind turbines can be divided into two main groups (Liu et al. 2022):

- Smaller models employ variable speed generators to convert mechanical energy into electrical power with adjustable voltage and frequency.
- Large variants use a gearbox to regulate the varying speed of the blades to match the generator's synchronous speed, ensuring a consistent output voltage and frequency. This type requires a controller to convert the variable frequency and voltage into direct current (DC) power for later usage. In multi-pole generating systems, a gearbox may be unnecessary.

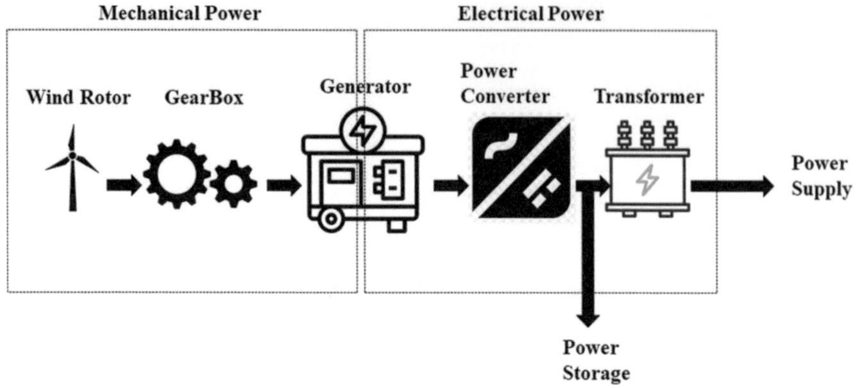

Fig. 2 Wind power system components

2.3 Biogas

Biogas systems produce clean, energy-efficient burnable gas through the natural interaction of microorganisms with organic wastes such as sewage, manure, and agricultural byproducts. Unlike natural gas, which is derived from subterranean deposits, biogas is a renewable source that can power the system as long as organic waste is available. Additionally, biogas generates methane-rich fertilizer that can be sold or used on farms to enhance crop yields. Biogas is produced through the anaerobic digestion of organic materials in oxygen-deficient environments, resulting in a mixture of carbon dioxide and methane. The main fuel source for biogas generation is biomass materials, which include municipal food waste, wastewater, agricultural waste, livestock waste, and urban wood. Worldwide, biogas is a popular energy source utilized for fuel consumption, energy production, and heating. Anaerobic digestion serves as an effective method for producing biogas and converting some waste into fertilizer. The effectiveness of the biogas system relies on careful planning and flawless execution (Farghali et al. 2022).

A typical setup of a biogas system is depicted at a high level in Fig. 3 (Nassereddine et al. 2024). Once received, the conveyor moves the organic waste to the input mixer to eliminate any unwanted elements. The waste is then directed into an anaerobic digestion system and a feedstock mixer/grinder, creating the final waste format that is prepared for use in the biogas system. Additionally, the resulting waste can be utilized in the composting process to produce fertilizer. Careful design and flawless implementation of essential components are necessary for the system's adequacy and high efficiency. The digestion system handles a significant volume of input and output feedstock every day. Therefore, it is crucial to ensure that the inlet and outlet designs meet the specific requirements of the facility. Various factors, including pressure feed and gravity feed, influence the design. The input must be capable of satisfying the hourly feed requirement for both feed types. This information is vital to ensuring

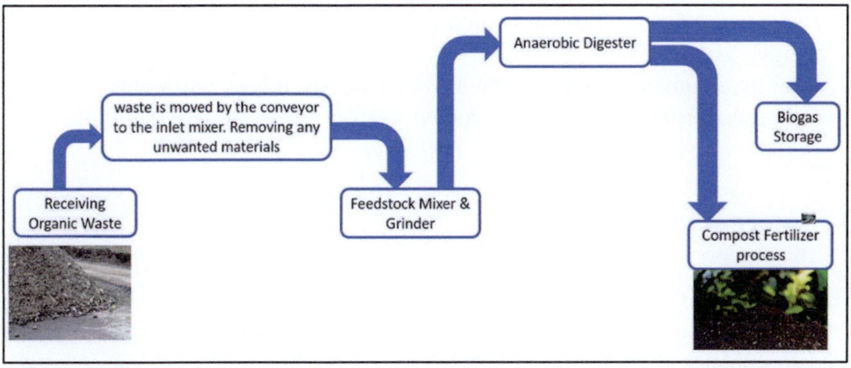

Fig. 3 Standard layout for biogas system setup; the process of transforming organic waste to composite fertilizer

the biogas system operates smoothly. Noncompliance with the digester requirements will result in system failure.

2.4 *Geothermal Power System*

The first nation in the world to produce energy using a geothermal power system is the United States of America. The country employs geothermal facilities to generate over 3.5 gigawatts annually, which can supply electricity to more than three and a half million homes (El Haj Assad et al. 2021). Figure 4 illustrates the major components of a geothermal power plant. The operation of this plant can be summarized in five steps (Chen et al. 2022):

- High pressure is used to force hot water from deep underground.
- As the pressure releases, the water becomes steam when it reaches the surface.

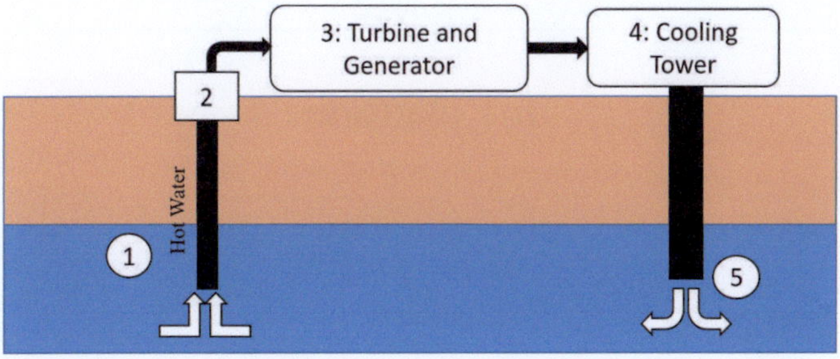

Fig. 4 Main components of geothermal power system

- The steam rotates a turbine, which is connected to a generator that produces energy.
- After the steam cools, the cooling tower condenses it back into water.
- The process is restarted after the cooled water is reinjected into the Earth.

3 Internet of Things

The Internet of Things (IoT) was first introduced by Kevin Ashton in 1999, describing a network of connected objects linked to the internet via sensors. Today, IoT is widely used to enhance the computing capabilities of machines, sensors, and everyday objects. It is a significant emerging technology that influences businesses and expands the world's boundaries by integrating physical and virtual aspects (Nassereddine et al. 2024). IoT technology has revolutionized the world by enabling devices to communicate without human interaction, creating a landscape where small items can connect and share information, ultimately enhancing human life. IoT devices detect human needs and preferences, providing innovative solutions to businesses, governments, and organizations worldwide (Sfar et al. 2017). This section explains the main components of IoT and its primary role in the renewable energy sector.

3.1 IoT Main Components

The main components of an IoT system are illustrated in Fig. 5. The first step in developing an IoT system is choosing the application and selecting the best components, such as sensors, storage devices, and communication tools (Ramamurthy and Jain 2017). The IoT device constitutes the second part of the IoT platform. These devices may include sensors, actuators, communication tools, data collection devices, processing devices, or any other gadgets that gather data. They can also function as IoT gateways, enabling two-way data exchange between the gateways and devices.

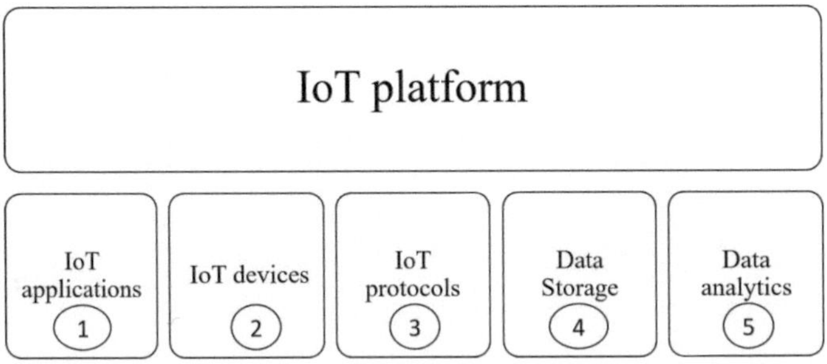

Fig. 5 Main components of IoT platform

The third part of the IoT platform consists of communication protocols. These protocols allow data and information to be shared with the controller and among the various components of the IoT platform, facilitating decision-making. IoT systems provide a vast array of communication techniques and technologies, each with unique characteristics. The choice of communication technology should be guided by the specific IoT application. These technologies include cellular options like Bluetooth, Wi-Fi, and LTE, as well as 4G and 5G networks (Karunarathne et al. 2018).

The fourth element of the IoT platform is data storage, which enables data collection from IoT devices. A significant amount of data is generated, necessitating effective storage solutions such as cloud servers. Data analytics can be performed either offline or in real-time once the data is stored, assisting in decision-making and helping to select the appropriate course of action. This constitutes the sixth aspect of the IoT platform (Nassreddine et al. 2024a).

3.2 *IoT in Renewable Energy Sector*

The emerging IoT technology has played a significant role in advancing the renewable energy sector for a more sustainable future. The IoT network deploys millions of sensors to collect various data to monitor the performance and reliability of RESs. By analysing this data, the energy suppliers can detect early any RESs issues or failures in real time, offering better infrastructure resiliency and potentially reducing maintenance costs (Mishra et al. 2022). Table 1 summarizes the main challenges of IoT integration in RES (Nižetić et al. 2020).

Due to the significant technological advancement, multiple solutions can be used to addressed the challenges illustrated in Fig. 6 (Mahale et al. 2023):

- Standardization: Compatibility and interoperability can be significantly improved by creating universal standards and protocols for solar panel systems and IoT

Table 1 The technical challenges of integration IoT in RES

Challenges	Description
Compatibility and interoperability	It may be difficult for devices made by various manufacturers to cooperate, which could result in inefficiencies and even system breakdowns. Moreover, the existence of many protocols and standards for IoT communications might raise interoperability issues and security vulnerabilities when integrating IoT devices with solar panel systems
Complexity of integration	It takes certain knowledge and experience to set up a smooth system that combines solar panels with IoT devices. Those homeowners who are not tech-savvy may find this intricacy to be a hurdle
Data management	It can be quite difficult to handle the enormous volumes of data produced by solar energy systems and IoT devices. For best results, data collection, storage, and analysis must be done effectively

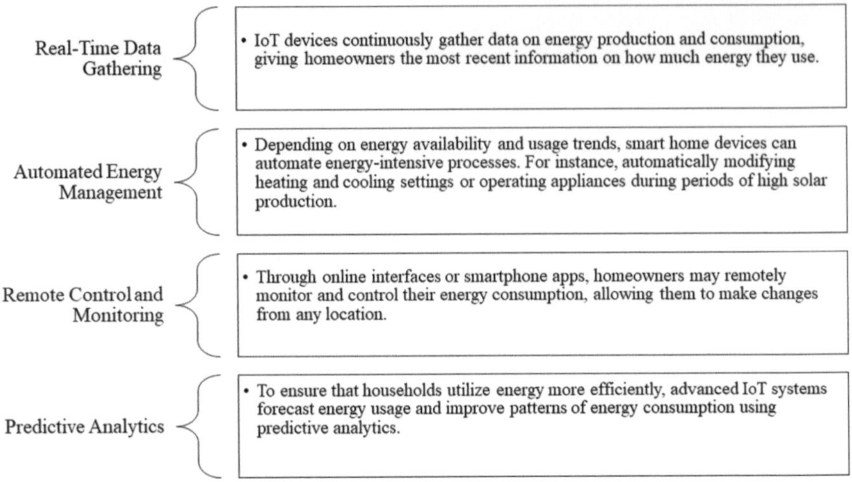

Fig. 6 Role of IoT in RES

devices. These standards are being established by organizations and leaders in the industry.

- Integrated Platforms: Businesses are creating integrated platforms that make it easier to link and control different RES and IoT devices. These solutions offer centralized control and an intuitive user interface.
- Edge Computing: By processing data closer to its source, at the network's edge, edge computing can improve data management efficiency and speed while decreasing the load on central data centers.
- AI and Machine Learning (ML): By anticipating patterns of energy use, controlling energy distribution, and spotting possible problems before they become serious ones, AI and ML technologies can help integrated systems operate more efficiently.
- AI and Machine Learning (ML): By utilizing AI and ML algorithms, the energy providers can analyze energy usage patterns for better demand forecasting, monitoring unusual energy consumption patterns due to failures or energy fraud, integrating effectively renewable energy resources into the grid, and price optimization.

Therefore, many advantages can be achieved via the integrations of IoT technology and renewable energy resources, such as real-time monitoring and control, increased reliability, environmental sustainability, and data-driven insights.

4 Artificial Intelligence

Today, Artificial Intelligence (AI) is considered as the key of success of any business or organization. It is acting as a prominent topic across different industries. AI based solution consists of creating intelligent systems able to present and simulate the human behavior. It is considered a major part for human everyday life application such as virtual assistants like Alexa, Chatbots, and many Gaming systems. AI systems is based on advanced statistical techniques with leveraging machine learning (ML). AI is a very interesting and hot topic that can identify patterns and generate smart for specific tasks (El Samad et al. 2023). Table 2 summarizes the usage of AI in some fields. As presented in Fig. 7, AI can be composed into seven main categories.

Table 2 Some of the AI application fields

Domain	Description
Healthcare (Johnson and Allen 2023)	AI can detect diseases, check and verify treatments, and perform better manage of health data
Transportation Systems (Carter and Lee 2022)	AI system optimizes road traffic and allow the usage of autonomous vehicles
Education (Smith and Kumar 2024)	AI can help in creating personalized learning system and enhancing teaching methodologies
Finance (Patel and Singh 2023)	AI is employed in the financial sector to automate customer interactions, optimize investment strategies, detect fraud, and analyse market trends
Marketing (Brown and White 2023)	Can be used for consumer behaviour analysis, personalized advertising, and optimizing customer engagement through data analysis
Business Operations (Walker et al. 2023)	AI streamlines business operations by automating workflows, improving decision-making through data analysis, and enhancing supply chain management
Agriculture (Greenfield and Jones 2022)	AI is used in agriculture for crop monitoring, yield prediction, pest detection, and optimizing irrigation systems to improve productivity
Retail (Harper and Singh 2023)	Retailers utilize AI for inventory management, dynamic pricing, customer behaviour analysis, and personalized shopping experiences
Energy (Torres and Lin 2023)	AI can apply algorithms to optimize power grids, to predict any failure, and finally to enhance the resource management

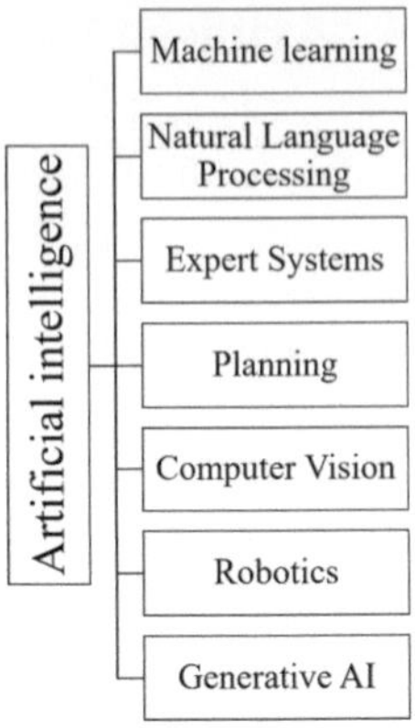

Fig. 7 AI main types

4.1 Machine Learning

Machine Learning (ML) is considered as a branch of AI. It is a very attractive research area (Choi et al. 2020; Sodhi et al. 2019). The power of ML comes from the ability to learn from data. More precisely, ML can analyse data, detect similar patterns and finally take decisions or prediction. For instance, in the healthcare field, given a set of features (gender, age, lifestyle, stress level) the ML algorithms can efficiently detect if a patient is more likely to have blood pressure or not. In the context of power systems, ML algorithms can predict the consumption of electricity for next winter and so on. In brief, the ML can learn, adapt and evolve thanks for data.

ML system can be classified in five main types:

- Supervised Learning: Algorithms are trained on what we call "labelled data" which means that each input has a specific output (Nasserddine et al. 2023).
- Unsupervised Learning: This subtype is based on "unlabelled data". Thus, ML model group these data according to some patterns. For example, cluster the energy consumption behaviour as normal or fraud consumption (Jaiswal et al. 2021).
- Semi-supervised Learning: This subtype combines the two previous subtypes, "labelled data" and "unlabelled data". This combination is efficient in certain domain like the prediction of fault in energy infrastructure (Erhueh et al. 2024).
- Reinforcement Learning: This subtypes is based on the usage of agents able to take cumulative decisions. This can be applied for the optimization of grid operations where balance between supply and demand is required (Sarker et al. 2021).
- Deep Learning (DL): It is a branch of ML that relies on artificial neural networks. It is based on a network of neural just like human brain. This kind of ML prove it efficiency in many applications especially in image processing and pattern recognition.

4.2 Natural Language Processing (NLP)

The Natural Language Processing (NLP) is an advanced application for ML since it is mainly used in chatbots, translation, and voice assistants (e.g., Siri, Alexa). The main goal of NLP is to understand and interpret human language. Moreover, it can also generate human language (speech, text). Many recent NLP systems can detect human sentiment. The NLP relies on advanced algorithms that can ensure the communication between humans and machines (Cai 2021).

4.3 Expert Systems

Expert Systems (Huda et al. 2024) are mainly employed for the decision-making abilities that are normally executed by human resource experts. This why, it is named Expert Systems. These systems rely on database that includes some rules. In a sample way, this database match for each problem a set of rules to be applied. Moreover, the expert systems include in addition to the database, an engine named "inference engine" that can efficiently apply the correct rule for a specific problem.

4.4 Planning

AI planning is mainly employed in many fields such as autonomous systems. The idea is that AI planning can handle complex decision systems. It is widely applied in areas like smart grid management and energy storage. The AI planning uses algorithms like heuristic search, Markov Decision Processes (MDPs) and finally reinforcement learning. Thanks to the efficient AI algorithms, AI planning can adjust to inconsistent energy demands by breaking down a complex process to small tasks (Omitaomu and Niu 2021). Another example of the usage of AI planning is to optimize energy routing, balancing supply and demand efficiently. AI planning systems are in progress and hopefully they allow real-time decision-making and enhanced resource management in renewable energy sectors.

4.5 Computer Vision

The Computer Vision subtype is an interesting subtype of AI since it enables the users, such as grid monitors, to visualize data. The monitoring of data can help to analyze data from the real world applications.

This subtype can be used in RES for solar panel monitoring and wind turbine inspection. The analysis of images can help for fault detections, to track the equipment performance. In addition, the Computer Vision can identify environmental conditions that affect energy generation (Manno et al. 2021). In wind farms, Computer Vision can help to detect turbine blade wear or malfunctions using real-time video footage (Zhang et al. 2021).

4.6 Generative AI

Generative AI is a new and trendy topic of AI. It is considered as a subtype and mainly dedicated for the creation of new contents. It can generate text, images and even coding. The Generative AI uses Large Language Models (LLMs), Generative Adversarial Networks (GANs), and Variational Autoencoders (VAEs).

The AI main focus is to do classification and analysis of data, in addition to prediction while Generative AI target is to generate new content (e.g., data, text, audio). In the context of power systems, Generative AI can be applied in smart grid optimization, urban planning, efficient management for the infrastructures particularly, in smart cities (Rane 2023).

4.7 Robotics

Robotics is a very interesting subtype of AI where intelligent algorithms are employed to enable a robot to be independent with minimum human intervention. Nowadays, robots can handle very delicate and specific task. They can achieve high accuracy where, sometimes human being can do errors. In smart cities, robots can help in the waste collection and sorting the wastes. As result (Moses et al. 2021), it is clear that robotics will play a major role for smart cities, urban management and energy production.

5 Collaboration of IoT and AI in Enhancing RES Reliability

Recently, the RES has become more efficient due to the integration of AI systems combined with IoT technology. IoT allows real-time data collection and systems monitoring using interconnected devices. Thus, energy management becomes possible with minimal human intervention. The data collected is used by the AI system to enhance decision-making and predict the amount of energy produced by the RES. This prediction allows consumption optimization and the balance between supply and demand, thus enhancing the grid's stability and reliability.

In addition, the combination of AI and IoT devices allows real-time monitoring and controlling of RES systems. This leads to smarter grid management, accurate maintenance prediction, and dynamic energy distribution. Thus, AI and IoT devices may improve the reliability and performance of RES by eventually contributing to more resilient and decentralized energy solutions.

In this section, the combination of IoT with AI will be explored according to the main parts: the role of IoT in monitoring and data collection, the role of AI in data analysis and optimization, and IoT and AI integration for ensuring system reliability. This section will end with a resume of the challenges and limitations of IoT and AI integration in RES.

5.1 Role of IoT in Monitoring and Data Collection

The use of RES makes the reduction of climate change easier, enhances the adaptation of sustainable energy sources, and improves energy efficiency. Recent technology such as IoT plays an essential function in the energy field, including supply, demand, transmission, and distribution parts. It can minimize the environmental influence of energy use, allow the exchange between RES, and boost energy productivity. Indeed, the growth demand of smart devices, fast information processing, and fast data communication has driven IoT to a higher stage. Recently, many organizations have embraced IoT-based solutions for information sharing, analyzing, and additionally enhanced yield.

The use of IoT devices makes the real-time data collection in RES feasible. Connected sensors and smart devices are used to monitor many principal factors in RES, such as energy production, weather conditions, equipment performance, and environmental features. They constantly send data to the central systems and offer the providers real-time information about the solar panels, wind turbines, and other renewable energy plants. This process allows the detection of any malfunction in RES. IT also follows energy generation, optimizes resource allocation, permits more robust energy management, and improves decision-making in RES.

In He et al. (2024), the authors suggests a new SCADA system for monitoring photovoltaic (PV) system based on dual IoT platforms. In this system, three sensors are used to collect data related to the PV module, battery, and energy load consumption. In addition, two other voltage sensors gathered the voltage of the PV and battery. Using ESP32-E, the data will be compiled and then transmitted to the Arduino cloud for real-time preview. After that, it will be sent using HTTO to the ESP32-S3 that manages the load and allows the relay to switch from on to off in response to the battery voltage. In addition, the ESP32-S3 employed HTTP to transmit the collected data to ThingSpeak for extreme examination, data storage, and real-time preview. Furthermore, a camera web server created with the ESP32-S3 displays load images. This system succeeds in controlling and monitoring the PV plant for a period of 20 days, proving its flexibility and adjustability even during the Arduino Cloud's outage. For one day, voltage evaluation ranged up to 13 V, and current value changed from zero amperes to 4.42 amperes. Visual load monitoring features that were not accessible in old systems were integrated into the suggested one.

IoT integration with RES can facilitate energy management systems. Voltage and current data can be gathered using sensors and IoT devices like the ESP32 controller. The gathered data can be transmitted to the cloud for further investigation, such as tracking load behavior, controlling energy consumption, and reducing greenhouse emissions. An IoT-based Smart Energy Management System (SEMS) approach was suggested, implemented and evaluated in Katuri et al. (2023). In this approach connected IoT devices were employed to real-time monitoring of household energy demand. The authors showed that the IoT-based SEMS succeeds in addressing issues related to household energy management.

Recently, hydrogen became an essential element in the current energy landscape. It assists in minimizing the reliance on fossil fuels and reducing carbon emissions for sustainable development. Polymer Electrolyte Membrane (PEM) technology using electricity and water can produce hydrogen and oxygen without generating any pollutants. In Folgado et al. (2023), the authors suggested a new process for collecting data and monitoring the PEM hydrogen producers. This mechanism produces an industrial controller that records key PEM electrolyzer (PEMEL) parameters such as current, voltage, pressure, and hydrogen flow. The gathered data are pushed into a communication network and stored in a database using a middleware layer. Full real-time online access is provided to users through a web-based interface. An industrial IoT (IIoT) architecture was used to manage and control the operation of all components, integrating sensor devices, IoT software, and industrial machinery. The validation of the proposed mechanism was performed by experimental data from a PEMEL under actual working conditions.

A novel SCADA system application for controlling and surveying a hybrid system that contains photovoltaic panels, wind turbine, and battery energy storage devices was suggested in Qays et al. (2022). In this system, the ThingSpeak platform was employed to real-time observe metrics like power, voltage, and current. The proposed SCADA system remotely manage power system components through network operators. To connect the SCADA system to the MATLAB/Simulink, the KEPServerEX client was employed. Affordable electronic components and an Arduino Integrated Development Environment (IDE) with the ATMega2560 remote terminal unit were used to create a hardware prototype for testing purposes. Experiments and simulations proved the proposed system's feasibility.

5.2 *Role of AI in Data Analysis and Optimization*

AI-based systems play a critical role in improving the efficiency and performance of RESs. Indeed, ML and DL techniques offer better prediction, resource management, and decision-making to promote energy production and distribution. Due to their higher capability in analyzing and handling huge volumes of data, ML and DL systems can examine data gathered by IoT devices to predict system failures, performance degradation, or maintenance needs. To enhance the sustainability and capability of PV solar plants, the authors in Duranay (2023) studied the possibility of using infrared solar module photos integrated with DL techniques to detect the faults in the panels. Efficientb0 model was used to track and classify the faults from twelve different classes: cell, cell-multi, cracking, diode, diode-multi, hot spot, hot spot-multi, no-anomaly, offline-module, shadowing, soiling, and vegetation. A dataset of 20,000 images delivered by infrared solar modules was used. The feature selection technique based on Neighborhood component analysis was employed to choose 17,000 best features. The classification process was performed using SVM. The suggested approach offers good performance in terms of accuracy, F1-score, precision, and sensitivity measures.

The role of DL models in detecting faults in solar PV panels was also examined in Rudro et al. (2024). The authors suggest a DL model that combines the InceptionV3-Net and U-Net techniques. Due to additional convolutional layers, squeeze-and-excitation (SE) blocks, residual connections, and global average pooling, the proposed method outperforms the InceptionV3 base with ImageNet weights. The performance metrics shows the effectiveness of the proposed approach with an F1 score of 0.94, a precision of 0.94, a recall of 0.94, a test accuracy of 94.35%. In order to identify fault types, the authors in Allal et al. (2024) analyses data gathered using an Irish wind turbine SCADA system. For fault detection two solutions were considered by the authors: a naive solution and a two-tiered technique. These solutions were adapted to test the performance of different ML techniques. Ensemble learners, neural networks, tree-based, and classical models were tested in this study. The proposed two-tier method succeeds in detecting fault types with 91% accuracy and separated defective samples with 95% accuracy. This study concluded that Ensemble learners gave more accurate results in identifying and detecting faults.

Furthermore, AI techniques is employed in forecasting energy generated by RES. In Akhter et al. (2022), a hybrid DL approach that combined SSA (Singular Spectrum Analysis), RNN (Recurrent Neural Network), and LSTM (Long Short term memory) was proposed to predict in hourly based the power generated by solar PV. The RMSE, MSE, MAE, and coefficient of determination (R2) were used to assess the forecasting performance of the proposed approach. The results and the comparison to other approaches illustrated that SSA-RNN-LSTM shows superior forecasting accuracy, higher R2 value, and the lowest (RMSE and MSE). In Zafar et al. (2022), the authors used data delivered by SCADA in a hybrid variational decomposition model for accurately prediction of the power production in microgrid farms. The proposed model employed an advanced dynamic group-based approach combined with a search mechanism and IDGC-RBFNN to improve short-term power forecasting. The proposed approach showed high prediction accuracy with a lower computational burden. It is also compared with various Neural Network topologies and three heuristic algorithms. This comparison demonstrated the effectiveness of the proposed approach. To estimate generated power in wind turbines under yaw control, the study in He et al. (2022) proposed a ML-based prediction method. The approach considered wake effects and the whole spectrum of yaw and inflow situations. The support Vector regression model was used to identify the relationships between inputs and outputs. The results of applying the proposed approach on a real dataset showed the effectiveness and the accuracy of the predicted outcomes.

Recently, biogas plant becomes one of the most famous sustainable and renewable source for energy production. It is considered as an emerged and viable alternative fuel source that assists in lowering greenhouse gas emissions. In Jamei et al. (2024), the authors suggested a ML based approach to predict the power generated by a biogas plant. They used extreme gradient boosting (XGB) with the SHapley Additive exPlanations (SHAP) tool to represent the five efficiency parameters of biogas-powered engines: peak pressure (PP), hydrocarbon (HC), carbon monoxide (CO), nitrogen oxides (NOx), and brake thermal efficiency (BTE). Four situations were considered in this study: engine load, compression ratio, fuel injection timing, and fuel injection pressure. In addition, the authors used the SHAP framework to show the impact of the input features on the engine performance and emissions.

5.3 *IoT and AI Integration for Ensuring System Reliability*

The combination of IoT and AI is changing the reliability of RES by integrating real-time data collection with intelligent analytics. This synergy allows proactive maintenance, fault detection, and optimization, ensuring enhanced system performance and minimal downtime. There are currently a lot of solar plants placed all over the world, and for them to be safe and dependable for the duration of their lives, they need to be constantly monitored and safeguarded. Fault detection technology is primarily used to protect and isolate PV plants from various faults, including arc faults, line-to-line, line-to-ground, and ground faults. However, solar plants are susceptible to a variety of defects and failures. Despite the existence of numerous international standards, solar plants still face significant issues due to invisible flaws. The limitations of current techniques and commercialized equipment will therefore be significantly addressed by building smart equipment, incorporating AI and the IoT, for remote sensing, problem detection, and diagnostics of photovoltaic plants. An overview of the use of AI and the IoT in photovoltaic plants was provided by the authors in Mellit and Kalogirou (2021). The most cutting-edge algorithms, including machine and deep learning, were also described in this study, along with their implementation costs, complexity, accuracy, software appropriateness, and viability for real-time applications. For PV plants situated in remote locations with expensive and difficult maintenance accessibility, it may be cost-effective and technically possible to integrate AI and IoT methods for fault detection and diagnosis into basic hardware, such as inexpensive chips.

In Kayalvizhi et al. (2024), the authors examine the integration of IoT and predictive maintenance in solar PV plants. This study shows that prediction performance was enhanced due to the integration of IoT real-time monitoring. Indeed, IoT devices could detect faults, and collect and assess data. Thus, they permit maintenance to be done on time. Furthermore, the use of IoT devices produced more efficient power distribution and storage. Thus, more stable and dependable grids became more possible.

Nowadays, the biogas plant plays an essential role in providing the required electricity power to meet daily energy demands for residential and commercial operations. To ensure the proper function of the anaerobic digestion of biodegradable organic material, real-time monitoring and evaluation is required. This monitoring becomes possible due to AI-based systems and IoT devices. In Swami et al. (2023), the author proposed an AI-based system that offered visual representation, and process monitoring tools. Modern power infrastructure, called also Smart Grid (SG), allows data and electricity to go back and forth between peers in electricity system networks (ESN). In Kumar et al. (2020), the authors combine blockchain, AI, and the IoT devices to substitute conventional grids that heavily deped on fossil fuels with distributed energy resources (DER). Blockchain and IoT play an essential role in data detection and storage. They also provide safe and secure peer-to-peer transactions. However, AI approaches such as fuzzy logic and neural networks could manage and controlDER-based SG. Hence, these recent technologies enhanced the sustainability, stability, protection, availability, and trustworthiness of power within smart grids.

The authors in Noviati et al. (2024) investigate the role AI in enhancing the efficacy and dependability of SGs. Using a mixed-approaches strategy, the authors applied MLtechniques like Linear Regression, Support Vector Regression, Recurrent Neural Networks, and Long Short-Term Memory to historical and real-time data to predict energy management accuracy. For grid balancing and resource scheduling, optimization methods such as Particle Swarm Optimization and Genetic Algorithms were also used. When compared to traditional systems, the results showed notable gains in grid stability and energy efficiency, as well as a decrease in prediction errors and operating expenses. These improvements demonstrated AI's revolutionary potential in smart grids, encouraging more sustainable and effective energy use.

5.4 Challenges and Limitations of IoT and AI Integration

Table 3 summarizes the main limitation and challenges of IoT and AI integration in RES (Esenogho et al. 2022).

6 Conclusion

The surge in global energy demands due to economic growth and industrialization in recent years has significantly impacted the existing grid infrastructure raising concerns about energy sustainability. Renewable energy resources, such as wind, solar, and hydropower, represent a viable solution to provide clean and reliable energy that, reducing carbon emissions and address the current challenges in the energy sector. However, the integration of these renewable resources with existing grid is a

Table 3 Challenges of IoT-AI integration in RES

Challenge/limitations	Description
Data quality and security	Guaranteeing the precision and good quality of data gathered from IoT devices is important as bad data quality can result in incorrect AI predictions.
Integration complexity	In RES exist, the integration of IoT and AI may be technically complex. This integration requires compatibility between different systems and devices
Scalability	Expanding IoT and AI solutions to large-scale RES can be costly and challenging
High initial cost	The costs for deploying IoT devices, AI infrastructure, and maintenance systems can be prohibitive, especially for smaller operations
Data privacy concerns	As data may be share across different users
Reliability of AI models	AI model accuracy depend on historical data quality
Interoperability issues	The need of standardization across various IoT devices and AI platforms becomes an essential requirement to guarantee easy communication and integration in mixed energy environments
Energy consumption of IoT devices	Continued gathering of data by IoT devices can boost energy consumption, thus, complexity in energy-efficient systems
Regulatory and compliance challenges	Dealing with industry regulations and standards can be challenging, especially when handling data security and operational protocols across regions

critical issue since these resources are intermittent highlighting the need for energy storage solutions, infrastructure upgrades and grid modernization.

The emergence of IoT and AI has played a major role in addressing many of these issues. IoT platforms play a crucial role in monitoring and controlling renewable energy systems, enabling a real-time data collection for predictive maintenance, asset management, and energy efficiency and optimization. In contrast, AI algorithms can analyze vast amount of data collected from millions of IoT sensors for energy demand forecasting, improved grid stability, seamless integrations of renewable energy resources into the grid, resource planning, and optimized energy management.

This chapter has investigated the role of IoT and AI in enhancing the stability and reliability of RES. First, the various types of RES were described. Next, the IoT architecture and AI techniques were explored. Moreover, the role of IoT and AI in managing and controlling RES was investigated, including their role in real-time monitoring and data collection, analysis and optimization, improved grid management, and predictive maintenance. However, there are many challenges yet to be addressed for the integration of IoT and AI with RES for efficient, secure, and sustainable energy management and distribution. These include privacy, data quality, interoperability, scalability, and regulatory compliance.

References

Akhter MN, Mekhilef S, Mokhlis H, Ali R, Usama M, Muhammad MA, Khairuddin ASM (2022) A Hybrid deep learning method for an hour ahead power output forecasting of three different photovoltaic systems. Appl Energy 307:118185

Allal Z, Noura HN, Vernier F, Salman O, Chahine K (2024) Wind turbine fault detection and identification using a two-tier machine learning framework. Intell Syst Appl 22:200372

Amiri AF, Oudira H, Chouder A, Kichou S (2024) Faults detection and diagnosis of pv systems based on machine learning approach using random forest classifier. Energy Convers Manag 301:118076

Anika OC, Nnabuife SG, Bello A, Okoroafor ER, Kuang B, Villa R (2022) Prospects of low and zero-carbon renewable fuels in 1.5-degree net zero emission actualisation by 2050: a critical review. Carbon Capture Sci Technol 5:100072

Bojek P (2023) IEA. https://www.iea.org/energy-system/renewables

Brown A, White P (2023) Generative AI: transforming creativity. J Artif Intell Innov 17(2):45–58

Cai M (2021) Natural language processing for urban research: a systematic review. Heliyon 7(3)

Carter L, Lee S (2022) Smart transportation systems powered by AI: enhancing safety and efficiency. In: Technological advancements in smart cities, pp 78–102. Springer

Chen H, Wang Y, Li J, Xu G, Lei J, Liu T (2022) Thermodynamic analysis and economic assessment of an improved geothermal power system integrated with a biomass-fired cogeneration plant. Energy 240:122477

Choi R, Coyner A, Kalpathy-Cramer J, Chiang M, Campbell J (2020) Introduction to machine learning, neural networks, and deep learning. Transl Vis Sci Technol 9(2)

Duranay ZB (2023) Fault detection in solar energy systems: a deep learning approach. Electronics 12(21):4397

El Haj Assad M, Ahmadi MH, Sadeghzadeh M, Yassin A, Issakhov A (2021) Renewable hybrid energy systems using geothermal energy: hybrid solar thermal-geothermal power plant. Int J Low-Carbon Technol 16(2):518–530

El Samad M, Nasserddine G, Kheir A (2023) Introduction to artificial intelligence. In: Artificial intelligence and knowledge processing, pp 1–14. CRC Press

Elsaraiti M, Merabet A (2022) Solar power forecasting using deep learning techniques. IEEE Access 10:31692–31698

Erhueh OV, Elete T, Akano OA, Nwakile C, Hanson E (2024) Application of internet of things (IoT) in energy infrastructure: lessons for the future of operations and maintenance. Compr Res Rev Sci Technol 2(2):28–54

Esenogho E, Djouani K, Kurien AM (2022) Integrating artificial intelligence internet of things and 5g for next-generation smartgrid: a survey of trends challenges and prospect. IEEE Access 10:4794–4831

Farghali M, Osman AI, Umetsu K, Rooney DW (2022) Integration of biogas systems into a carbon zero and hydrogen economy: a review. Environ Chem Lett 20(5):2853–2927

Folgado FJ, González I, Calderón AJ (2023) Data acquisition and monitoring system framed in industrial internet of things for PEM Hydrogen generators. Internet Things 22:100795

Greenfield J, Jones R (2022) Artificial intelligence in sustainable agriculture. J Agric Technol 16(3):45–62

Harper S, Singh R (2023) AI-Driven retail: improving customer experience and operational efficiency. Retail Innov Rev 9(2):77–94

He R, Yang H, Sun S, Lu L, Sun H, Gao X (2022) A machine learning-based fatigue loads and power prediction method for wind turbines under yaw control. Appl Energy 326:120013

He W, Baig MJA, Iqbal MT (2024) An Internet of things—supervisory control and data acquisition (IoT-SCADA) architecture for photovoltaic system monitoring, control, and inspection in real time. Electronics 14(1):42

He X, Geng H, Mu G (2021) Modeling of wind turbine generators for power system stability studies: a review. Renew Sustain Energy Rev 143:110865

Huda NU, Ahmed I, Adnan M, Ali M, Naeem F (2024) Experts and intelligent systems for smart homes' transformation to sustainable smart cities: a comprehensive review. Expert Syst Appl 238:122380

Jaiswal D, Kaushal V, Singh PK, Biswas A (2021) Green market segmentation and consumer profiling: a cluster approach to an emerging consumer market. Benchmarking Int J 28(3):792–812

Jamei M, Sharma P, Ali M, Bora BJ, Malik A, Paramasivam P, Abdulla S (2024) Application of an explainable glass-box machine learning approach for prognostic analysis of a biogas-powered small agriculture engine. Energy 288:129862

Johnson R, Allen P (2023) Artificial Intelligence in Healthcare: current trends and future possibilities. J Med Innov 12(4):23–45

Karunarathne, GGKWMSIR, Kulawansa KADT, Firdhous MFM (2018) Wireless communication technologies in internet of things: a critical evaluation. In: 2018 international conference on intelligent and innovative computing applications (ICONIC), pp 1–5. IEEE

Katuri K, Somalal J, Kumar RS, Kumar BH, Chandrika VS, Kishore YK (2023) Monitoring and controlling of solar driven real time smart energy management system integrated with IoT devices. In: 2023 second international conference on trends in electrical, electronics, and computer engineering (TEECCON), pp 347–52. IEEE

Kayalvizhi N, Santhosh M, Thamodharan R, Dhileep M (2024) Iot-enabled real-time monitoring and predictive maintenance for solar systems: maximizing efficiency and minimizing downtime. In: 2024 international conference on smart systems for applications in electrical sciences (ICSSES), pp 1–5

Kumar NM, Chand AA, Malvoni M, Prasad KA, Mamun KA, Islam FR, Chopra SS (2020) Distributed energy resources and the application of AI, IoT, and blockchain in smart grids. Energies 13(21):5739

Liu X, Yue Y, Huang X, Xu W, Lu X (2022) A review of wind energy output simulation for new power system planning. Front Energy Res 10:942450

Mahale MD, Patra BB, Mahale SM (2023) Applications of IoT in renewable energy sector: smart energy management, opportunities, challenges and applications. Highly Effic Therm Renew Energy Syst 62–82

Manno D, Cipriani G, Ciulla G, Di Dio V, Guarino S, Brano VL (2021) Deep learning strategies for automatic fault diagnosis in photovoltaic systems by thermographic images. Energy Convers Manag 241:114303

Mazzeo D, Matera N, De Luca P, Baglivo C, Congedo PM, Oliveti G (2021) A literature review and statistical analysis of photovoltaic-wind hybrid renewable system research by considering the most relevant 550 articles: an upgradable matrix literature database. J Clean Prod 295:126070

Mellit A, Kalogirou S (2021) Artificial intelligence and internet of things to improve efficacy of diagnosis and remote sensing of solar photovoltaic systems: challenges, recommendations and future directions. Renew Sustain Energy Rev 143:110889

Mishra R, Naik BKR, Raut RD, Kumar M (2022) Internet of things (IoT) adoption challenges in renewable energy: a case study from a developing economy. J Clean Prod 371:133595

Moses OA, Chen W, Adam ML, Wang Z, Liu K, Shao J, Yu X (2021) Integration of data-intensive, machine learning and robotic experimental approaches for accelerated discovery of catalysts in renewable energy-related reactions. Mater Rep Energy 1(3):100049

Nasserddine G, Nassereddine M, El Arid AA (2023) Internet of things integration in renewable energy systems. In: Handbook of research on applications of AI, digital twin, and internet of things for sustainable development, pp 159–85. IGI Global

Nassereddine M, Khang A (2024) Applications of internet of things (IoT) in smart cities. In: Advanced IoT technologies and applications in the industry 4.0 digital economy, pp 109–36. CRC Press

Nassereddine M, Nassreddine G, El Arid A (2024) Hybrid photovoltaic and biogas system for stable power system. Next Energy 5:100172

Nassreddine G, Arid AE, Nassereddine M, Barbulescu E, Ahmed H (2024a) The performance effect of seasonality feature in solar photovoltaic power prediction using machine learning. In: International conference on electrical and electronics engineering, pp 167–183. Singapore: Springer Nature Singapore

Nassreddine G, El Arid A, Nassereddine M (2024b) Internet of things in intelligent transportation systems. In: IoT edge intelligence, pp 291–314. Springer Nature Switzerland

Nassreddine G, El Arid A, Nasseredine M (2023) Solar PV power prediction system based on machine learning approach. In: 2023 IEEE international conference on energy technologies for future grids (ETFG), pp 1–7. IEEE

Nižetić S, Šolić P, Gonzalez-De DLDI, Patrono L (2020) Internet of things (IoT): opportunities, issues and challenges towards a smart and sustainable future. J Clean Prod 274:122877

Noviati ND, Maulina SD, Smith S (2024) Smart grids: integrating AI for efficient renewable energy utilization. Int Trans Artif Intell 3(1):1–10

Omitaomu OA, Niu H (2021) Artificial intelligence techniques in smart grid: a survey. Smart Cities 4(2):548–568

Patel K, Singh R (2023) Advances in computer vision for image recognition. Comput vis Res 15(3):112–124

Qays MO, Ahmed MM, Parvez Mahmud MA, Abu-Siada A, Muyeen SM, Hossain ML, Rahman MM (2022) Monitoring of renewable energy systems by IoT-aided SCADA system. Energy Sci Eng 10(6):1874–1885

Ramamurthy A, Jain P (2017) The internet of things in the power sector opportunities in Asia and the Pacific. Mandaluyong, Philippines: Asian Development Bank

Rane N (2023) Contribution of ChatGPT and other generative artificial intelligence (AI) in renewable and sustainable energy

Rudro RAM, Nur K, Al Sohan MFA, Mridha MF, Alfarhood S, Safran M, Kanagarathinam K (2024) SPF-net: solar panel fault detection using U-net based deep learning image classification. Energy Rep 12:1580–1594

Sarker E, Halder P, Seyedmahmoudian M, Jamei E, Horan B, Mekhilef S, Stojcevski A (2021) Progress on the demand side management in smart grid and optimization approaches. Int J Energy Res 45(1):36–64

Şerban AC, Lytras MD (2020) Artificial intelligence for smart renewable energy sector in Europe—smart energy infrastructures for next-generation smart cities. IEEE Access 8:77364–77377

Sfar AR, Chtourou Z, Challal Y (2017) A systemic and cognitive vision for IoT security: a case study of military live simulation and security challenges. In: 2017 international conference on smart, monitored and controlled cities (SM2C), pp 101–5. IEEE

Smith J, Kumar V (2024) AI in personalized education: adapting learning for the modern classroom. In: Innovative educational technologies, pp 145–72. Wiley

Sodhi P, Awasthi N, Sharma V (2019) Introduction to machine learning and its basic application in python. In: Proceedings of the 10th international conference on digital strategies for organizational success

Sohail A, Islam NU, Haq AU, Islam SU, Shafi I, Park J (2023) Fault detection and computation of power in PV cells under faulty conditions using deep-learning. Energy Rep 9:4325–36

Swami S, Suthar S, Singh R, Thakur AK, Gupta LR, Sikarwar VS (2023) Integration of anaerobic digestion with artificial intelligence to optimise biogas plant operation. Environ Dev Sustain 1–31

Torres L, Lin X (2023) Optimizing renewable energy systems with artificial intelligence. Adv Energy Manag Syst 98–121. Academic Press

Twidell J (2021) Renewable energy resources. Routledge

Ullah Z, Rehman AU, Wang S, Hasanien HM, Luo P, Elkadeem MR, Abido MA (2023) IoT-based monitoring and control of substations and smart grids with renewables and electric vehicles integration. Energy 282:128924

Walker M, Taylor J, Fields S (2023) AI in business operations: automation and decision-making. Bus Innov Q 15(4):27–45

Zafar MH, Khan NM, Mansoor M, Mirza AF, Moosavi SKR, Sanfilippo F (2022) Adaptive ML-based technique for renewable energy system power forecasting in hybrid PV-wind farms power conversion systems. Energy Convers Manag 258:115564

Zazoum B (2022) Solar photovoltaic power prediction using different machine learning methods. Energy Rep 8:19–25

Zhang C, Lu Y (2021) Study on artificial intelligence: the state of the art and future prospects. J Ind Inf Integr 23:100224

Zhang J, Cosma G, Watkins J (2021) Image enhanced mask R-CNN: a deep learning pipeline with new evaluation measures for wind turbine blade defect detection and classification. J Imaging 7(3):46

Three New Considerations Applied to a Transportation Model: Accelerated Convergence, Ordered Optimal Solutions, and Analysis of a Protean Transportation System

Santosh Kumar, Elias Munapo, Trust Tawanda, and Philimon Nyamugure

Abstract The transportation model was recognized as an assignment model and was solved by the Hungarian method of assignment. This aspect of viewing a transportation model as an assignment model opens a few more new possibilities. This chapter investigates three such possibilities in context of a transportation model, which are: (1) accelerated convergence, (2) Kth best solution, and (3) analysis of a protean transportation system.

Keywords Transportation model · Accelerated convergence · Kth best solution · Protean transportation model

1 Introduction

The assignment and transportation models in OR are well-known for their industrial applications, details are documented in the OR literature, for example, see, Eppen et al. (1963), Hillier and Lieberman (2015), Munapo and Kumar (2022), Sharma (2018), Taha (2017) and Winston (2004). Many approaches were developed

S. Kumar
Department of Mathematical and Geospatial Sciences, School of Sciences, RMIT University, Melbourne, Vic, Australia
e-mail: Santosh.Kumar@rmit.edu.au

E. Munapo (✉)
School of Economics and Decision Sciences, North West University, Mafikeng, South Africa
e-mail: emunapo@gmail.com

T. Tawanda · P. Nyamugure
Department of Statistics and Operations Research, National University of Science and Technology, Bulawayo, Zimbabwe
e-mail: trust.tawanda@nust.ac.zw

P. Nyamugure
e-mail: philimon.nyamugure@nust.ac.zw

M. Ram et al. (eds.), *Advances in Mathematics for Engineering Sciences*, Mathematical Engineering, https://doi.org/10.1007/978-3-031-95693-5_2

for analysis of these special degenerate linear programming models, and various methods developed for analysis have also been documented in the above cited literature. Munapo et al. (2012) identified a relationship between transportation and an assignment model. They established that a balanced transportation model can be viewed as a MXM assignment model, where M represents the total demand and supply for a given balanced transportation model. Recently Munapo (2020) considered accelerated convergence for an assignment problem, and that idea has been extended for the transportation model in this chapter. Recently Kumar et al. (2024a) developed an algorithm to solve the transportation model by the Hungarian method of assignment. Kumar and Munapo (2024b) also discussed a few innovative ways for creating an alternative solution for a few problems in the mathematics of OR and they used those alternative solutions for solving many hard combinatorial optimization problems. One of these problems was to find the Kth best optimal solution for an assignment model. We find the Kth best optimal solution to the transportation model in this chapter. Finally, we consider analysis of a protean transportation model, which simply means that the same model keeps appearing again and again with slightly different input data. Kumar (1995) has reviewed analysis of these protean systems and identified several such OR models in the category of protean systems. In the case of protean systems, information concerning the previous solution may still be useful for identifying the new optimal solution to the changed problem. For information recycling, see Kumar (2005, 2006). In view of the above, this chapter revisits the well-known transportation model to study the following:

(1) Accelerating convergence to reach an optimal solution of a transportation model.
(2) Find the Kth optimal solution for a transportation model.
(3) Apply the information recycling concept from the waste management to a transportation model to identify the new solution to the changed problem from the previous solution to the changed transportation model.

For each case, numerical illustrations have also been presented.

The chapter has been organized in 6 sections, mathematical model and some relevant information about the transportation model have been discussed in Sect. 2. The accelerated convergence for the transportation model is discussed in Sect. 3. The ordered optimum for any mathematical programming problem is computationally a difficult combinatorial problem. However, a simple procedure to find a few ordered optimal solutions for a given transportation model have been discussed in Sect. 4. Information recycling concept in context of a protean transportation model is presented in Sect. 5. In each section, some numerical illustrations have been presented. The chapter is concluded with some concluding remarks in Sect. 6.

2 Mathematical Model of the Transportation Problem and Its Relevant Properties

Consider a mathematical model of a balanced transportation model given by expressions (1), (2), and (3), where $c_{ij} is$ a unit transportation cost from the origin i to the destination j. The variable x_{ij} represents the number of units to be sent from the origin i to the destination j. The problem is to find the optimal values of the variables x_{ij}. The supply and demand constraints are represented by (2), and the condition (3) indicates that the total demand and supply are equal for a balanced transportation model. All variables to be determined will be non-negative integer values.

$$\text{Minimize}\, Z = \sum_{i=1}^{m} \sum_{j=1}^{n} c_{ij}\, x_{ij} \tag{1}$$

$$\sum_{j=1}^{n} x_{ij} = a_i,\, i = 1, 2, \ldots, m$$

$$\sum_{i=1}^{m} x_{ij} = b_j, j = 1, 2, \ldots, n. \tag{2}$$

$$\sum_{i=1}^{m} a_i = \sum_{j=1}^{n} b_j = M\,,\, x_{ij} \geq 0\, and\, an\, integer\, \forall (i, j) \tag{3}$$

The a_i and b_j represent the supply at the origin points i and demand at the point j, respectively. Kumar et al. (2024a) have considered the above transportation model as an assignment model of size MXM by replacing the row i for a_i number of times and similarly replacing the column j for b_j number of times, where, $i = 1, 2, \ldots, m$, and, $j = 1, 2, \ldots, n$.

A balanced transportation model (1), (2), and (3) when viewed as an assignment model of MXM size has some interesting properties. Some of these properties were observed and stated in Kumar et al. (2024a). Briefly these properties were:

1. The Hungarian method of assignment transforms the cost matrix in an equivalent cost matrix without changing relative merits of various cost elements.
2. The transformation of the cost elements is applied to create zero elements in the transportation cost matrix. If the number of zero elements in the modified cost matrix is denoted by p, then p will satisfy

$$\text{Max}(m, n) \leq p \leq (m + n - 1) \tag{4}$$

3. The Hungarian method of assignment can easily deal with all degenerate transportation problems, as the assignment problem is a degenerate transportation model, were $a_i = b_j = 1, i = 1, 2, \ldots, m$, and, $j = 1, 2, \ldots, n$, and $m = n$.

For non-degenerate transportation models, the maximum number of independent zero elements that can be created in an m × n matrix is $(m+n-1)$, which is the required number of independent zero elements for a feasible solution of a non-degenerate transportation model. A cell (i, j) in the ith row and the jth column is an independent cell, if it satisfies fully the requirement at the jth column or the supply from the ith point. It means that each allocation satisfies either the demand or the supply through that cell, consequently that column or row is crossed out.

4. For the Hungarian method of assignment, at any time maximum number of rows crossed out is $\leq (m-1)$ and similarly, the maximum number of columns crossed out is $\leq (n-1)$.

a. ***The transportation-assignment theorem***

The Hungarian approach, while maintaining relative merit of various cost elements, transforms some of the cost elements c_{ij} to zero values and those cells have an additional significance for analysis of the transportation problem, which leads to the following theorem:

Transportation-Assignment theorem: Although the total cost in the case of a transportation model is given by $\sum_{i=1}^{m}\sum_{j=1}^{n} c_{ij}x_{ij}$, which is a function of the transportation cost rate c_{ij} and the number of units x_{ij} that are transported through that combination of the demand and supply points. However, when the cost-matrix is modified by using the Hungarian approach of assignment, i.e., some elements in the cost-matrix get modified to zero cost, while maintaining their relative merits. These zero cost elements become independent of the number x_{ij}, as the sum-product $\sum_{i=1}^{m}\sum_{j=1}^{n} c_{ij}x_{ij}$ remains zero. It simply means that one of the two quantities in each term of the sum function representing the objective function of the transportation model is maintained at a zero level, i.e., when the modified $C_{ij} > 0$, *the corresponding* $x_{ij} = 0$ and similarly when the modified $c_{ij} = 0$, the corresponding x_{ij} *is allowed to assume values* $>$ 0. This property is not only useful for solving the transportation by the method of assignment, but it is useful for advancing the transportation model to two related cases, which are:

(1) Determination of the ordered optimal solution for a transportation problem, and for
(2) Solving a protean transportation model.

These cases will be discussed in Sects. 4 and 5.

Proof of the transportation-assignment theorem: For developing a feasible solution, positive allocations are made i.e., $x_{ij} > 0$, where the modifies relative cost $c_{ij} = 0$ and make $x_{ij} = 0$, where $c_{ij} > 0$. In other words the value of the objective function with transformed cost matrix will be $\sum_{i=1}^{m}\sum_{j=1}^{n} c_{ij}x_{ij} = 0$.

Effectively, it means the solution in the transformed cost matrix is independent of the values of the variable $x_{ij} > 0$. Therefore there is no difference between the transportation model, where $x_{ij} > 0$ *are integers values and* for an assignment model, these values were 1 as all $a_i = b_j = 1$. For the transportation model where these supply and demand values are known integer values >=1 and the overall sum satisfies the condition $\sum_{i=1}^{m} x_{ij} = b_j, \sum_{j=1}^{n} x_{ij} = a_i, and\ x_{ij} \geq 0\ integer\ quantities\ \forall(i,j)$.

3 Accelerating Convergence for the Optimal Solution of a Transportation Model

The acceleration concept in context of an assignment model was developed by Munapo (2020). Since we are solving the transportation model by the Hungarian method of assignment, it is natural to apply the acceleration idea when solving the transportation model by the Hungarian method of assignment. Instead of developing a zero in each row and in each column by subtracting the minimum value from each element in that row or the column, Munapo (2020) suggested to create a row or a column with the same value and thereby when the minimum is subtracted, in that row or the column, subtraction resulted in all zero entries. These zero cells help accelerating convergence to reach a feasible solution confined to zero cost cells and the optimal solution. This is discussed in Sect. 3.1.

3.1 The Steps of the Accelerated Hungarian Method for the Transportation Model

The steps of the accelerated transportation algorithm are as follows:

Step 1. Using the accelerating Hungarian method by Munapo (2020), modify elements in any one column or a row to have the same value and then create zeros in each column and each row. Suppose we wish to create column 1 of zero elements. For this we ensure that all elements in column 1 have the same value by adding appropriate integer valued quantities without changing the relative merits of the cost elements. Therefore, we bring all elements equal to the maximum value element in the selected row or the column. Any column or row can be changed to zero valued cells; however, we may like to create the column or the row with a maximum value of demand or supply for the transportation model.

Step 2. Ensure that all other columns/rows have at least a zero. The chosen row or the column with equal values will be transformed to all zero value cells in the transformed matrix.

Step 3. Each allocation in a zero cell results in covering either a row or a column. Cover all zeros with a minimum number of lines and obtain a feasible solution to the transportation model. Remember maximum number of crossed out rows to remain less than (m − 1) and similarly number of crossed out columns should also be less than (n − 1). If a feasible solution can be identified in zero cost cells, it is an optimal solution. If the solution is not feasible, check if there is a zero-cost cell which has not been crossed out. Go to Step 4.

Step 4: If the solution is infeasible and an uncrossed zero cell exist, develop a transportation loop with the intension of filling the zero-cell with some positive allocation and move towards feasibility of the solution. If feasibility of the solution has improved, find the new solution. However, if feasibility of the new solution reduces, try other combination of covering lines so that feasibility is improved compared to the previous solution. If the solution is still infeasible and the number of lines is less than (n + m − 1) then go to Step 5. At no stage we cross all rows or all columns for an application of the Hungarian method of assignment to be able to develop a greater number of effective zero value cells.

Step 5. Select the smallest uncovered element and update the cost elements as is done in the Hungarian method of assignment. Ensure once again that each row and each column have a zero element and feasibility is improved compared to the previous solution. Return to Step 3.

3.2 *Numerical Illustrations*

3.2.1 Illustration 3.1

Let us reconsider a transportation problem taken from Eppen et al. (1963) p. 283, which was also solved earlier by Kumar et al. (2024a) by the Hungarian method of assignment. However, here we solve the same problem by using the acceleration concept by Munapo (2020). The cost, demand and supply data are given in Table 1.

Since the relative merit of various costs in a row or a column will not change when a constant is added to all elements in that row or the column, we can easily achieve a row or a column having the same values and when the minimum element is subtracted from all other elements in that row or column, it will generate all zero entries in that chosen row or column. Since maximum demand corresponds to column 2, we arbitrarily select column 2 to transform to zero values, for that first we make all elements in this column to attain the same value. The maximum in column 2 is c_{i2} with a value 13 in the cell c_{12}, we can add 9 to all elements in row 2 and 4 to all elements in row 3. This will not alter the relative merits of various elements;

Table 1 The transport problem

	1	2	3	4	Supply
1	12	13	4	6	5
2	6	4	10	11	7
3	10	9	12	4	8
Demand	4	9	2	5	20

however, it will transform the cost data as shown in Table 2. Creating a zero in each column by subtracting the minimum, the relative cost elements will become as given in Table 3, where we subtract 12 from each element in column 1, 13 from each element in column 2, 4 is subtracted from each element in column 3 and 6 is subtracted from each element in column 4. Each row and column have a zero-cost cell in Table 3. Making allocations in independent cells, we do not get a feasible solution, hence more zero elements must be created.

Step 1: **Initial data transformation without changing relative merits**

Iteration 1

The above solution in Table 3 is not a feasible solution, hence it will require a greater number of independent zero cells. For the solution in Table 3, the measure of infeasibility is 6 units, which is reflected in the last column. Lines are drawn through columns 1 and 2 and rows 1 and 2. The minimum of the element not crossed out is 2 in the cell (3,4). Updating the cost values by the Hungarian approach, we have Table 4. Note that column 1 and row 2 resulted in non-zero cells, hence once again the minimum has been subtracted from all elements in this column.

Table 2 With all elements same (i.e., 13) in column 2

	1	2	3	4	Supply
1	12	13	4	6	5
2	$6+9=15$	$4+9=13$	$10+9=19$	$11+9=20$	7
3	$10+4=14$	$9+4=13$	$12+4=16$	$4+4=8$	8
Demand	4	9	2	5	20

Table 3 With zero elements in each column

	$1 \uparrow 1$	$2 \uparrow 4$	3	4	Supply
$1 \rightarrow 2$	0 (4)	0	0	0(1)	5
$2 \rightarrow 3$	3	0 (7)	15	14	7
3	2	0 (2)	12	2	8 (6)
Demand	4	9	2 (2)	5 (4)	20 (6)

Table 4 Obtained from Table 3

	$1 \uparrow 3$	2	$3 \uparrow 1$	$4 \uparrow 5$	Supply
$1 \rightarrow 4$	~~2~~0	2	0 (2)	0 (3)	5
$2 \rightarrow 2$	~~5~~1	0 (7)	13	12	7
$3 \rightarrow 6$	~~2~~0 (4)	0 (2)	10	0 (2)	8
Demand	4	9	2	5	20 (0)

Table 5 The alternative solution

	1	2	3	4	Supply
1	~~2~~0 (3)	2	0 (2)	0	5
2	~~3~~ 1	0 (7)	13	12	7
3	~~2~~ 0 (1)	0 (2)	10	0 (5)	8
Demand	4	9	2	5	20

The solution in Table 4 is feasible, hence optimal. There is a zero cell in the location (1,1), indicating one more optimal solution exists for the problem, which is shown in Table 5.

The same optimal solutions were obtained earlier in Kumar et al. (2024a), as was expected, however, the optimal solution has been attained in one iteration only.

3.2.2 Illustration 3.2

Let us reconsider the transportation model taken from Munapo and Kumar (2022, p. 132), as given in Table 6.

Solution

Initial Transformed Data Without Changing the Relative Merits

For the application of the accelerated algorithm, first we create a row or a column of all zero elements. Let us select row 5, where the supply is maximum. The maximum element in row 5 is 15. We wish to make all elements in row 5 as 15. For that we add 13 to all elements in column 1. Similarly add 10 to all elements in column 3, 3 to all elements in column 4, 7 to all elements in column 5 and 8 to all elements in column 6. Nothing has been added to elements in column 2. This results in Table 7.

Subtracting minimum elements in each row from all the elements in that row, results in Table 8.

Each column and each row in Table 8 has a zero element, where 5 number of allocations are possible, and five lines are drawn in columns 1, 2 and 4 and rows 2

Table 6 Cost matrix and demand and supply for a balanced problem

	1	2	3	4	5	6	Supply
1	8	9	20	4	4	6	2
2	2	1	1	1	1	1	3
3	2	8	19	7	3	14	4
4	2	6	6	4	11	8	5
5	2	15	5	12	8	7	6
Demand	3	4	5	2	2	4	20

Table 7 Making all elements in row 5 equal to 15

	1	2	3	4	5	6	Supply
1	21	9	30	7	11	14	2
2	15	1	11	4	8	9	3
3	15	8	29	10	10	22	4
4	15	6	16	7	18	16	5
5	15	15	15	15	15	15	6
Demand	3	4	5	2	2	4	20

Table 8 Creating a zero element in each row

	1 ↑ 4	2 ↑ 3	3	4 ↑ 1	5	6	Supply
1	14	2	23	0 (2)	4	7	2
2 → 2	14	0 (3)	10	3	7	8	3
3	7	0 (1)	21	2	2	14	4 (3)
4	9	0	10	1	12	10	5 (5)
5 → 5	0 (3)	0	0 (3)	0	0	0	6
Demand	3	4	5 (2)	2	2 (2)	4 (4)	20 (8)

and 5. These allocations are shown in Table 8 and lines are indicated by a vertical arrow ↑ to cross out a column and a horizontal arrow → for crossing a row. A number besides an arrow just indicates the order of allocation. The minimum element that has not been crossed out is 2 in the cell (3,5). Updating the cost matrix by the Hungarian approach, we have, Table 9.

Iteration 1

Note column 1 row 2 has no zero element, and to create a zero element, we subtract the minimum element, which is 2. This results in Table 10. New allocations are once again shown in the parenthesis. Although, feasibility of the solution has improved compared to the solution we had in Table 8, it remains infeasible as 6 more units need to be allocated.

Table 9 Updated values of the cost elements

	1	2	3	4	5	6	Supply
1	14	2	21	0	2	5	2
2	16	2	10	5	7	8	3
3	7	0	19	2	0	12	4
4	9	0	8	1	10	8	5
5	2	2	0	2	0	0	6
Demand	3	4	5	2	2	4	20

Table 10 Updated values of the cost elements from Table 9

	1 ↑ 4	2 ↑ 3	3	4 ↑ 1	5 ↑ 6	6	Supply
1	12	2	21	0 (2)	2	5	2
2 → 2	12	0 (3)	8	3	5	6	3
3	5	0 (1)	19	2	0 (2)	12	4 (1)
4	7	0	8	1	10	8	5 (5)
5 → 5	0 (3)	2	0 (3)	2	0	0	6
Demand	3	4	5 (2)	2	2	4 (4)	20 (6)

Note that in Table 10, we made a total number of 6 allocations in 6 cells and have crossed 4 vertical lines in columns 1, 2, 4, 5 and two horizontal lines in rows 2 and 5. The minimum element among those which have not been crossed out is 5 in the cell (1,6). The updated values are given in Table 11.

Iteration 2

Note that in Table 11, we have no zero element in column 1 and row 2. The minimum is 5, which is subtracted from all other elements, and we obtain Table 12. Infeasibility has reduced, however, the solution is not feasible, as is reflected in Table 12.

Table 11 Updated values of the cost elements from Table 10

	1	2	3	4	5	6	Supply
1	12	2	16	0	2	0	2
2	17	5	8	8	10	6	3
3	5	0	14	2	0	7	4
4	7	0	3	1	10	3	5
5	5	7	0	7	5	0	6
Demand	3	4	5	2	2	4	20

Table 12 Updated values of the cost elements from Table 11

	1 ↑ 5	2 ↑ 7	3	4	5 ↑ 3	6	Supply
1 → 1	7	2	16	0 (2)	2	0	2
2 → 2	7	0 (3)	3	3	5	1	3
3 → 4	0 (2)	0	14	2	0 (2)	7	4
4	2	0 (1)	3	1	10	3	5 (4)
5 → 6	0 (1)	7	0 (5)	7	5	0	6
Demand	3	4	5	2	2	4 (4)	20 (4)

Table 13 Updated values of the cost elements from Table 12

	1	2	3	4	5	6	Supply
1	8	3	16	0	3	0	2
2	8	1	3	3	6	1	3
3	1	1	14	2	1	7	4
4	2	0	2	0	10	2	5
5	1	8	0	7	6	0	6
Demand	3	4	5	2	2	4	20

Table 14 Updated values of the cost elements from Table 13

	$1 \uparrow 3$	$2 \uparrow 5$	$3 \uparrow 1$	$4 \uparrow 6$	$5 \uparrow 7$	6	Supply
$1 \rightarrow 8$	7	3	16	0	2	0 (2)	2
$2 \rightarrow 4$	6	0 (1)	2	2	4	0 (2)	3
3	0 (2)	1	14	2	0 (2)	7	4
4	1	0 (3)	2	0 (2)	9	2	5
$5 \rightarrow 2$	0 (1)	8	0 (5)	7	5	0	6
Demand	3	4	5	2	2	4	20 (0)

Iteration 3

In Table 12, only uncrossed row is 4 and column 3, 4 and 6. The minimum element is 1 in the cell (4,4). This results in Table 13.

Note that zero element is missing in column 1 and 5 and in row 2. Subtracting the minimum, we obtain Table 14.

The above is a feasible solution and the total cost associated with the above solution is as follows:

Cost for the allocations in cell (1,6) is $(6 \times 2) = 12$

Cost for the allocations in cell (2,2) and (2,6) is $(1 \times 1 + 1 \times 2) = 3$

Cost for the allocations in cell (3,2) and (3,5) is $(2 \times 2 + 3 \times 2) = 10$

Cost of allocations in cell (4,2) and (4,4) is $(6 \times 3 + 4 \times 2) = 26$

Cost of allocations in cell (5,1) and (5,3) is $(2 \times 1 + 5 \times 5) = 27$, Total cost $= 78$

The above example took 4 iterations but when attempted by the accelerated Hungarian approach, the optimal solution has been obtained in 3 iterations.

3.2.3 Illustration 3.3

Consider a degenerate transportation model given in Table 15.

Solution

Initial Preparation

We create each element in column 1 as 15 without altering the relative merits of various coast elements. This can be achieved by adding 7 to each element in row 1, and 12 to each element in row 3. We get Table 16.

Developing zero elements in each row and column, we have Table 17.

After crossing rows 3 and 2 and column 1, the minimum element is 1 from the uncrossed elements in the cell (1,3). The upgraded matrix is given in Table 18.

Row 3 has no zero, subtracting the minimum, we obtain Table 19.

The feasible solution in Table 19 is an optimal solution.

Table 15 Cost, demand and supply data for a balanced transportation model

Demand/supply	D_1	D_2	D_3	Supply
S_1	8	5	6	12
S_2	15	10	12	8
S_3	3	9	10	8
Demand	15	8	5	28

Table 16 Transformed cost elements from Table 15

Demand/supply	D_1	D_2	D_3	Supply
S_1	15	12	13	12
S_2	15	10	12	8
S_3	15	21	22	8
Demand	15	8	5	28

Table 17 Zero element in each row and column

Demand/supply	$D_1 \uparrow 3$	D_2	D_3	Supply
S_1	0 (7)	2	1	12 (5)
$S_2 \rightarrow 2$	0	0 (8)	0	8
$S_3 \rightarrow 1$	0 (8)	11	10	8
Demand	15	8	5 (5)	28 (5)

Table 18 Upgraded cost matrix from Table 17

Demand/supply	D_1	D_2	D_3	Supply
S_1	0	1	0	12
S_2	1	0	0	8
S_3	1	11	10	8
Demand	15	8	5	28

Table 19 Optimal solution

Demand/supply	D_1	D_2	$D_3 \uparrow 3$	Supply
S_1	0 (7)	1	0 (5)	12
$S_2 \rightarrow 2$	1	0 (8)	0	8
$S_3 \rightarrow 1$	0 (8)	10	9	8
Demand	15	8	5	28 (0)

4 Determination of the Kth Best Solution in a Transportation Model

Determination of the Kth best solution is computationally a demanding problem, as the Kth best is defined to be the best solution after excluding the (K − 1) number best solutions, K > 1. In other words, there is no direct test to identify the Kth best solution for a specific value of K, other than when K = 1. Another mathematical difficulty associated with the Kth best solution is that it is associated to the previous (K − 1) best solutions. Mathematical requirement for determination of the Kth best was identified when Kumar and Wagner (1979) developed a method for solving an extreme point mathematical programming problem, which is given by the model (5).

4.1 Mathematical Justification for the Kth Best Optimal Solution

Consider a mathematical model given by (5)

$$Max z = Cx$$

Subject to

$$AX \leq b$$

where X is an extreme point of

$$DX \leq f, X \geq 0 \tag{5}$$

The mathematical model (5) was developed by Kirby et al. (1972), Kirby and Scobey (1970) while dealing with a problem that needed to be processed on 'n' number of identical machines. Approach proposed by Kumar and Wagner (1979) for solving the model (5) requires determination of ordered optimal solutions of the linear programming model (6):

$$\text{Max } z = Cx,$$

Subject to:

$$DX \leq f, X \geq 0. \tag{6}$$

The problem of the Kth best has been studied by many authors, for example, see, Murthy (1968) studied it for an assignment model, Kumar et al. (2020), Kumar, Munapo and Jones (2007) studied it for a pure integer programming model, Munapo and Kumar (2022) and Kumar and Munapo (2024b) for an assignment model. In this section, we discuss a method to find the 2nd and 3rd best solution for the transportation model.

The basic idea is to find the optimal solution to a given transportation model by using the Hungarian method of assignment and strategically increase the length of the optimal solution in a controlled way to create an alternative optimal solution. This leads to two optimal solutions. One of the solutions that was optimal earlier represents the optimal solution and the newly created optimal solution becomes the 2nd best solution. Extension of this idea can explore the 3rd best solution; however, generalization of this idea may become unwieldy and unmanageable.

4.2 The Algorithm to Find the Kth Best Solution to the Transportation Model

Step 1: Apply the accelerated Hungarian method of assignment to find an optimal solution to the given balanced transportation model.

Step 2: Find the minimum transformed non-zero cost element in the final cost matrix, where the optimal solution has been identified.

Step 3: Increase an appropriate zero cost element equal to the minimum non-zero cost value, and make sure to have a zero entry in each row and column. Find the new optimal solution. Go to the next step.

Step 4: We must obtain alternative solutions, one of them must be the previous solution and the other one will be using the minimum non-zero cost cell created by us.

Step 5: With respect to the original cost data, these alternative solutions will be the best and the 2nd best optimal solutions. If the next best solution is required, we may repeat the same idea for the next best solution, however, implementation will get more and more cumbersome.

4.3 *Numerical Illustration for the Kth Best Solution for the Transportation Model*

To illustrate determination of the 2nd best solution of a transportation model, let us reconsider the transportation model illustration 3.2, for which the original cost data (Table 6) and the final solution (Table 14) is reproduced below as Tables 20 and 21.

Note that the minimum cost element in the optimal solution in Table 19 appears in two cells (3,2) and the cell (4,1).

Possibility 1

Consider the minimum value 1 in the cell (4,1).

Table 20 Balanced transportation cost matrix with demand and supply

	1	2	3	4	5	6	Supply
1	8	9	20	4	4	6	2
2	2	1	1	1	1	1	3
3	2	8	19	7	3	14	4
4	2	6	6	4	11	8	5
5	2	15	5	12	8	7	6
Demand	3	4	5	2	2	4	20

Table 21 Optimal solution to the problem in Table 20

	1 ↑ 3	2 ↑ 5	3 ↑ 1	4 ↑ 6	5 ↑ 7	6	Supply
1 → 8	7	3	16	0	2	0 (2)	2
2 → 4	6	0 (1)	2	2	4	0 (2)	3
3	0 (2)	1	14	2	0 (2)	7	4
4	1	0 (3)	2	0 (2)	9	2	5
5 → 2	0 (1)	8	0 (5)	7	5	0	6
Demand	3	4	5	2	2	4	20 (0)

To increase the cost associated with the optimal solution in Table 14a. In this column, there are two allocations in the cell, (3,1) and (5,1). We change these cost elements associated with these cells (3,1) and (5,1) equal to the cost associated with the cell (4,1). This will result in Table 22.

Note there is no zero element in the column 1 in Table 22. Subtracting the minimum, we obtain Table 23.

Note that allocation in the cell (4,1) is possible by moving one unit along the transportation loop shown in Fig. 1. This results in Table 24.

Table 22 Modified cost associated in column 1

	1 ↑ 3	2 ↑ 5	3 ↑ 1	4 ↑ 6	5 ↑ 7	6	Supply
1 → 8	7	3	16	0	2	0 (2)	2
2 → 4	6	0 (1)	2	2	4	0 (2)	3
3	1 (2)	1	14	2	0 (2)	7	4
4	1	0 (3)	2	0 (2)	9	2	5
5 → 2	1 (1)	8	0 (5)	7	5	0	6
Demand	3	4	5	2	2	4	20 (0)

Table 23 Table with modified cost values from Table 22

	1 ↑ 3	2 ↑ 5	3 ↑ 1	4 ↑ 6	5 ↑ 7	6	Supply
1 → 8	6	3	16	0	2	0 (2)	2
2 → 4	5	0 (1)	2	2	4	0 (2)	3
3	0 (2)	1	14	2	0 (2)	7	4
4	0	0 (3)	2	0 (2)	9	2	5
5 → 2	0 (1)	8	0 (5)	7	5	0	6
Demand	3	4	5	2	2	4	20 (0)

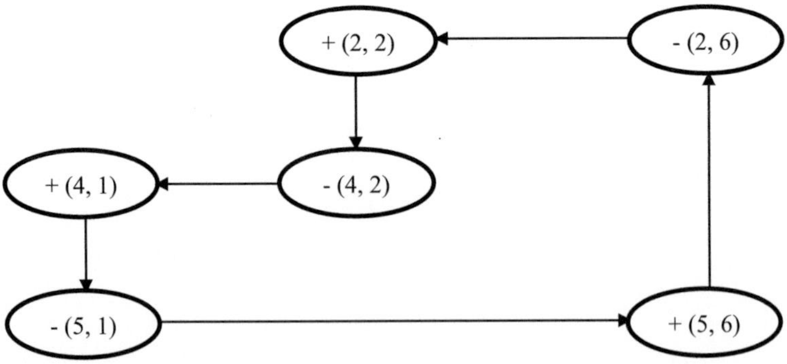

Fig. 1 Transportation loop

Table 24 Table with modified cost values from Table 23

	$1\uparrow 3$	$2\uparrow 5$	$3\uparrow 1$	$4\uparrow 6$	$5\uparrow 7$	6	Supply
$1\rightarrow 8$	6	3	16	0	2	0 (2)	2
$2\rightarrow 4$	5	0 (2)	2	2	4	0 (1)	3
3	0 (2)	1	14	2	0 (2)	7	4
4	0 (1)	0 (2)	2	0 (2)	9	2	5
$5\rightarrow 2$	0	8	0 (5)	7	5	0 (1)	6
Demand	3	4	5	2	2	4	20 (0)

There are two optimal solutions, one of them is displayed in Table 23 and the other in Table 24. Total cost associated with these solutions are given by: $\sum_{i=1}^{5}\sum_{j=1}^{6} c_{ij}x_{ij} =$

For the solution in Table 19 and the cost matrix in Table 18, the total cost is =

$$2\times 2+2\times 1+1\times 1+6\times 3+5\times 5+4\times 2+3\times 2+6\times 2+1\times 2=78$$

Similarly calculating the cost for the solution in Table 22 using the cost matrix from Table 18, the total cost is

$$\begin{aligned}&=2\times 2+2\times 1+1\times 2+6\times 2+5\times 5\\&\quad+4\times 2+3\times 2+6\times 2+1X1+7\times 1=79\end{aligned}$$

Thus, resulting in the best and the second-best solutions.

Possibility 2

Alternatively, if we consider row 4 in Table 14a, we must increase the cost in cells (4,2) and (4,4) and that will make row 4 without a zero element. After subtracting the minimum, we get back to Table 23, leading once again to the same solution as obtained in Table 24. Therefore, we can conclude that the total cost associated with the optimal and the 2nd best optimal solutions to Illustration 2.2 are 78 and 79 respectively.

Possibility 3

There is one more minimum element 1 in the cell (3,2). This case will once again lead to two similar considerations with respect to column 2 and row 3. However, no allocation can be made to the cell (3,2), therefore it leads to no other solution. Finally, we conclude that Table 22 gives the optimal and the 2nd best solutions to the transportation model for the illustration 3.2.

The 3rd Best Solution to the Transportation Model

From Table 24, we are now investigating if we can identify the next best solution. Making necessary changes we get from Tables 24 to 25. Note that the minimum cost element in Table 23 is still 1 in the cell (3,2). To increase the cost of the cells, we

Table 25 Table with modified cost values from Table 24

	$1 \uparrow 3$	$2 \uparrow 5$	$3 \uparrow 1$	$4 \uparrow 6$	$5 \uparrow 7$	6	Supply
$1 \rightarrow 8$	6	3	16	0	2	0 (2)	2
$2 \rightarrow 4$	5	0 (2)	2	2	4	0 (1)	3
3	0 (2)	1	14	2	0 (2)	7	4
4	0 (1)	0 (2)	2	0 (2)	9	2	5
$5 \rightarrow 2$	0	8	0 (5)	7	5	0 (1)	6
Demand	3	4	5	2	2	4	20 (0)

Table 26 Table with modified cost values from Table 25

	$1 \uparrow 3$	$2 \uparrow 5$	$3 \uparrow 1$	$4 \uparrow 6$	$5 \uparrow 7$	6	Supply
$1 \rightarrow 8$	6	3	16	0	2	0 (2)	2
$2 \rightarrow 4$	5	0 (2)	2	2	4	0 (1)	3
3	$0\ (2-2-0)$	$0\ (0+2=2)$	14	2	0 (2)	7	4
4	$0\ (1+2=3)$	0 (2–2 = 0)	2	0 (2)	9	2	5
$5 \rightarrow 2$	0	8	0 (5)	7	5	0 (1)	6
Demand	3	4	5	2	2	4	20 (0)

can either increase the cost associated with zero cost elements in column 2 or row 3. This will eliminate zero cost cells in this row or the column. When subtracted, we can allocate in the cells (3,2). This will result in Table 26.

Total cost for this solution is given by: $2c_{16} + 2c_{22} + 1c_{26} + 2c_{32} + 2c_{35} + 3c_{41} + 2c_{44} + 5c_{53} + 1c_{56} = 12 + 2 + 1 + 16 + 6 + 6 + 8 + 25 + 7 = 83$.

It is certainly true that implementation of this approach for higher value of K will become difficult to execute.

5 Protean Transportation Model

Protean systems change from time to time, and they often arise in many real-life situations, when the same mathematical model is applicable but values of input data changes due to external control. Here we are considering a transportation model where a given cost matrix remains unchanged but demand and supply, which is influenced by some external situations change from one period to another period. In such situations, information recycling is possible for that part of the model which is invariant with respect to changes. Here we consider that in a transportation model, cost matrix is unchanged but values of the a_i and b_j are changing. The Information recycling concept that was developed by Kumar in context of a linear programming problem and geometric programming problems by Kumar (2005, 2006), is applied to the transportation model.

Note that the cost matrix of the given transportation model is transformed with a zero element in each row and column, in other words the transformed matrix is independent of the values of a_i and b_j. Therefore, zero cost cells are independent of the values of a_i and b_j. It leads to the fact that optimal solution to the transportation model does not have to be approached from the beginning, instead new optimal solution can be derived from the previous solution. Since the maximum number of independent zero cells in a balanced (m × n) transportation model will be at most (m + n − 1), additional zero cost cells when required will always form a loop with the existing independent cells. Therefore, when values of a_i and b_j change, they will either fit in the existing independent zero cells or they will form a loop with existing zero cells. For a degenerate solution, if changed problem has a non-degenerate solution, additional zero cost cells will be developed by an application of the Hungarian method of assignment. Hence to deal with a protean transportation model, it can be solved by using the following steps.

Algorithm for protean transportation system

Step 1: Rewrite the problem in the form of the previous transportation model for which optimal solution is known.

Step 2: Since the zero elements in the cost matrix are independent of the number of units transported through that combination, adjust in these cells as much as is possible from the leftover demand and supply points. Each demand or supply point when fully satisfied will lead to deletion of either a row or a column.

Step 3: If a feasible solution can be identified that solution will becomes an optimal solution to the changed problem. The search process stops.

Step 4: If a feasible solution is not possible among the existing cells, apply the Hungarian approach to increase the number of independent zero elements to find a feasible solution among the zero cost cells and return to Step 3.

5.1 *Numerical Illustration*

Consider illustration 3.1, which is reproduced here as Table 27. The solution is given here as Table 28.

Table 27 The transport problem i.e. the given problem

	1	2	3	4	Supply
1	12	13	4	6	5
2	6	4	10	11	7
3	10	9	12	4	8
Demand	4	9	2	5	20

Table 28 Optimal solution to the problem in Table 25

	1↑ 3	2↑5	3↑1	4↑6	5↑7	6	Old/new/change
1 → 8	7	3	16	0	2	0 (2) (3)	2/3 /(−1)
2→ 4	6	0 (1) (1)	2	2	4	0 (2)(1)	3/2/(+1)
3	0 (2) (2)	1	14	2	0 (2) (3)	7	4/5/ (−1)
4	1	0 (3) (3)	2	0 (2) (3)	9	2	5/6/(−1)
5 → 2	0 (1) (2)	8	0 (5) (4)	7	5	0	6/6/(0)
Old/new	3/4/ (−1)	4/4/(0)	5/4/(+1)	2/3/ (−1)	2/3/(−1)	4/4/(0)	20 /22 (0)

The optimal solution was obtained earlier and is given by Table 26. We try and allocate the zero cost cells by the new values shown in red.

The solution to the old problem is shown in back ink. Now consider the situation as follows:

Remark 1 Previous supply at a_1 was 2 and now it is 3. Hence one more unit is ready to go from supply point 1. Similarly, the supply a_2 has gone down from 3 to 2. These changes can be accommodated at the cell (1,6) and adjusted from the cell (2,6) and the supply point a_2.

Remark 2 The supply from the a_3 has gone up from 4 to 5. The demand at the demand point b_5 has also gone up. Hence the extra supply from row 3 has been adjusted with the demand at column 5.

Remark 3 Finally, Column 1 demand is satisfied from the cell (5,1) and adjusted through the cell (5,3).

This completes the solution to the new problem. New solution is shown in red. It is optimal as it is confined in the zero cells obtained earlier by the Hungarian method of assignment.

Alternatively, one can just go the Table with modified cost matrix and new demand and supply values and fill the demand and supply in the zero cells as shown in Table 29.

Note that in Table 29, we have for every entry a crossed row, or the column exist as shown in Table 29. This is a degenerate optimal solution.

Table 29 Alternative approach to reach the optimal solution for changed demand and supply

	1↑ 8*	2↑7*	3	4↑6*	5↑4*	6↑ 2 *	new
1 → 1*	7	3	16	0	2	0 (3)	3
2 → 3*	6	0 (1)	2	2	4	0 (1)	2
3 → 5 *	0 (2)	1	14	2	0 (3)	7	5
4	1	0 (3)	2	0 (3)	9	2	6
5 → 9*	0 (2)	8	0 (4)	7	5	0	6
new	4	4	4	3	3	4	22 (0)

6 Some Concluding Remarks

6.1 The Hungarian method of assignment when applied to the transportation model has opened new possibilities for investigations, and some of them have been discussed in this chapter.

6.2 In this chapter, we have applied accelerated convergence idea to the transportation model and observed that the optimal solution was attained in a smaller number of iterations. More computational investigations are needed for assurance.

6.3 In the case of a transportation model, there is no unique way of covering the independent zero elements. It is important that at each iteration, the feasibility of the solution should improve. This is possible by changing the order of covering the rows and columns.

6.4 The Kth best solution to the transportation model is possible by creating an alternative solution, however, the problem of Kth best solution remains computationally demanding for larger values of K, for example, when $K \geq 4$. More work is needed for solving higher ordered optimization problems. Ordered solutions have been attempted by Chan et al. (1992), Murthy (1968), Kumar et al. (2020). The problem of ordered optimal solutions for an integer program was attempted by Kumar and Munapo (2012), Kumar et al. (2007, 2021). More work is needed in this direction.

6.5 Information recycling in the transportation model, when approached by the Hungarian method of assignment is possible, which otherwise is not possible when solution to the transportation model is obtained by any conventional approaches discussed in the OR books.

6.6 In this chapter, we have discussed protean aspect with respect to the changes in the demand and supply at various points in a transportation model, however, changes can also arise in cost elements and in the values of m and n. Further studies are required for more detailed protean transportation models, when changes arise in the cost matrix and values of m and n. This aspect of a protean transportation model will be discussed in a subsequent publication.

References

Chan S, McPhee J, Kumar S (1992) Determination of the kth best in an assignment problem. In: Puha et al (ed) Optimization techniques and applications. World Scientific Publishers, pp 1110–18

Eppen GD, Gould FJ, Schmidt CP (1963) Introduction to management science. Prentice Hall

Kirby M, Love HR, Swarup K (1972) Extreme point mathematical programming problem. Manage Sci 18(9):540–549

Kirby M, Scobey PF (1970) Production scheduling on ‘n’ identical machines. J Can Oper Res Soc 8:14–27

Kumar S (1995) Optimization of protean systems: a review. In: Fushimi M, Tone K (eds) APORS’94 Fakuoka, Japan. World Scientific Publishers, pp 139–146

Kumar S (2005) Information recycling mathematical methods for protean systems: a path-way approach. S Afr J Ind Eng 16(2):81–101
Kumar S (2006) Information recycling mathematical methods: a path-way approach to a geometric program. S Afr J Ind Eng 17(2):127–43. ISSN 1012-277X
Kumar S, Al-Hasani A, Al-Rabeeah M, Ebrhard A (2020) A random search method for finding k >= 2 number of ranked optimal solutions to an assignment problem. In: Journal of physics: conference series, vol 1490, pp 1–3. https://doi.org/10.1088/1742-6596/1490/1/012063
Kumar S, Wagner D (1979) Some algorithms for solving the extreme point mathematical programming problem. N Z Oper Res J 7(2):127–149
Kumar S, Munapo E (2012) Some lateral ideas and their applications for developing new solution procedure. In: Mariappan S, Amritraj (eds) Chapter in intelligent solutions through mathematics and statistics. Excell Indian Publishers, pp 13–21
Kumar S, Munapo E (2024b) Innovative ways of developing and using specific purpose alternatives for solving hard combinatorial network routing and ordered optimization problems. AllpiedMath 4:791–805. https://doi.org/10.3390/appliedmath4020042
Kumar S, Munapo E, Jones BC (2007) An integer equation controlled descending plane to a pure integer program. Indian J Math 49(2):211–237
Kumar S, Munapo E, Nyamugure P (2021) An insight into the characteristic's equation for an integer program. Int J Math Eng Manag Sci 6(2):611–620
Kumar S, Tawanda T, Munapo E, Nyamugure P (2024a) A note on solving the transportation model by the Hungarian method of assignment: unification of the transportation and assignment models. Lecture notes in Networks and systems. Intelligent computing and optimization. In: Vasant P et al (ed) ICO 2023, LNNS, vol 855, pp 301–313. https://doi.org/10.1007/978-3-031-50158-6_30
Munapo E (2020) Development of an accelerating Hungarian method for assignment problems. East-Eur J Enterp Technol. ISSN 1729-3774
Munapo E, Kumar S (2022) Linear integer programming: theory, applications, recent developments. De Gruyter series on the application of mathematics in engineering and information sciences. ISBN 978-3-11-070292-7
Munapo E, Nyamugure P, Lesaoana M, Kumar S (2012) A note on unified approach to solving transportation and assignment models in operations research. Int J Math Model Simul Appl 5:140–149
Murthy KG (1968) An algorithm for ranking all assignments in order of increasing costs. Oper Res 16:682–687
Sharma JK (2018) Operations research: theory and applications. Trinity Press
Taha HA (2017) Operations research: an introduction, Pearson educators, 10th ed
Winston WL (2004) Operations research applications and algorithms, 4th edn. Duxbury Press

Specified Number of Nodes Travelling Salesman Problem

Trust Tawanda, Santosh Kumar, Elias Munapo, and Philimon Nyamugure

Abstract This paper introduces the specified number of nodes travelling salesman problem (SNN-TSP), a variant of the travelling salesman problem. In the SNN-TSP number of nodes to be visited P are specified. The problem is to determine a minimum cost tour that visits specified number of nodes exactly once. This paper provides an integer programming model for the problem. TSP based algorithm is modified to solve the optimally for all values of P. Numerical illustrations have been used to prove that modified TSP bases algorithm can solve SNN-TSP optimally.

Keywords Specified number of nodes travelling salesman problem · Travelling salesman problem · Minimum cost tour · Algorithm

1 Introduction

The routing problem is a classical problem in graph theory, and it has attracted lot of attention from researchers, for example, see (Bellman 1958; Bellman and Dreyfus 1968; Dantzig 1960; Dijkstra 1959; Pollock and Wiebenson 1960; Munapo et al. 2008). Many variants of this problem have been proposed and new approaches were developed to find answers to many real-life applications, see Saksena and Kumar (1966), Arora and Kumar (1994), Kumar and Bappoo (1995), Kumar et al. (1999). For industrial applications of shortest routes in networks, see Hastings (1973) and Saksena (2007), Saksena (1970), Kumar and Bappoo (1999). New applications and challenges will continue to grow. Consequently, the shortest route solution procedures have remained as an active area for research. New variants of this problem

T. Tawanda (✉) · P. Nyamugure
Department of Statistics and Operations Research, Bulawayo, Zimbabwe
e-mail: trustawanda@gmail.com

S. Kumar
School of Mathematical and Geospatial Sciences, Melbourne, Australia

E. Munapo
School of Economics and Decision Sciences, Mafikeng, South Africa

M. Ram et al. (eds.), *Advances in Mathematics for Engineering Sciences*, Mathematical Engineering, https://doi.org/10.1007/978-3-031-95693-5_3

have been realized from time to time and solution procedures have been developed, for example, shortest path between two nodes, which is constrained to pass through K-specified nodes, see Saksena and Kumar (1966), Kumar et al. (2013, 2020). The travelling salesman problem is a generalization, when the set of specified node contains all nodes in the set. These problems have attracted lot of attention, for example, Dantzig et al. (1954), Gonzalez (1962), Kumar et al. (2020, 2022). These methods are refined every day. For example, as more information is available, more challenges are faced by the researchers. A good example is the modern navigation system, where navigation provides alternative routes when one is going from the place to the place B. The system provides routes, which are shortest, or fastest, or a route that avoids congestion etc. The race of generalization is a never-ending journey. Here in this paper, we propose a variants of the travelling salesman problem. Travelling salesman problem is one of the useful combinatorial optimizations that have several applications in areas such as engineering, military, health, agriculture, transport, among others. TSP can be defined as a routing problem that arises when there is a set of locations (airports, buildings, cities, warehouses) interconnected by links (roads, wires, rivers, rail roads) to form a network with link weights (distance, cost, probability).

The salesman is then required to start from a certain location (origin) and visit all the other nodes exactly once and return back to the origin while minimizing total circuit length. TSP is somehow related to shortest path problem in that the optimal tour is a shortest path that start and ends at the origin node, while SPP the path the path starts at the source node and ends at the destination node (Tawanda 2013; Maposa 2014). Several solution approaches have been proposed to solve the TSP thus the Labelling method (Tawanda et al. 2023a), Network reconstruction approach (Munapo 2020), Minimum spanning tree solution to TSP (Kumar et al. 2018) and Branch and Bound (Munapo 2013). Over the time TSP has branched to form variants due to technology changes. The K specified nodes TSP was solved in Srivastava et al. (1969) and Garg et al. (1970). Laporte and Martello (1990) proposed the Selective travelling salesman problem (STSP). The STSP is a variant of the TSP. The objective in this variant is to select a tour that gives the maximum profit while within specified bound. Laporte and Martello (1990) formulated and solved integer linear programming models for the STSP. Chisman (1975) proposed a variant of the TSP namely, the Clustered travelling salesman problem (CTSP). In CTSP cities are grouped to form clusters and cities in each group must be visited contiguously while giving the optimal solution. Ahmed (2013) solved the CTSP using the Lexi-search algorithm and the algorithms computed optimal solutions to the problem. Laporte and Palekar (2002) listed some applications of CTSP which include disk defragmentation, examination timetabling, computer program restructuring, integrated circuit testing, and cytology and embroidery. Moran-Mirabal (2014) introduced the Family travelling salesman problem (FTSP). FTSP can be applied to oder picking in warehouses (Moran-Mirabal, 2014). Bernado and Paias (2018) defined FTSP as follows, the vertices are grouped into clusters called families. The aim is to visit all predefined number of vertices from each family while using optimal circuit. Fischer et al (2014) proposed the Quadratic travelling salesman problem (QTSP) where the objective is a quadratic function. The problem was solved using Branch and Bound (BB) algorithm and Branch and Cut (BC) algorithm. BC algorithm proved to be the best algorithm on real world instances. QTSP can be applied to telecommunications, bioinformatics, and robotics (Rostami et al. 2016). Flying sidekick travelling salesman problem (FS-TSP) was introduced by Murray and Chu (2015). This variant of TSP arise when there is a set of customers that are supposed to be saved once either by driver operated vehicle or unmanned aerial vehicle (UAV) working in coordination with each other. UAV cannot serve the customers in situations such as parcel weight exceeds the capacity of UAV, no landing space for UAV and when signature is required after delivery. Several solution approaches have been developed for FS-TSP such as Mixed integer linear programming (MILP), Branch and Bound, and Iterative algorithm (See, Dell'Amico 2021). Price collecting travelling salesman problem (PCTSP) was proposed by Balas (1989). In this problem, the salesman gets a prize for every city visited and a penalty for every city that remains unvisited. The objective is to minimize travelling cost and penalties, while collecting the prize from as many as cities as possible. Applications of PCTSP include but not limited to helicopter tour planning for offshore oil platforms and tourist routes design (Climaco et al. 2021). Several TSP variants have been proposed over years which include but not limited to the Balanced travelling salesman problem (BTSP) in Larusic and Punnen (2011), the Maximum scatter travelling salesman problem (MSTSP) considered in Arkin et al. (1999) and the Equality generalized travelling salesman problem (E-GTSP) solved by Tawanda et al. (2022).

2 Problem Description

Given $G = (V \cup \{0\}, A)$ be a complete TSP network with vertices set given by $V = \{v_1, v_{2,\ldots,} v_n\}$ and 0 is the start and finish node. The cost of visiting j from i is given by $c_{ij} > 0 \forall i \neq j$ and $c_{ij} = \infty \forall i = j$. Specified number of vertices to be visited is given by K where $1 < P < V$. When $P = V$ thus, the normal travelling salesman problem and when $P = 1$ there is no problem to solve, the solution is obvious. The objective of the SNN-TSP is to start from the origin vertex 0 and visit P specified number of vertices exactly once and return to the origin node in such a way that the total circuit or tour cost is minimized.

2.1 Motivation

Several applications have motivated the proposal of the SNN-TSP variant of the TSP. (a) Delivery problem: suppose that there are V customers that require one unit of certain type to be delivered to them on daily basis. Due to circumstances beyond the manufacturer, the manufacturer has managed to produce P. The manufacturer must supply any P customers while minimizing total delivery cost. (b) Politician routing: suppose that the politician plans to visit P places for campaign purposes out of V places in the constituency, the problem of determining which places to visit while minimizing distance travelled can be modelled as SNN-TSP. (c) Order picking in warehouses: Same products are stored individually in separated and isolated spaces due to radio frequency identification technologies. In this set up these locations are treated as nodes V and the problem arises when a customer orders P items and the items should be picked for the customer in the shortest possible way.

3 SNN-TSP in Mathematical Form

Binary and non-binary variables are defined respectively

$$x_{ij} = \left\{ \begin{array}{l} 1,\ \textit{if link } (i,j) \textit{ part of the circuit} \\ 0,\ \textit{Otherwise} \end{array} \right\}$$

$$V = \text{Number of vertices in the problem}$$

$$\text{Minimize } C = \sum_{i \in V \cup \{0\}} \sum_{j \in V \cup \{0\}} c_{ij} x_{ij},\ c_{ij} = \infty\ \forall i = j \tag{1}$$

Subjected to the following constraints:

$$\sum_{i \in V} 0, i = 1 \tag{2}$$

$$\sum_{j \in V} j, 0 = 1 \tag{3}$$

$$\sum_{i \in V \cup \{0\}} x_{ij} \le 1 \quad \forall j \in V \tag{4}$$

$$\sum_{j \in V \cup \{0\}} x_{ij} \le 1 \quad \forall i \in V \tag{5}$$

$$\sum_{i \in V \cup \{0\}} \sum_{j \in V \cup \{0\}} x_{ij} = P + 1 \tag{6}$$

$$\sum_{i \in V \cup \{0\}} x_{ij} - \sum_{i \in V \cup \{0\}} x_{ji} \quad \forall j \in V \tag{7}$$

$$\text{Solution forms a roundtrip } P + 1 \text{vertices tour} \tag{8}$$

Source: Taha (2017)

$$x_{ij} = \{0, 1\} \quad \forall i, j \in V \tag{9}$$

The objective function (1) minimizes total travelled circuit distance while visiting the specified number of nodes P, constraints (2) and (3) ensure that the tour starts and ends at the origin node. Similarly, constraints (4) and (5) ensure that at most one arc enters and leaves each node. Constraint (6) establishes the total number of arcs in any feasible solution, constraint (7) guarantees flow conservation, while constraint (8) restricts sub-tours in the solution and finally, constraint (9) imposes binary nature restrictions on decision variables.

4 TANYAKUMU Labelling Method for SNN-TSP

The TANYAKUMU algorithm was applied successfully to compute exact solution to the Travelling salesman problem (TSP), the Equality generalized travelling salesman problem (E-GTSP) and the Shortest path problem (SPP) as demonstrated in (Tawanda et al. 2022, 2023a, 2023b). In this section, we modify the TANYAKUMU labelling method to solve SNN-TSP. Table 1 contains the notations used by the method and the modified algorithm for SNN-TSP is presented as Algorithm 1.

Table 1 Algorithm notations

Notation	Description
V	Vertices in the TSP
P	Specified vertices
$\mathbf{L_0}$	Visited vertices set
$\mathbf{L_0^*}$	Origin vertex set
$\mathbf{w_{(s,j)^z}}$	Label for node j showing partial circuit from origin s to j through z intermediate nodes and w is the partial circuit total cost
$\mathbf{w^*_{(s,k)^z}}$	Minimum weight label for the currently added vertex k
∞	Restriction node label

Algorithm 1: TANYAKUMU labelling method for SNN-TSP

Step 1.	Connect source node s to all other nodes j $w_{(s,j)^z}$ and $z = 0$.
Step 2.	Determine the minimum weight node label $w^*_{(s,k)^z}$ for all nodes j using the formula $w^*_{(s,j)^z} = \mathrm{Min}\{w_{(s,j)^z}\}$
Step 3.	If there are $z < 1$ intermediate nodes in the partial tours, Go to Step 4, Else go to Step 5.
Step 4.	Set $j \rightarrow k$ and label all nodes in the particular partial tour with node label (∞) and also label other nodes j using node label computed as follows $w_{(s,j)^{z+1}} = w^*_{(s,k)^z} + c(k, j)$ and go to Step 2
Step 5.	Set $j \rightarrow k$, $j = s$ and connect the minimum weight partial tours to s, the origin node.
Step 6.	$P = 1$, Set $P = P + 1$, and select the best tour as the SNN-TSP solution with P specified number of nodes.
Step 7.	While $z \leq V - 1$, go to Step 2. Else go to Step 8.
Step 8.	Stop all solutions for $1 < P < V$ have been found.

Flow chart for Algorithm 1 is given in Fig. 1. This flow chart gives a better picture of how the algorithms flows, how decisions are made and how the algorithm iterates as it computes all the required solutions.

4.1 *Illustration of the Algorithm*

We explain how the algorithm computes the optimal tours for all values of P using a 5×5 symmetric TSP and a 7×7 asymmetric TSP.

Example 1 Consider the following 5×5 symmetric travelling salesman problem presented in Table 2. Node 1 is the origin node and it is required to determine the circuits that minimizes total cost for all values of P such that $1 < P \leq 3$.

Solution

Initialization: To initialize the search, connect origin node 1 to all the nodes in the network to set the node labels $w_{(1,j)^0}$ and $w^*_{(1,j)^0} = w_{(1,j)^0}$. Set $j \rightarrow k$ and then consider Iteration 1.

Iteration 1 Modify node labels $w^*_{(1,k)^0}$ to get new node labels $w_{(1,j)^1}$ by connecting k to all j not part of the partial tour. The best partial tours to node j through $z = 1$ intermediate nodes is given by $w^*_{(1,j)^1}$ the minimum weight node label (Table 3).

Solution for P = 2 specified number of nodes tour is determined by connecting the minimum weight node label $w^*_{(1,j)^1}$ back to the origin node 1. The best tour becomes the optimal solution.

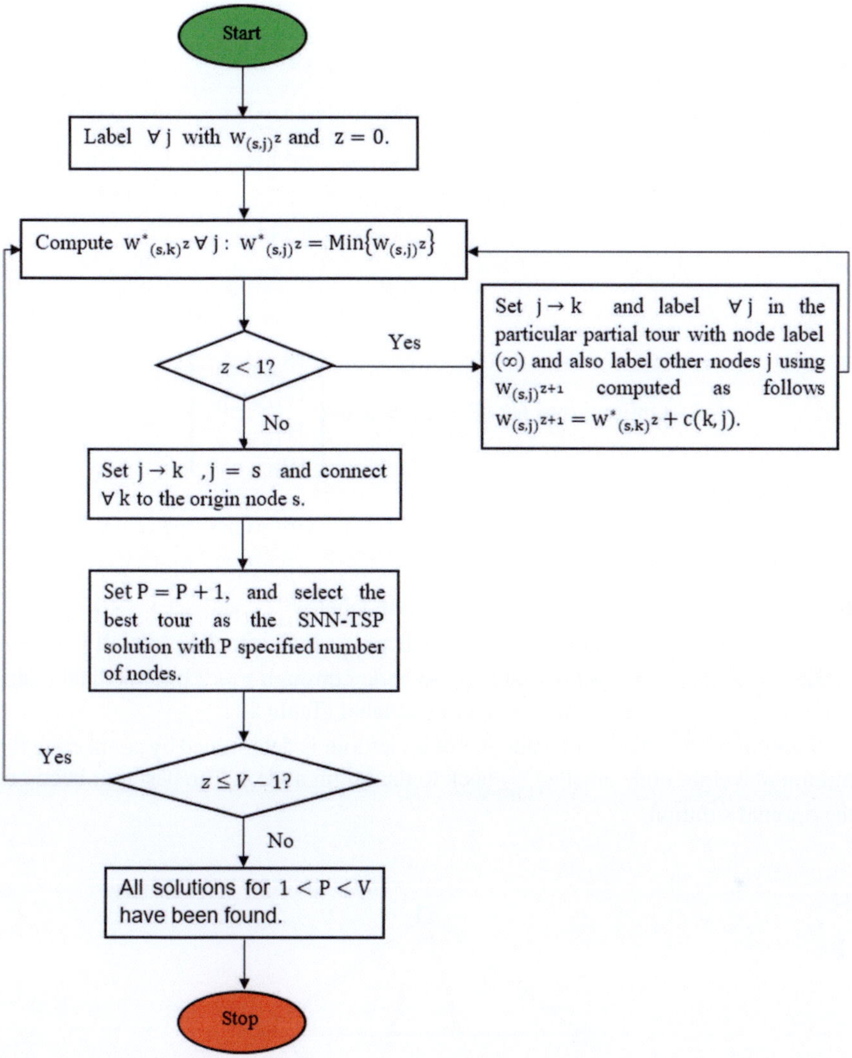

Fig. 1 Algorithm 1 flow chart

Table 2 Cost matrix

	1	2	3	4	5
1	INF	10	3	6	9
2	10	INF	5	4	2
3	3	5	INF	7	8
4	6	4	7	INF	4
5	9	2	8	4	INF

Table 3 Iteration 1 node labels

$w^*_{(1,k)^0}$	j = 2	j = 3	j = 4	j = 5
$12_{(1,2)^0}$	∞	$15_{(1,2,3)^1}$	$14_{(1,2,4)^1}$	$12_{(1,2,5)^1}$
$3_{(1,3)^0}$	$8_{(1,3,2)^1}$	∞	$10_{(1,3,4)^1}$	$11_{(1,3,5)^1}$
$6_{(1,4)^0}$	$10_{(1,4,2)^1}$	$13_{(1,4,3)^1}$	∞	$10_{(1,4,5)^1}$
$9_{(1,5)^0}$	$11_{(1,5,2)^1}$	$17_{(1,5,3)^1}$	$13_{(1,5,4)^1}$	∞
$w^*_{(1,j)^1}$	$8_{(1,3,2)^1}$	$13_{(1,4,3)^1}$	$10_{(1,3,4)^1}$	$10_{(1,4,5)^1}$

$$\text{Optimal tour for (P} = 2) = \text{Min}\left\{\begin{matrix} 18_{(1,3,2,1)^2} \\ 16_{(1,4,3,1)^2} \\ 16_{(1,3,4,1)^2} \\ 19_{(1,4,5,1)^2} \end{matrix}\right\} \tag{10}$$

From Eq. (10) the best tour is $\{1 - 4 - 3 - 1\}$ with tour value = 16. Figure 2 shows the solution on the network.

Iteration 2 Using the minimum weight node label $w^*_{(1,j)^1}$ and set j → k. Modify node labels $w^*_{(1,k)^1}$ to obtain new node labels $w_{(1,j)^2}$ by connecting k to all j not part of the partial tour. The best partial tours to node j through $z = 2$ intermediate nodes is given by $w^*_{(1,j)^2}$ the minimum weight node label (Table 4).

Solution for P = 3 specified number of nodes tour is determined by connecting the minimum weight node label $w^*_{(1,j)^2}$ back to the origin node 1. The best tour becomes the optimal solution.

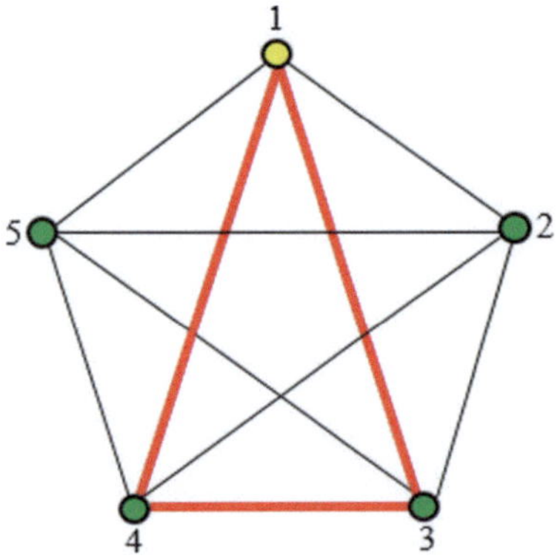

Fig. 2 Optimal tour for P = 2

Table 4 Iteration 2 node labels

$w^*_{(1,k)^1}$	j = 2	j = 3	j = 4	j = 5
$8_{(1,3,2)^1}$	∞	∞	$12_{(1,3,2,4)^2}$	$10_{(1,3,2,5)^2}$
$13_{(1,4,3)^1}$	$18_{(1,4,3,2)^2}$	∞	∞	$21_{(1,4,3,5)^2}$
$10_{(1,3,4)^1}$	$14_{(1,3,4,2)^2}$	∞	∞	$14_{(1,3,4,5)^2}$
$10_{(1,4,5)^1}$	$12_{(1,4,5,2)^2}$	$18_{(1,4,5,3)^2}$	∞	∞
$w^*_{(1,j)^2}$	$12_{(1,4,5,2)^2}$	$18_{(1,4,5,3)^2}$	$12_{(1,3,2,4)^2}$	$10_{(1,3,2,5)^2}$

$$\text{Optimal tour for } (P = 3) = \text{Min}\left\{\begin{matrix} 22_{(1,4,5,2,1)^3} \\ 21_{(1,4,5,3,1)^3} \\ 18_{(1,3,2,4,1)^3} \\ 19_{(1,3,2,5,1)^3} \end{matrix}\right\} \tag{11}$$

From (11) the best tour is {1 − 3 − 2 − 4 − 1} with tour value = 18. Figure 3 shows the solution on the network.

Example 2 Consider the following 7 × 7 asymmetric travelling salesman problem (Ahmed 2013) presented in Table 5. Node 1 is the origin node and it is required to determine the circuits that minimizes total cost for all values of P such that $1 < P \leq 5$.

Solution

Initialization: To initialize the search, connect origin node 1 to all the nodes in the network to set the node labels $w_{(1,j)^0}$ and $w^*_{(1,j)^0} = w_{(1,j)^0}$. Set j → k and go to Iteration 1.

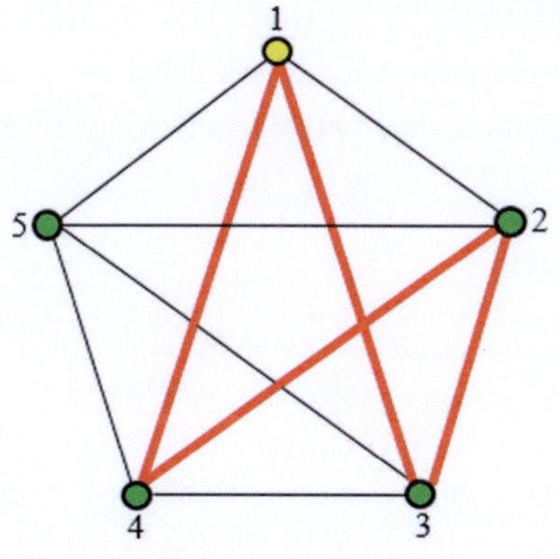

Fig. 3 Optimal tour for P = 3

Table 5 Cost matrix

	1	2	3	4	5	6	7
1	INF	75	99	9	35	63	8
2	51	INF	86	46	88	29	20
3	100	5	INF	16	28	35	28
4	20	45	11	INF	59	53	49
5	86	63	33	65	INF	76	72
6	36	53	89	31	21	INF	52
7	58	31	43	67	52	60	INF

Iteration 1 Modify node labels $w^*_{(1,k)^0}$ to get new node labels $\mathrm{w}_{(1,j)^1}$ by connecting k to all j not part of the partial tour. The best partial tours to node j through $z = 1$ intermediate nodes is given by $w^*_{(1,j)^1}$ the minimum weight node label (Table 6).

Solution for P = 2 specified number of nodes tour is determined by connecting the minimum weight node label $w^*_{(1,j)^1}$ back to the origin node 1 as shown below. The best tour becomes the optimal solution.

$$\text{Optimal tour for } (\mathrm{P} = 2) = \mathrm{Min}\left\{\begin{array}{l} 90_{(1,7,2,1)^2} \\ 120_{(1,4,3,1)^2} \\ 95_{(1,9,4,1)^2} \\ 146_{(1,9,5,1)^2} \\ 98_{(1,4,6,1)^2} \\ 116_{(1,4,7,1)^2} \end{array}\right\} \tag{12}$$

From (12) the best tour is $\{1 \to 7 \to 2 \to 1\}$ with tour value $= 90$. Figure 4 shows the solution on the network.

Iteration 2 Using the minimum weight node label $w^*_{(1,j)^1}$ and set j $\to$ k. Modify node labels $w^*_{(1,k)^1}$ to obtain updated node labels $\mathrm{w}_{(1,j)^2}$ by connecting k to all j not

Table 6 Iteration 1 node labels

$\mathbf{w}^*_{(1,\mathbf{k})^0}$	j = 2	j = 3	j = 4	j = 5	j = 6	j = 7
$75_{(1,2)^0}$	∞	$161_{(1,2,3)^1}$	$121_{(1,2,4)^1}$	$163_{(1,2,5)^1}$	$104_{(1,2,6)^1}$	$95_{(1,2,7)^1}$
$99_{(1,3)^0}$	$104_{(1,3,2)^1}$	∞	$115_{(1,3,4)^1}$	$127_{(1,3,5)^1}$	$134_{(1,3,6)^1}$	$127_{(1,3,7)^1}$
$9_{(1,4)^0}$	$54_{(1,4,2)^1}$	$20_{(1,4,3)^1}$	∞	$68_{(1,4,5)^1}$	$62_{(1,4,6)^1}$	$58_{(1,4,7)^1}$
$35_{(1,5)^0}$	$98_{(1,5,2)^1}$	$68_{(1,5,3)^1}$	$100_{(1,5,4)^1}$	∞	$111_{(1,5,6)^1}$	$107_{(1,5,7)^1}$
$63_{(1,6)^0}$	$116_{(1,6,2)^1}$	$152_{(1,6,3)^1}$	$94_{(1,6,4)^1}$	$84_{(1,6,5)^1}$	∞	$115_{(1,6,7)^1}$
$8_{(1,7)^0}$	$39_{(1,7,2)^1}$	$51_{(1,7,3)^1}$	$75_{(1,7,4)^1}$	$60_{(1,7,5)^1}$	$68_{(1,7,6)^1}$	∞
$\mathbf{w}^*_{(1,\mathbf{j})^1}$	$39_{(1,7,2)^1}$	$20_{(1,4,3)^1}$	$75_{(1,7,4)^1}$	$60_{(1,7,5)^1}$	$62_{(1,4,6)^1}$	$58_{(1,4,7)^1}$

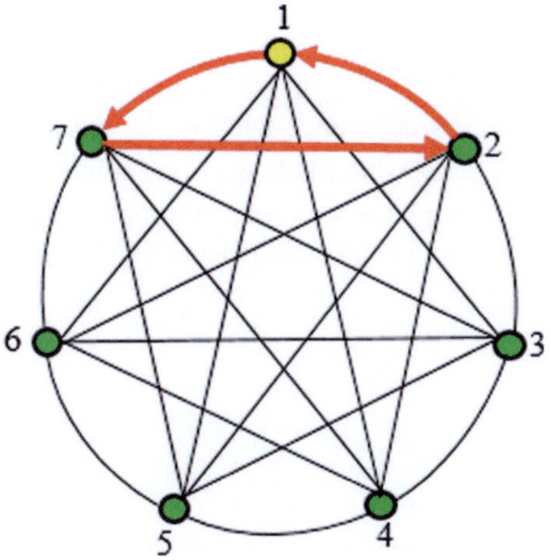

Fig. 4 Optimal tour for P = 2

part of the partial tour. The best partial tours to node j through $z = 2$ intermediate nodes is given by $w^*_{(1,j)^2}$ the minimum weight node label (Table 7).

Solution for P = 3 specified number of nodes tour is determined by connecting the minimum weight node label $w^*_{(1,j)^2}$ back to the origin node 1. The best tour becomes the optimal solution.

$$\text{Optimal tour for } (P = 3) = \text{Min}\left\{\begin{array}{c} 76_{(1,4,3,2,1)^3} \\ 186_{(1,7,4,3,1)^3} \\ 105_{(1,7,2,4,1)^3} \\ 136_{(1,4,3,5,1)^3} \\ 91_{(1,4,3,6,1)^3} \\ 106_{(1,4,3,7,1)^3} \end{array}\right\} \quad (13)$$

Table 7 Iteration 2 node labels

$\mathbf{w}^*_{(1,\mathbf{k})^1}$	**j** = 2	**j** = 3	**j** = 4	**j** = 5	**j** = 6	**j** = 7
$39_{(1,7,2)^1}$	∞	$125_{(1,7,2,3)^2}$	$85_{(1,7,2,4)^2}$	$127_{(1,7,2,5)^2}$	$68_{(1,7,2,6)^2}$	∞
$20_{(1,4,3)^1}$	$25_{(1,4,3,2)^2}$	∞	∞	$48_{(1,4,3,5)^2}$	$55_{(1,4,3,6)^2}$	$48_{(1,4,3,7)^2}$
$75_{(1,7,4)^1}$	$120_{(1,7,4,2)^2}$	$86_{(1,7,4,3)^2}$	∞	$134_{(1,7,4,5)^2}$	$128_{(1,7,4,6)^2}$	∞
$60_{(1,7,5)^1}$	$123_{(1,7,5,2)^2}$	$93_{(1,7,5,3)^2}$	$125_{(1,7,5,4)^2}$	∞	$136_{(1,7,5,6)^2}$	∞
$62_{(1,4,6)^1}$	$115_{(1,4,6,2)^2}$	$151_{(1,4,6,3)^2}$	∞	$83_{(1,4,6,5)^2}$	∞	$114_{(1,4,6,7)^2}$
$58_{(1,4,7)^1}$	$89_{(1,4,7,2)^2}$	$101_{(1,4,7,3)^2}$	∞	$110_{(1,4,7,5)^2}$	$118_{(1,4,7,6)^2}$	∞
$\mathbf{w}^*_{(1,\mathbf{j})^2}$	$25_{(1,4,3,2)^2}$	$86_{(1,7,4,3)^2}$	$85_{(1,7,2,4)^2}$	$48_{(1,4,3,5)^2}$	$55_{(1,4,3,6)^2}$	$48_{(1,4,3,7)^2}$

From (13) the best tour is $\{1 \to 4 \to 3 \to 2 \to 1\}$ with tour value = 76. Figure 5 shows the solution on the network.

Iteration 3 Using the least weight node label $w^*_{(1,j)^2}$ and set $j \to k$. Modify node labels $w^*_{(1,k)^2}$ to obtain updated node labels $w_{(1,j)^3}$ by connecting k to all j not part of the partial tour. The best partial tours to node j through $z = 3$ intermediate nodes is given by $w^*_{(1,j)^2}$ the minimum weight node label (Table 8).

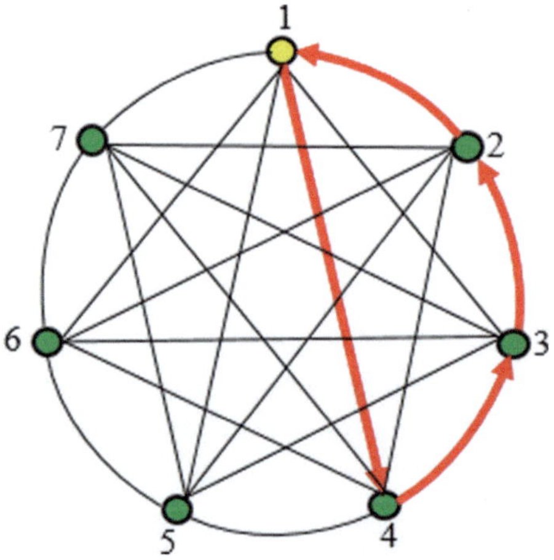

Fig. 5 Optimal tour for P = 3

Table 8 Iteration 3 node labels

$\mathbf{w}^*_{(1,\mathbf{k})^2}$	j = 2	j = 3	j = 4	j = 5	j = 6	j = 7
$25_{(1,4,3,2)^2}$	∞	∞	∞	$113_{(1,4,3,2,5)^3}$	$54_{(1,4,3,2,6)^3}$	$45_{(1,4,3,2,7)^3}$
$86_{(1,7,4,3)^2}$	$91_{(1,7,4,3,2)^3}$	∞	∞	$114_{(1,7,4,3,5)^3}$	$121_{(1,7,4,3,6)^3}$	∞
$85_{(1,7,2,4)^2}$	∞	$96_{(1,7,2,4,3)^3}$	∞	$144_{(1,7,2,4,5)^3}$	$138_{(1,7,2,4,6)^3}$	∞
$48_{(1,4,3,5)^2}$	$111_{(1,4,3,5,2)^3}$	∞	∞	∞	$124_{(1,4,3,5,6)^3}$	$120_{(1,4,3,5,7)^3}$
$55_{(1,4,3,6)^2}$	$108_{(1,4,3,6,2)^3}$	∞	∞	$76_{(1,4,3,6,5)^3}$	∞	$107_{(1,4,3,6,7)^3}$
$48_{(1,4,3,7)^2}$	$79_{(1,4,3,7,2)^3}$	∞	∞	$100_{(1,4,3,7,5)^3}$	$108_{(1,4,3,7,6)^3}$	∞
$\mathbf{w}^*_{(1,\mathbf{j})^3}$	$79_{(1,4,3,7,2)^3}$	$96_{(1,7,2,4,3)^3}$		$76_{(1,4,3,6,5)^3}$	$54_{(1,4,3,2,6)^3}$	$45_{(1,4,3,2,7)^3}$

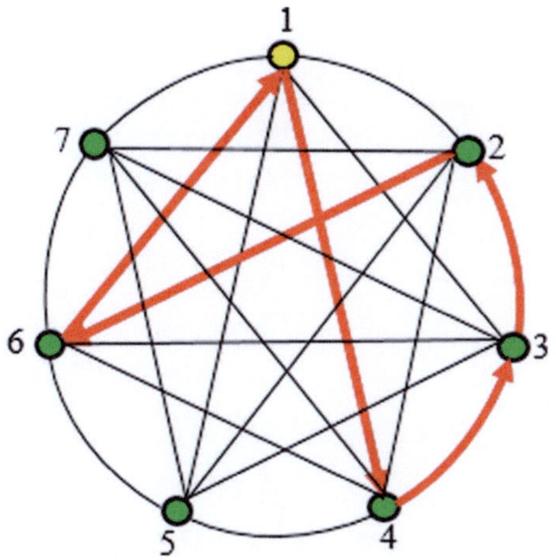

Fig. 6 Optimal tour for P = 4

Solution for P = 4 specified number of nodes tour is determined by connecting the minimum weight node label $w^*_{(1,j)^3}$ back to the origin node 1. The best tour becomes the optimal solution.

$$\text{Optimal tour for (P = 4)} = \text{Min}\begin{Bmatrix} 130_{(1,4,3,7,2,1)^4} \\ 196_{(1,7,2,4,3,1)^4} \\ 162_{(1,4,3,6,5,1)^4} \\ 90_{(1,4,3,2,6,1)^4} \\ 103_{(1,4,3,2,7,1)^4} \end{Bmatrix} \tag{14}$$

From (14) the best tour is {1 → 4 → 3 → 2 → 6 → 1} with tour value = 90. Figure 6 shows the solution on the network.

Iteration 4 Using the minimum weight node label $w^*_{(1,j)^3}$ and set j → k. Modify node labels $w^*_{(1,k)^3}$ to obtain updated node labels $w_{(1,j)^4}$ by connecting k to all j not part of the partial tour. The best partial tours to node j through $z = 4$ intermediate nodes is given by $w^*_{(1,j)^4}$ the minimum weight node label (Table 9).

Solution for P = 5 specified number of nodes tour is determined by connecting the minimum weight node label $w^*_{(1,j)^4}$ back to the origin node 1. The best tour becomes the optimal solution.

$$\text{Optimal tour for (P = 5)} = \text{Min}\begin{Bmatrix} 190_{(1,4,3,6,5,2,1)^5} \\ 161_{(1,4,3,2,6,5,1)^5} \\ 141_{(1,4,3,2,7,6,1)^5} \\ 164_{(1,4,3,2,6,7,1)^5} \end{Bmatrix} \tag{15}$$

Table 9 Iteration 4 node labels

$w^*_{(1,k)^3}$	j = 2	j = 3	j = 4	j = 5	j = 6	j = 7
$79_{(1,4,3,7,2)^3}$	∞	∞	∞	$167_{(1,4,3,7,2,5)^4}$	$108_{(1,4,3,7,2,6)^4}$	∞
$96_{(1,7,2,4,3)^3}$	∞	∞	∞	$124_{(1,7,2,4,3,5)^4}$	$131_{(1,7,2,4,3,6)^3}$	∞
$76_{(1,4,3,6,5)^3}$	$139_{(1,4,3,6,5,2)^4}$	∞	∞	∞	∞	$148_{(1,4,3,6,5,7)^4}$
$54_{(1,4,3,2,6)^3}$	∞	∞	∞	$75_{(1,4,3,2,6,5)^4}$	∞	$106_{(1,4,3,2,6,7)^4}$
$45_{(1,4,3,2,7)^3}$	∞	∞	∞	$97_{(1,4,3,2,7,5)^4}$	$105_{(1,4,3,2,7,6)^4}$	∞
$w^*_{(1,j)^4}$	$139_{(1,4,3,6,5,2)^4}$	-	-	$75_{(1,4,3,2,6,5)^4}$	$105_{(1,4,3,2,7,6)^4}$	$106_{(1,4,3,2,6,7)^4}$

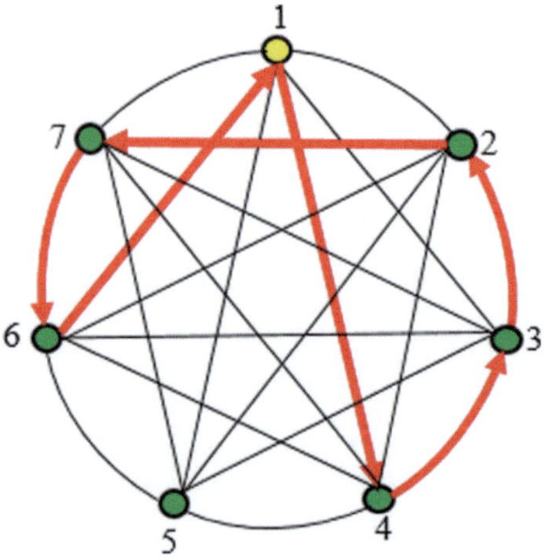

Fig. 7 Optimal Tour for P = 5

From (15) the best tour is $\{1 \rightarrow 4 \rightarrow 3 \rightarrow 2 \rightarrow 7 \rightarrow 6 \rightarrow 1\}$ with tour value = 141. Figure 7 shows the solution on the network.

5 Worst-Case Time Complexity

The algorithm requires $O(V)$ time to compute initial labels thus, connected origin node to V vertices. Modifying node labels requires $O(V^2)$ time and $O(V)$ time is required to determine minimum weight partial tours. Another $O(V)$ is requires at each iteration to connect partial tours to the origin. The algorithm takes $V - 1$ iterations to compute all solutions for P. As a result the overall time complexity is given by $O(V^3)$.

6 Conclusions

In this chapter, we proposed a variant of the Travelling salesman problem (TSP) namely the Specified number of nodes travelling salesman problem (SNN-TSP). This variant can be applied to order picking in warehouses, delivery routing, and politician routing. We presented integer programming model formulation for the problem. TANYAKUMU labelling method for TSP was modified to compute optimal solutions for the SNN-TSP. Two small problems one symmetric and the other asymmetric were used to demonstrate how the proposed algorithm computes solutions for all values of P specified number of nodes. Computational complexity reduces as the number of iterations increase due to a decrease in total number of operations as iterations increase. As a result the proposed algorithm has overall worst case time complexity of $O(V^3)$ where V is the number of vertices. There is need to develop a code or software for the algorithm so that large problem instances can be solved. Further research directions include combining SNN-TSP with other variants of TSP to form hybrid variants. For instance SNN-TSP can be combined with FSTP to form SNN-FTSP, combined with PCTSP to form SNN-PCTSP, combined with FS-TSP to form SNN-FS-TSP and lastly can be combined with MSTSP to form SNN-MSTSP. This hybrid variants can address several problems that are emerging due to the fact that technology changes every time and bring about complex problems that can be modelled as hybrid TSP variants.

Declaration of competing interest The authors declare no conflict of interest.

Institutional Review Board Statement Not applicable.

Informed Consent Statement Not applicable.

Data Availability Statement The data supporting this research are contained in this paper.

Funding No funding received.

References

Ahmed ZH (2013) An exact algorithm for the clustered travelling salesman problem. Opsearch 50:215–228

Arkin EM, Chiang YJ, Mitchell JS, Skiena SS, Yang TC (1999) On the maximum scatter traveling salesperson problem. SIAM J Comput 29(2):515–544

Arora H, Kumar S (1994) (K+1)th shortest path from the Kth shortest path in a network. Math Today 12A:75–84

Balas E (1989) The prize collecting traveling salesman problem. Networks 19(6):621–636

Bellman R (1958) On a routing problem. Quart Appl Maths 16

Bellman R, Dreyfus S (1968) Applied dynamic programming. Princeton University Press

Bernardino R, Paias A (2018) Solving the family traveling salesman problem. Eur J Oper Res 267(2):453–466
Chisman JA (1975) The clustered traveling salesman problem. Comput Oper Res 2(2):115–119
Clímaco G, Simonetti L, Rosseti I (2021) A branch-and-cut and MIP-based heuristics for the prize-collecting travelling salesman problem. RAIRO Oper Res 55:S719–S726
Dantzig GB (1960) On the shortest route through a network. Manage Sci 6(2):187–190
Dantzig G, Fulkerson R, Johnson S (1954) Solution of a large-scale traveling-salesman problem. J Oper Res Soc Am 2(4):393–410
Dell'Amico M, Montemanni R, Novellani S (2021) Algorithms based on branch and bound for the flying sidekick traveling salesman problem. Omega 104:102493
Dijkstra EW (1959) A note on two problems in connection with graph theory. Numer Math 1:269–271
Fischer A, Fischer F, Jäger G, Keilwagen J, Molitor P, Grosse I (2014) Exact algorithms and heuristics for the quadratic traveling salesman problem with an application in bioinformatics. Discret Appl Math 166:97–114
Garg RC, Kumar S, Dass P, Sen P (1970) Generalized travelling salesman problem through N sets of nodes in a competitive market. AKOR Ahlanf und Planningforschung 2(2):116–120
Gonzalez RH (1962) Solution of the travelling salesman problem by dynamic programming on the Hypercube. Interim Technical Report No. 18, MIT
Hastings NAJ (1973) Dynamic programming with management applications. Butterworh Publications
Kumar S, Bapoo R (1995) Shortest routes in protean networks. In: Operations research and its applications. World Publishing Corporation, Beijing, pp 228–237
Kumar S, Bappoo R (1999) Deployment of refuse vehicles in a protean environment. Opsearch 36:242–259
Kumar S, Arora H, Chiera C, Raviganesh G (1999) Paths for self-healing protean networks. Bull Allahabad Math Soc 14:79–94
Kumar S, Munapo E, Lesaoana M, Nyamugure (2018) A minimum spanning tree based heuristic for the travelling salesman tour. Opsearch 55:150–164
Kumar S, Munapo E, Lesaoana M, Nyamugure P, Dikgale P (2020) Link-weight modification for network optimization: is it a tool, philosophy, or an art? In: Recent advances in mathematics for engineering, pp 221–235. CRC Press
Kumar S, Munapo E, Ncube O, Sigauke C, Nyamugure P (2013) A minimum weight labelling method for determination of a shortest route in a non-directed network. Int J Syst Assur Eng Manag 4:13–18
Kumar S, Munapo E, Nyamugure P, Tawanda T (2022) Mathematics of OR: significance and application of virtual directions in reducing computational complexity in network optimization. In: Chauhan IS (ed) Emerging trends in applied research, vol 3, Ch 3, pp 33–47. Integrated Publications, New Delhi
Laporte G, Martello S (1990) The selective travelling salesman problem. Discret Appl Math 26(2–3):193–207
Laporte G, Palekar U (2002) Some applications of the clustered travelling salesman problem. J Oper Res Soc 53(9):972–976
Larusic J, Punnen AP (2011) The balanced traveling salesmanproblem. Comput Oper Res 38(5):868–875
Maposa D, Mupondo NC, Tawanda T (2014) Non-iterative algorithm for finding shortest route. Int J Logist Econ Glob 6(1):56–77
Morán-Mirabal LF, González-Velarde JL, Resende MG (2014) Randomized heuristics for the family traveling salesperson problem. Int Trans Oper Res 21(1):41–57
Munapo E (2013) A network branch and bound approach for the traveling salesman model. S Afr J Econ Manag Sci 16(1):52–63
Munapo E (2020) Network reconstruction–a new approach to the traveling salesman problem and complexity. In: Intelligent computing and optimization: proceedings of the 2nd international

conference on intelligent computing and optimization 2019 (ICO 2019), pp 260–272. Springer International Publishing

Munapo E, Jones BC, Kumar S (2008) A minimum incoming weight label method and its application in CPM networks. ORiON 24(1):37–48

Murray CC, Chu AG (2015) The flying sidekick traveling salesman problem: optimization of drone-assisted parcel delivery. Transp Res Part C Emerg Technol 54:86–109

Pollack M, Wiebenson W (1960) Solutions of the shortest-route problem—a review. Oper Res 8(2):224–230

Rostami B, Malucelli F, Belotti P, Gualandi S (2016) Lower bounding procedure for the asymmetric quadratic traveling salesman problem. Eur J Oper Res 253(3):584–592

Saskena JP (1970) Mathematical model of scheduling clients through welfare agencies. J Can Oper Res Soc 8:185–200

Saksena JP (2007) Dynamic programming applications. Siddharth Publications, Delhi

Saksena JP, Kumar S (1966) Path through K specified nodes, Opns Res 14(5)

Srivastava SS, Kumar S, Garg RC, Sen P (1969) Generalized traveling salesman problem through n sets of nodes. CORS J 7(2):97

Taha AH (2017) Operations research: an introduction; Pearson Prentice Hall: Upper Saddle River. NJ, USA

Tawanda T (2013) Tawanda's non-iterative optimal tree algorithm for shortest route problems. Sci J Pure Appl Sci 2(2):87–94

Tawanda T, Munapo E, Kumar S, Nyamugure P (2023a) Extended TANYAKUMU labelling method to compute shortest paths in directed networks. Int J Math Eng Manag Sci 8(5):991

Tawanda T, Nyamugure P, Kumar S, Munapo E (2022) Modified TANYAKUMU labelling method to solve equality generalized travelling salesman problem. In: International conference on intelligent computing & optimization, pp 936–947. Cham: Springer International Publishing

Tawanda T, Nyamugure P, Kumar S, Munapo E (2023b) A labelling method for the travelling salesman problem. Appl Sci 13(11):6417

Mathematical Analysis of Real Gas Effects on the Accuracy of Microfluidic Gas Pressure Profiles

Vishal Ramnath

Abstract Fluid mechanics simulations modelled in terms of the Navier–Stokes partial differential equations require equations of states for mathematical completeness. In general, equations of states for ideal fluids include the incompressibility assumption for liquids where the density is constant, the polytropic assumption where the pressure depends on a known power of the density, and the ideal gas assumption for gases. The ideal gas equation is frequently combined with the assumptions of a specific heat and thermal conductivity that are independent of the pressure and only depend on the temperature. Real gases on the other hand, exhibit non-ideal behaviour with complex algebraic mathematical models for microfluidic flows where different physics of flow phenomena are present. In order to address current equations of states limitations, microfluidic behaviour in this study is modelled with a new mathematical non-linear model that incorporates a coupled pressure-density-viscosity equation of state. A numerical method is then developed for solving a nonlinear Reynolds integral equation that incorporates real gas effects, and is used to perform a mathematical analysis in order to determine real gas accuracy effects. Results from this study demonstrate that ideal gas assumptions can have a detrimental effect on accuracies for microfluidic simulations.

1 Introduction

Fluid mechanics may be studied from an experimental, computational or theoretical perspective as a branch of applied science. Theoretical studies in fluid mechanics are usually approached either from a discrete particle basis using statistical mechanics as a branch of computational physics using a mixture of mathematical and statistical techniques to solve approximations of the integro-differential Boltzmann equation as reported by Sharipov (2016), or from a continuum basis modelled in terms of

V. Ramnath (✉)
Department of Mechanical, Bioresources and Biomedical Engineering, University of South Africa, Gauteng, Republic of South Africa
e-mail: ramnav@unisa.ac.za

M. Ram et al. (eds.), *Advances in Mathematics for Engineering Sciences*, Mathematical Engineering, https://doi.org/10.1007/978-3-031-95693-5_4

the Navier–Stokes partial differential equations (PDEs) for incompressible fluids as reported by Lions (1996) and subsequently for compressible fluids as reported by Lions (1998). The system of Navier–Stokes PDEs consist of the three conservation equations for mass, momentum and energy. The three unknowns in this coupled system of PDEs to solve for are the fluid's mass density $\rho(\mathbf{x})$, hydrostatic pressure $p(\mathbf{x})$, and velocity vector $\mathbf{v}(\mathbf{x})$ for all points $\mathbf{x} \in \Omega \subset \mathbb{R}^3$ for a domain Ω and boundary $\Gamma = \partial\Omega$, from which all fluid behaviour may be uniquely characterized. Although the existence, smoothness, and uniqueness of solutions for the Navier–Stokes PDEs for the special case of incompressible and homogeneous fluids is still an open mathematical problem that is yet to yield to a formal proof as summarized by Fefferman (2025), real engineering fluids problems frequently incorporate compressible fluids where the viscosity may also exhibit complex behaviour. As a result, in studies of mathematical fluid mechanics problems, either in terms of the classical Navier–Stokes equations for macroscopic length scales at smaller Knudsen numbers or in extensions such as the generalized Burnett hydrodynamics equations as reported by Bobylev and Meleshko (2020) that extend the validity for microscopic length scales at larger Knudsen numbers, it is advantageous to understand the potential influences of real fluid properties effects. These real fluid effects which depend on the invariant primitive variables of the pressure p and temperature T, may exhibit variations which may not be immediately apparent from ideal fluid properties assumptions of the density ρ, viscosity μ, and enthalpy h constituent equations of states models that are typically applied in many engineering studies for simplicity and convenience with standardized Computational Fluid Dynamics (CFD) solvers.

Typical equation of states for ideal fluids include the incompressibility assumption for liquids where the density is constant such that $\frac{\partial \rho}{\partial p} = 0$ and $\frac{\partial \rho}{\partial T} = 0$, the polytropic assumption $p = K\rho^{\left(1+\frac{1}{n}\right)}$ where the pressure depends on a known power of the density where K is a constant and n is the polytropic index, and the ideal gas assumption $pV = n_m RT$ where V is the volume of the fluid, n_m is the number of moles, and R is the universal gas constant for gases. The ideal gas equation is frequently combined with the assumptions of a specific heat c_p and thermal conductivity k that are independent of the pressure and only depend on the temperature. Real gases on the other hand, exhibit non-ideal behaviour with complex algebraic mathematical models for microfluidic flows where different physics of flow phenomena are present. In order to address current equations of states limitations, microfluidic behaviour in this study is modelled with a new mathematical non-linear model that incorporates a coupled pressure-density-viscosity equation of state.

In this study a numerical method is then developed for solving a non-linear Reynolds integral equation that incorporates real gas effects, and is used to perform a mathematical analysis in order to determine real gas accuracy effects. Results from this study demonstrate that ideal gas assumptions can have a detrimental effect on accuracies for microfluidic simulations particularly for elevated ranges in high accuracy precision engineering instrumentation.

1.1 Fluids Theoretical Overview

In engineering the governing macroscopic Navier–Stokes equations may be considered from the vector forms as discussed by White (1991). The underlying mathematical basis of the derivation of the Navier–Stokes partial differential equations which exhibit a Lagrangian nature is usually formulated in terms of the conservation laws of mass, momentum and energy, either by considering a balance of fluxes approach in an elemental volume $\mathcal{V}$, or through the application of the Reynolds transport theorem. A balance of fluxes derivation approach involves accounting for the conservation of momentum and energy within an elemental volume, as the mass conservation equation is straightforward to derive with the aid of the particle derivative operator. On the other hand, the derivation of the Navier–Stokes equations through the application of the Reynolds transport theorem approach, which may be considered as a 3-dimensional generalization of the Leibniz integral rule, and which in the particular case of continuum mechanics takes the special form $\frac{\mathrm{d}}{\mathrm{d}t}\left(\int_{\Omega(t)} \mathbf{f}\,\mathrm{d}\mathcal{V}\right) = \int_{\Omega(t)} \frac{\partial \mathbf{f}}{\partial t}\,\mathrm{d}\mathcal{V} + \int_{\partial\Omega(t)} (\mathbf{v}\cdot\mathbf{n})\,\mathbf{f}\,\mathrm{d}\mathcal{A}$, where $\mathbf{f}$ is some quantity within the control volume $\Omega(t)$ that may vary with time t, $\mathbf{v}$ is the velocity, and $\mathbf{n}$ is the outwards normal vector, involves the sequential application of the Reynolds transport theorem to scalar/vector terms that are representative of the mass, momentum and energy. When this formulation is utilized a generic conservation equation of the form $\frac{\partial \varphi}{\partial t} + \nabla\cdot(\varphi\mathbf{v}) + s = 0$ may first be derived where s represents sources/sinks into the control volume. This can then be sequentially applied to mass, momentum, and energy terms, to ultimately derive the Navier–Stokes equations representing the conservation of mass, momentum and energy. The final mathematical form is then

$$\frac{D\rho}{Dt} + \rho\,\mathrm{div}\,\mathbf{v} = 0 \tag{1}$$

$$\rho\frac{D\mathbf{v}}{Dt} = \rho\mathbf{g} - \nabla p + \frac{\partial}{\partial x_j}\left[\mu\left(\frac{\partial v_i}{\partial x_j} + \frac{\partial v_j}{\partial x_i}\right) + \delta_{ij}\lambda\mathrm{div}\,\mathbf{v}\right] \tag{2}$$

$$\rho\frac{Dh}{Dt} = \frac{Dp}{Dt} + \mathrm{div}\,(k\nabla T) + \Phi \tag{3}$$

In the above system of equations, $\frac{D}{Dt} = \frac{\partial}{\partial t} + (\mathbf{v}\cdot\nabla)$ represents the particle or Lagrangian derivative operator, ρ the fluid mass density, $\mathbf{g}$ the gravitational acceleration to account for body forces, p the fluid pressure, $\mathbf{v} = [u, v, w]^{\mathsf{T}} \in \mathbb{R}^3$ the fluid velocity, δ_{ij} the Kronecker delta-function, μ the dynamic viscosity or first viscosity coefficient, λ the second viscosity coefficient, k the fluid thermal conductivity, T the fluid temperature, and Φ the dissipation function.

An equivalent form of the momentum conservation equation may be written as $\frac{\partial}{\partial t}(\rho\mathbf{v}) + \nabla\cdot(\rho\mathbf{v}\mathbf{v}) = \rho\mathbf{f} + \nabla\cdot\Pi_{ij}$, and that of the energy conservation equation which may be written as $\frac{\partial E_t}{\partial t} + \nabla\cdot(E_t\mathbf{v}) = \frac{\partial Q}{\partial t} - \nabla\cdot\mathbf{q} + \rho\mathbf{f}\cdot\mathbf{v} + \nabla\cdot\left(\Pi_{ij}\cdot\mathbf{v}\right)$, where Π_{ij} is a fluid stress tensor, from which Tannehill et al. (1997) remarked that the Navier–Stokes equations in these particular forms are sufficiently general and equally valid for both continuum as well as non-continuum flows, and that differences only arise when particular choices for approximations of the shear stress tensor Π_{ij} are chosen.

The distinction between continuum flows and non-continuum flows is formally delineated through the use of the Knudsen number Kn following the approach by Sharipov (2016) where the Knudsen number is defined as $Kn = \ell/a$ where ℓ is a Mean Free Path (MFP) and a is a characteristic length. Usually the reciprocal of the Knudsen number is specified as $\delta = \frac{a}{\ell}$, where $\ell = \frac{\mu v_m}{p}$ and $v_m = \sqrt{\frac{2k_B T}{m}}$ is a most probable molecular speed, k_B is the Boltzmann constant, and m is the mass of the molecule of the fluid, such that free molecular flows are defined as the limit $\delta \to 0$ and continuum flows are defined as the limit $\delta \to \infty$. Under these definitions the corresponding limits for flow regimes may be specified following Dongari et al. (2009) as continuum flows for $0.0001 \leqslant Kn \leqslant 0.001$, slip flows for $0.001 \leqslant Kn \leqslant 0.1$, transition flows for $0.1 \leqslant Kn \leqslant 10$, and free molecular flows for $10 \leqslant Kn$. Corresponding mathematical models for these flow regimes which exhibit unique mathematical characteristics are the traditional Navier–Stokes equations for continuum flows with standard Dirichlet or Newmann boundary conditions, the extended Navier–Stokes equations also known as the ENS equations for slip flows reported by Guo et al. (2007), higher order and/or non-linear boundary conditions for transitional flows using the extended Navier–Stokes equations as reported by Akhlaghi et al. (2023), and the Direct Simulation Monte Carlo (DSMC) method for molecular flows as reported by Bird (1994).

Restricting the analysis in this study to continuum flows with a Newtonian fluid model so that the fluid exhibits isotropic behaviour for simplicity, the coupled mass, momentum and energy partial differential equations arising from the Navier–Stokes equations may be reduced to solving for the three unknowns of the fluid velocity $\mathbf{v}$, the thermodynamic pressure p, and the absolute temperature T. In order to ensure mathematical closure four auxiliary thermodynamic relations for the density, enthalpy, viscosity and thermal conductivity for ideal or real fluids must be specified as

$$\rho = \rho(p, T) \tag{4}$$

$$h = h(p, T) \tag{5}$$

$$\mu = \mu(p, T) \tag{6}$$

$$k = k(p, T) \tag{7}$$

In many CFD studies the above auxiliary thermodynamic relations are usually utilized with ideal equation of state properties in both introductory texts such as that of Versteeg and Malalasekera (1995) as well as in advanced texts such as that of Blazek (2001). Typical approximations for the equation of state include the dilute gas approximation $p = \rho R_g T$, the constant specific heat approximation $i = C_V T$ for the internal energy, and the absence of pressure effects on both the thermal conductivity as well as the viscosity. Unfortunately, real gases exhibit non-ideal equations of states which cannot be disregarded and which detrimentally effects the accuracy of any underlying mathematical analysis in precision engineering equipment and instruments.

A particular example of a real world engineering problem in which real gas effects cannot be disregarded is with the introduction of newly developed quantum metrology standards as discussed by Thakur et al. (2023) for an optical based pascal as a unit of pressure. This new measurement concept can now in principle be realized by linking the thermodynamic pressure to optical refractometry standards, which are in turn based on and ultimately traceable to quantum electrodynamic standards. This linking of mechanical based pressure measurements to electrical based optical measurements is thus essentially achieved by the utilization of the ideal gas law $pV = n_m RT$ for dilute pressures only where the pressure of a gas depends on the temperature T, volume V, number of moles n_m, number density ρ_N, and associated physical constants of nature such as the Boltzmann constant k_B, the universal gas constant R, and Avogadro number N_A amongst others that are intrinsically defined in terms of quantum metrology standards. If the pressure is not sufficiently low then the ideal gas equation will become inaccurate and real gas effects would have to be considered in order to refine the validity and accuracy of the mathematical analysis. Although the optical pascal approach as originally outlined by Jousten et al. (2017) is conceptually straightforward using the ideal gas equation $p = \rho_N k_B T$, where optical refractory measurements specify ρ_N and electrical based measurements specify T, the implementation of this new optical pascal method due to practical constraints is envisioned to be mainly limited to national measurement laboratories that can develop, maintain and disseminate quantum metrology standards.

More recent investigations on the optical pascal method by Forssen et al. (2022) has utilized an equation of the form $p = RT\rho[1 + B_\rho(T)\rho]$ where $\rho = \frac{2}{3A_R}(n - 1)[1 + b_{n-1}(n - 1)]$ is the expression for the gas molar density in terms of the optical refractivity n, A_R is the molar polarizability, $b_{n-1} = -(1 + 4B_R/A_R^2)/6$ is an algebraic term, B_R is the second refractivity virial coefficient in the Lorentz-Lorenz equation, and $B_\rho(T)$ is the second density virial coefficient which can quantified with *ab initio* simulations for common pressure balance working fluids such as helium or nitrogen gas, in order to incorporate real gas effects for greater accuracy. This approach has thus demonstrated the technical feasibility of potentially extending the validity for elevated pressures that are not dilute gas pressures. Current experimental and theoretical limitations for the optical pascal technique constrain the method direct validity for $1 \leqslant p/[\text{Pa}] \lesssim 3 \times 10^5$ at laboratories such as NIST (2023). The practical implication of this constraint is that a mathematical analysis of real gas effects is unavoidable for pressures above 300 kPa at accuracy levels of parts-per-million which is the typical accuracy of national measurement laboratory pressure standards.

Related engineering work that also requires a mathematical analysis of real gas effects, includes indirect pressure measurements outside the direct validity measurement for the optical pascal range below $p = 1$ Pa. These physical measurements may conveniently be achieved with a Static Expansion System (SES) under the simplifying assumptions of conservation of mass and isothermal conditions. Under these conditions the ideal gas equation $pV = n_m RT$ may be modified to the equivalent approximate form $p_1 V_1 = p_2 V_2$ for a known reference chamber volume V_1 and a larger evacuated expansion chamber volume V_2, as outlined in more technical detail by Bock et al. (2021) accounting for imperfections such as pressure valve flow conductances between chamber volumes with a system of coupled ordinary differential equations. More detailed refinements in vacuum pressure measurements by Cheng etal. (2021) replace the ideal gas equation $pV = n_m RT$ with a real gas Van der Waals equation of state of the form $\left(p + a\frac{n_m^2}{V^2}\right)\left(\frac{V}{n_m} - b\right) = RT$ where a and b are the Van der Waals constants for the particular choice of gas molecular species in order to adequately account for real gas effects as the ideal gas equation is only strictly valid in the limit as $p \to 0$. Atmospheric pressure $(1 \leqslant p/[\mathrm{Pa}] \leqslant 1 \times 10^5)$ and vacuum pressure $(1 \times 10^{-6} \leqslant p/[\mathrm{Pa}] \leqslant 1)$ measurements with a SES are therefore known to only be accurate in the limit of dilute gases. A similar indirect principle applying a SES in reverse may also be utilized so that instead of a lower pressure $p_2 = \left(\frac{V_1}{V_2}\right) p_1$ being generated by varying p_1, higher pressures in the kilo-pascal range for $p \gtrsim 3 \times 10^5$ Pa may instead also be generated and measured with a pressure amplification system if $V_2 < V_1$ is considered as a smaller compressed volume. Both of these fluid mechanics approaches, for both elevated atmospheric pressures as well as for reduced vacuum pressures, therefore implicitly require either the validity of the ideal gas equation of state at hard vacuum pressures $(1 \times 10^{-9} \leqslant p/[\mathrm{Pa}] \leqslant 1 \times 10^{-6})$, or a refinement such as the Van der Waals equation at weak vacuum pressures $(1 \times 10^{-6} \leqslant p/[\mathrm{Pa}] \leqslant 1 \times 10^5)$ in order to accommodate real gas non-ideal characteristics, in order for the underlying mathematical model to meaningfully reflect the fluid physical equation of state for an adequate level of measurement accuracy.

It may thus be observed that real gas effects may significantly influence achievable measurement accuracies not just in traditional pressure balance measurements, but also in related vacuum measurement systems work as well. This physical observation therefore provides the engineering basis and motivation to perform a mathematical analysis of the influence of real gas effects, as the results may have practical and tangible effects on the performance, reliability, accuracy and validity of physical instruments and equipment.

1.2 Research Objective

This study seeks to examine the underlying validity assumptions of real gas equations of states, and to perform a mathematical analysis in order to develop an appropriate numerical technique that is able to incorporate complex algebraic model of real non-ideal equations of states for practical real world fluid mechanics numerical simulations. An additional research objective is to investigate a mathematical approach to incorporate the coupling effect of the auxiliary thermodynamic variables based on the underlying primitive variables, and to investigate a new numerical scheme for performing the mathematical analysis of real gas effects.

2 Mathematical Models

In the mathematical study of fluid mechanics under the assumption of a continuum fluid, the Navier–Stokes equations may be used without loss of generality to model any practical macroscopic flow for a wide variety of engineering flow behaviours, with traditional Dirichlet and Neumann boundary conditions for the underlying system of PDEs. The system of PDEs with traditional Dirichlet/Neumann boundary conditions are usually solved either with finite differences for simple geometries as outlined by Tannehill et al. (1997), the finite element method as outlined by Zienkiewicz et al. (2005), the finite volume method as outlined by Blazek (2001), or occasionally the spectral element method as outlined by Kopriva (2009) when high accuracy resolutions must be obtained. In special case of nonlinear boundary conditions such as the Extended Navier–Stokes equations which extend the validity of the model to varying flow regimes, alternative techniques such as either the Radial Basis Function (RBF) approach by Ramnath (2012) or the Generalized Finite Difference Method (GFDM) approach by Song et al. (2020) may instead be utilized for low Reynolds numbers flows as special cases corresponding to subsonic conditions.

2.1 General Application of the Navier–Stokes Equations

For nozzles the theory of one dimensional flow may be considered as a fluid flow in the limit of a streamline for a flow within a fluid domain as developed by Shapiro (1953) under the assumption of isentropic conditions. In this study a one dimensional flow was then generalized by Shapiro to what is termed a quasi one dimensional flow that can incorporate area change, wall friction, drag within the internal body, external heat generation, chemical reaction, change of phase (e.g. evaporation/condensation), mixing of gases within the nozzle, and changes in molecular weight and specific heat of the working fluid within the nozzle caused by combustion, evaporation or gas injection. While the general method of quasi one dimensional flow is very powerful in performing an analysis of nozzle flow in many types of equipment such as pressure regulation devices and rocket exhaust nozzles, the original analysis by Shapiro is based on the three underlying assumptions that:

1. the flow is one dimensional and steady
2. the changes in stream-wise properties are continuous (thus there are no discontinuities in for example local shock-wave propagation within the nozzle)

3. the gas obeys the Boyles law and Charles law for ideal gases and has a specific heat c_p which varies only with the temperature T and has no pressure dependence.

These original historical assumptions were also utilized in later studies such as that by Sforza (2016) again by using the perfect gas equation $p = (\rho R_u T)/W$ where R_u is the universal gas constant and W is the molecular weight and the corresponding differential form is $\frac{\mathrm{d}p}{p} = \frac{\mathrm{d}\rho}{\rho} + \frac{\mathrm{d}T}{T} - \frac{\mathrm{d}W}{W}$ for aerospace engineering applications, and by Babu (2021) where the quasi one dimensional flow takes the mathematically equivalent form of three equations $\rho_1 u_1 A_1 = \rho_2 u_2 A_2$ which represents the knowledge that the mass flow rate $\dot{m} = \rho u A$ is constant, $(p_1 + \rho_1 u_1^2)A_1 + \int_1^2 (p\mathrm{d}A)_x = (p_2 + \rho_2 u_2^2)A_2$ which represents the momentum equation, and $h_1 + \frac{1}{2}u_1^2 = h_2 + \frac{1}{2}u_2^2$ which represents the energy equation using the enthalpy h.

Later work by Atmaca et al. (2020) utilized a quasi one dimensional flow model formulated in terms of a coupled system of ordinary differential equations to model the flow for a convergent-divergent nozzle in a two phase ejector for a refrigeration cycle to decrease the throttling losses and improve the performance. By contrast Oliva et al. (2023) utilized instead an integral form of the governing equations with a calorically perfect ideal gas to derive analytical equations for the density and velocity profiles in the nozzle. As a result, the standard quasi one dimensional flow equations in earlier studies are not directly applicable to determining the pressure profiles for gas microflows due to a combination of their underlying modelling assumptions.

An earlier study by Sabuga et al. (2011) that was later refined in more technical detail by Sharipov et al. (2016) extended the mathematical analysis to validate the gas flow in an interface gap using rarefied gas dynamics for the transitional flow regime to the continuum flow regimes for very low to low pressures where the continuum modelling assumption breaks down at large Knudsen numbers in pressure balance interface gaps with a nominal interface gap of $h = \mathcal{O}(1\ \mu\mathrm{m})$ for pressures from $p = 10$ Pa to $p = 100$ kPa. The study by Sabuga et al. (2006) has outlined the analysis approach for liquids such as oils for pressures from $p = 50$ MPa to $p = 500$ MPa that may in principle be extended for oils down to a working pressure of 30 MPa.

Taking into account the above approaches and validity limitations, by starting from the vector form of the Navier–Stokes equations which is the conservation equations for mass, momentum and energy from Tannehill et al. (1997) the vector form may equivalently be written as

$$\frac{\partial \mathbf{U}}{\partial t} + \frac{\partial \mathbf{E}}{\partial x} + \frac{\partial \mathbf{F}}{\partial y} + \frac{\partial \mathbf{G}}{\partial z} = \mathbf{0} \tag{8}$$

In the above multi-dimensional system **U**, **E**, **F**, and **G** are 5×1 dimensional vectors that represent aspects of the continuum fluid state, where $\mathbf{0}$ is zero vector as per standard CFD practise. The first component in the above vectors corresponds to the mass conservation equation, the second to fourth components correspond to the momentum conservation equations, and the fifth component corresponds to the

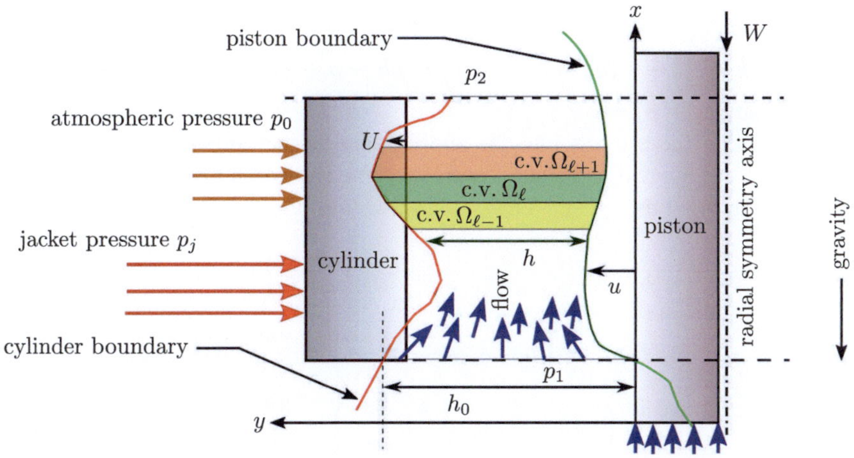

Fig. 1 Schematic of a pressure balance Dadson model geometry for fluid domain and boundary conditions

energy conservation equation. Modelling the fluid flow within crevices and channels as a quasi one dimensional real gas in steady state, it immediately follows that $\frac{\partial \mathbf{U}}{\partial t} = \frac{\partial \mathbf{F}}{\partial y} = \frac{\partial \mathbf{G}}{\partial z} = \mathbf{0}$ so that there is only a variation in the stream-wise flow direction which may conveniently taken as x in a Cartesian Eulerian coordinate system as shown in Fig. 1.

The remaining vector in the system of PDEs is then

$$\mathbf{E} = \begin{bmatrix} \rho u \\ \rho u^2 + p - \tau_{xx} \\ \rho u v - \tau_{xy} \\ \rho u w - \tau_{xz} \\ (E_t + p)u - u\tau_{xx} - v\tau_{xy} - w\tau_{xz} + q_x \end{bmatrix} \tag{9}$$

In the above system ρ is the mass density which is not necessarily assumed to follow an ideal gas EoS, u, v, w are the x, y, z components of the fluid velocity vector $\mathbf{v} = [u, v, w]^\mathsf{T}$, τ_{ij} are the shear stresses for an appropriate choice of shear stress tensor Π_{ij}, q_x is a heat source term, and E_t is the fluid total energy per unit volume. The value of E_t may equivalently be written as $E_t = \rho(e + \frac{1}{2}|\mathbf{V}|^2 + \text{potential energy} + \cdots)$ where e is the internal energy per unit mass. Utilizing standard thermodynamic definitions, the internal energy per unit mass may in turn be defined in terms of the enthalpy h as $h = e + \frac{p}{\rho}$.

Noting that $\tau_{xx} = \frac{2}{3}\mu\left(2\frac{\partial u}{\partial x} - \frac{\partial v}{\partial y} - \frac{\partial w}{\partial z}\right)$, $\tau_{xy} = \mu\left(\frac{\partial u}{\partial y} + \frac{\partial v}{\partial x}\right)$ and $\tau_{xz} = \mu\left(\frac{\partial w}{\partial x} + \frac{\partial u}{\partial z}\right)$, by applying a one dimensional flow assumption within the interface gap by setting $u \neq 0$, $v = 0$ and $w = 0$, the shear stress components then reduces to $\tau_{xx} = \frac{4}{3}\mu\frac{\partial u}{\partial x}$, and $\tau_{xy} = 0$, $\tau_{xz} = 0$. Since $h = e + \frac{p}{\rho}$ via the

Maxwell thermodynamic equations it then follows that $e = h - \frac{p}{\rho}$ so that $E_t = \rho\left(\left(h - \frac{p}{\rho}\right) + \frac{1}{2}\left(u^2 + v^2 + w^2\right)\right) = \rho h - p + \frac{1}{2}\rho u^2$ from which it then follows that $(E_t + p) = \rho h + \frac{1}{2}\rho u^2$.

Substituting these terms into the above equation then produces a set of three PDEs for a quasi one dimensional control volume Ω_ℓ within the crevice or channel illustrated in Fig. 1 as

$$\frac{\partial}{\partial x}(\rho u) = \frac{\partial \Phi_1}{\partial x} = 0 \tag{10}$$

$$\frac{\partial}{\partial x}\left(\rho u^2 + p - \frac{4}{3}\mu\frac{\partial u}{\partial x}\right) = \frac{\partial \Phi_2}{\partial x} = 0 \tag{11}$$

$$\frac{\partial}{\partial x}\left(\rho u h + \frac{1}{2}\rho u^3 - \frac{4}{3}\mu\frac{\partial u}{\partial x}u\right) = \frac{\partial \Phi_3}{\partial x} = 0 \tag{12}$$

The above set of equations $\frac{\partial \Phi_\xi}{\partial x} = 0,\ \xi \in [1, 2, 3]$ are not strictly ordinary differential equations as the control volume Ω_ℓ which is three dimensional in space is not taken in the limit where the control volume converges to a stream line, and as a result the above system of equations is not a true mathematical one dimensional flow. A limitation of a pure one dimensional flow in many mathematical nozzle flow studies is that it does not represent the physical non-zero cross-sectional area in a real nozzle which varies in the stream-wise flow direction.

Letting the above system of mass, momentum and energy conservation equations apply for an interior control sub-volume Ω_ℓ within the channel/crevice of a pressurized fluid flow which has a total volume from the inlet to the outlet of $\Omega = \cup_{\ell=1}^{N}\Omega_\ell$ for a sequence of control volumes (c.v's) $\Omega_1, \ldots, \Omega_{\ell-1}, \Omega_\ell, \Omega_{\ell+1}, \ldots, \Omega_N$ which are constructed in the stream-wise x direction, and applying the standard finite volume discretization methodology by integrating over the sub-volume Ω_ℓ so that for $\xi \in [1, 2, 3]$ corresponding to the each of the above three equations $\frac{\partial \Phi_\xi}{\partial x} = 0$, the above volume integrals may then be rewritten as an integral of a divergence of a dummy column vector built up of Φ_ξ that was previously defined so that

$$\int_{\Omega_\ell} \frac{\partial \Phi_\xi}{\partial x} \mathrm{d}\Omega_\ell = \int_{\Omega_\ell} \nabla \cdot \begin{bmatrix} \Phi_\xi \\ 0 \\ 0 \end{bmatrix} \mathrm{d}\Omega_\ell = 0,\ \xi \in [1, 2, 3] \tag{13}$$

Utilizing the divergence theorem $\iiint_V \nabla \cdot \mathbf{F}\mathrm{d}V = \oiint_S \mathbf{F} \cdot \mathrm{d}\mathbf{S}$ then converts the above triple volume integrals into equivalent closed surface integrals so that

$$\oiint_{\partial\Omega_\ell} (\Phi_\xi \hat{\boldsymbol{i}}) \cdot \hat{\boldsymbol{n}}\mathrm{d}S = 0,\ \ell \in [1, \ldots, N],\ \xi \in [1, 2, 3] \tag{14}$$

The above system of surface integrals may now be discretized in terms of the primitive variables of the density $\rho(x)$, the velocity $u(x)$, and the temperature $T(x)$ using standard quadrilaterals for the sub-volumes Ω_ℓ following the approach of Blazek (2001). The corresponding simultaneous system of equations in the interior sub-volumes $\Omega_2, \ldots, \Omega_{N-1}$ when combined along with the initial sub-volume Ω_1 for the inlet flow and the final sub-volume Ω_N may then be solved simultaneously to recover the density, velocity and temperature field within the interface gap using the previously defined real gas EoS formula for the pressure $p(\rho, T)$, thermal conductivity $k(\rho, T)$, viscosity $\mu(\rho, T)$ and enthalpy $h(\rho, T)$ in terms of the primitive variables where the velocity gradient $\frac{\partial u}{\partial x}$ may be approximated using central difference from the values of the centroids of the control volumes in the interior.

2.2 *Models of Thermophysical Properties*

The ideal gas Equation of State (EoS) $pV = n_m RT$ may equivalently be written as $p = \rho R_g T$ for pure gases where p is the pressure in Pa, V the volume in m^3, $n_m = \frac{m}{M}$ is the number of moles, m the mass in kg, M the molar mass in kg/mol with $M_{\text{air}} \approx 28.97 \times 10^{-3}$ kg mol^{-1} and $M_{N_2} = 28.014 \times 10^{-3}$ kg mol^{-1}, $R = 8.3144958$ J mol^{-1} K^{-1} is the universal gas constant, T the absolute temperature in K, and where $R_g = \frac{R}{M}$ is the particular molecular gas species constant in units of J kg^{-1} K^{-1} respectively. This equation that is derived from statistical mechanics considerations is only strictly valid for dilute gases in the limit $p \to 0$ where the pressure tends to zero.

Typical refinements to the ideal gas equation are the well known Van der Waals EoS which includes higher order virial coefficients and the Soave-Redlich-Kwong cubic EoS. Currently the most accurate up to date EoS for gases is the Peng–Robinson (PR) EoS by Peng and Robinson (1976) which was originally developed in 1976. Various refinements have been applied to the PR EoS over the intervening years and new alpha functions to refine the PR EoS have been reported by Sun et al. (2022). In this paper for simplicity an earlier PR with an alpha function correction by Saffari and Zahedi (2013) takes the form

$$p = \frac{RT}{V_m - b} - \frac{a(T)}{V_m(V_m + b) + b(V_m - b)} \tag{15}$$

$$b = 0.07780 \frac{RT_c}{p_c} \tag{16}$$

$$a(T) = \left[0.45724 \frac{R^2 T_c^2}{p_c^2}\right] \alpha(T) \tag{17}$$

In the above equation $V_m = \frac{M}{\rho}$ is the molar volume in units of m^3 mol^{-1}, M is the molar mass in units of kg mol^{-1}, ρ is the mass density in kg m^{-3}, T_c is the critical temperature in K and p_c is the critical pressure in Pa where $T_{cN_2} = 126.2$ K and

$p_{cN_2} = 3.390 \times 10^6$ Pa, while b and $a(T)$ are the corresponding PR EoS parameters where $\alpha(T)$ is the alpha function that refines the PR EoS prediction. The form originally proposed by Peng and Robinson is

$$\alpha(T) = \left[1 + \kappa\left(1 - \sqrt{T_r}\right)\right]^2 \tag{18}$$

In the above equation $T_r = \frac{T}{T_c}$ is the reduced temperature which is dimensionless, $\kappa = 0.37464 + 1.54226\omega - 0.26992\omega^2$ is a parameter, and ω is the acentric factor where $\omega_{N_2} = 0.039$ and $\omega_{air} \approx 0.0335$. The above particular choice of alpha function corresponds to the well-known Redlich-Kwong (RK) EoS.

Viscosity of gases denoted by μ in units of Pa.s is usually modelled by the Sutherland equation of the form $\frac{\mu}{\mu_0} = \left(\frac{T}{T_0}\right)^{\frac{3}{2}} \frac{T_0+S}{T+S}$ where S is an effective temperature called the Sutherland constant and T_0 in K is a reference temperature. Representative values are $T_{0air} = 273$ K, $\mu_{0air} = 1.716 \times 10^{-5}$ N s m^{-2} and $T_{0N_2} = 273$ K, $\mu_{0N_2} = 1.663 \times 10^{-5}$ N s m^{-2} by White (1991) however this choice of EoS is technically only valid for a dilute gas approximation. Formally the gas viscosity depends on a combination of both the pressure and the temperature and a very early approximation by James Clerk Maxwell estimated the viscosity as $\mu \approx 0.67\rho\ell a$ where ℓ is a mean free path of the gas which depends on the temperature and a is the local speed of sound within the gas which depends on a combination of the temperature and pressure. More refined statistical mechanics analysis with a dilute gas approximation by Chapman and Cowling (1970) using a generalized kinetic theory formula specify that the gas viscosity takes the form $\mu = \frac{2.68\times10^{-6}\sqrt{MT}}{\sigma^2\Omega_v}$ where σ is a collision diameter in Angstroms, M the molecular weight of the gas, μ the gas viscosity in units of kg m^{-1} s^{-1}, T the absolute temperature, and Ω_v is a dimensionless collision integral quantity to account for molecular collisions.

An approximation of the collision integral Ω_v using a Stockmayer potential function for molecular dynamics simulations is $\Omega_v \approx 1.147\left(\frac{T}{T_\epsilon}\right)^{-0.145} + \left(\frac{T}{T_\epsilon+0.5}\right)^{-2.0}$ where T_ϵ is an effective temperature of the force potential for the molecular dynamics simulation. Typical values reported by White (1991) are $(\sigma)_{air} = 3.711$ Å, $(T_\epsilon)_{air} = 78.6$ K, $(\sigma)_{N_2} = 3.798$ Å, and $(T_\epsilon)_{N_2} = 71.4$ Å. Although the generalized kinetic approximation for a gas viscosity allows for useful physical insights it is only valid limit in the dilute gas limit, and can therefore not be applied at atmospheric pressures in the kilo-pascal range.

A formula for the gas viscosity for nitrogen gas that incorporates both density and temperature was developed by Latto and Saunders (1972) valid from 1 atmosphere to 150 atmospheres i.e. for $101325 \leqslant p/[\mathrm{Pa}] \leqslant 15.199 \times 10^6$ and $90 \leqslant T/[\mathrm{K}] \leqslant 400$ and takes the form

$$\eta = \frac{5.5811\times10^{-4}T^{1.0432}}{(T+9.0067\times10^2)} + 9.8426 \times 10^{-5}\rho_\star + 6.9817 \times 10^{-4}\rho_\star^2 \tag{19}$$

In the above formula $\rho_\star$ is the density of the gas in units of g cm^{-3}, T is the absolute temperature in kelvin, and the viscosity η is in units of poise. As a result the corresponding gas viscosity may be calculated as

$$(\mu)_{N_2} = 0.10 \times \eta(\rho, T) \tag{20}$$

Although the EoS formally apply only for pure substances an approximate non-ideal EoS for air in terms of the EoS for nitrogen was developed by Lemmon et al. (2000) where equivalent air temperature T_{air}, air density ρ_{air}, and air pressure p_{air} may be modelled as

$$T_{air} = (1.038128 + 0.000054933 T_{N_2}) T_{N_2} \tag{21}$$

$$\rho_{air} = 1.043492 \rho_{N_2} \tag{22}$$

$$p_{air} = \frac{T_{air}}{T_{N_2}} \times \frac{\rho_{air}}{\rho_{N_2}} \times p_{N_2} \tag{23}$$

Formulae for the ideal gas and an air mixture were also developed by Lemmon et al. (2000) where the specific heat capacity at constant pressure (isobaric heat capacity) c_p is in units of J mol^{-1} K^{-1}. In this study the critical points are $(T_c)_{air} = 132.5306$ K, $(p_c)_{air} = 3.7860$ MPa, $(\rho_c)_{air} = 11.8308$ mol dm^{-3}, $(T_c)_{N_2} = 126.192$ K, $(p_c)_{N_2} =$ 3.3958 MPa, and $(\rho_c)_{N_2} = 11.1839$ mol dm^{-3}. It may be noted that the critical density is in units of moles per cubic decimeter and not in kilograms per cubic meter.

The results for nitrogen gas specific heat is given by

$$\left(\frac{c_p^0}{R}\right)_{N_2} = \sum_{i=1}^{4} N_i T^{(i-1)} + \frac{N_5 u^2 e^u}{(e^u - 1)^2} \tag{24}$$

whilst the results for an air mixture specific heat is given by

$$\left(\frac{c_p^0}{R}\right)_{air} = \sum_{i=1}^{4} N_i T^{(i-1)} + \frac{N_5}{T^{1.5}} + \frac{N_6 u^2 e^u}{(e^u - 1)^2} \tag{25}$$

where the values of the various parameters in the above two equations are documented in the work by Lemmon et al. (2000). Using the above formulae for the ideal gas specific heat of nitrogen and air $(c_p^0)_X$ with $X \in [N_2, \text{air}]$ the corresponding ideal gas enthalpy in units of J mol^{-1} may then be defined as

$$h_i^0 = h_{0i}^0 + \int_{T_0}^{T} c_{pi}^0 \, dT \tag{26}$$

In the above expression h_{0i}^0 is the enthalpy data value at a specified reference temperature, noting that the enthalpy is defined as a difference in values between a working temperature and reference temperature. The reference temperature is usually taken as the ambient temperature for convenience in many studies.

The predictions with the formulae for $(c_p^0)_X,\ X \in [\mathrm{N_2}, \text{air}]$ may be used to calculate the ideal gas estimates for the corresponding ideal gas estimate of the specific enthalpy h_i^0 however this must be adjusted for real gas effects. An investigation by Dai et al. (2012) gives the formulae using the compressibility factor $Z = \frac{pv}{R_g T}$ where the specific volume $\nu = \frac{V}{m} = \frac{1}{\rho}$ is the reciprocal of the mass density using the thermodynamic equations of state to calculate the real gas enthalpy as

$$h = h^{ig} + (Z-1)R_g T + \int_\infty^\nu \left[T\left(\frac{\partial p}{\partial T}\right)_\nu - p\right] \mathrm{d}\nu \tag{27}$$

The partial derivatives in the above expression will use an appropriate choice of EoS such as the Peng–Robinson equation.

An equivalent expression to calculate the real gas specific heat is

$$c_p(T,\nu) = T\int_\infty^\nu \left(\frac{\partial^2 p}{\partial T^2}\right)_\nu \mathrm{d}\nu - \frac{T\left(\frac{\partial p}{\partial T}\right)_\nu^2}{\left(\frac{\partial p}{\partial \nu}\right)_\nu} - R_g + c_p^{ig}(T,\nu) \tag{28}$$

Models for the thermal conductivity EoS models for nitrogen and air were investigated in a later study by Lemmon and Jacobsen (2014) where the thermal conductivity λ in units of $10^{-3}\mathrm{W\ m^{-1}\ K^{-1}}$ (milli-watts per meter per kelvin) take the general form

$$\lambda = \lambda^0(T) + \lambda^r(\tau,\delta) + \lambda^c(\tau,\delta) \tag{29}$$

$$\tau = \frac{T_c}{T} \tag{30}$$

$$\delta = \frac{\rho}{\rho_c} \tag{31}$$

Expressions for the dilute gas thermal conductivity λ^0, the residual fluid thermal conductivity λ^r, and the thermal conductivity critical enhancement λ^c are calculated as

$$\lambda^0 = N_1\left[\frac{\eta^0(T)}{1\ \mu\mathrm{Pa\cdot s}}\right] + N_2\tau^{(t_2)} + N_3\tau^{(t_3)} \tag{32}$$

$$\lambda^r = \sum_{i=1}^{n} N_i \tau^{(t_i)}\delta^{(d_i)} \exp\left[-\gamma_i \delta^{(\ell_i)}\right] \tag{33}$$

$$\lambda^c = \rho c_p \frac{k R_0 T}{6\pi\xi\eta(T,\rho)}(\tilde{\Omega} - \tilde{\Omega}_0) \tag{34}$$

2.3 Reynolds Equation as a Specific Navier–Stokes Model

In certain earlier pressure balance studies such as that by Sabuga et al. (2006) the fluid flow in the interface gap for liquids has incorporated both integral equation as well as differential equation formulations for the pressure profile. An integral equation for the unknown pressure $p_z \equiv p(z)$ along an engagement length for a piston-cylinder circular annulus, assuming a Newtonian fluid, steady state, single phase, one dimensional, viscous fluid, laminar, and isothermal conditions then takes the form

$$p_z = p - \frac{6Q_m}{\pi r_0}\int_0^z \frac{\eta(p_z)}{\rho(p_z)}\frac{1}{h_p^3}\mathrm{d}z : \text{general integral equation model} \tag{35}$$

$$Q_m = \frac{\pi r_0 p}{6\int_0^L \frac{\eta(p_z)}{\rho(p_z)}\frac{1}{h_p^3}\mathrm{d}z} \tag{36}$$

The above Volterra integral formulation is usually solved iteratively from specified formula for the viscosity $\eta(p)$ and density $\rho(p)$ relationship in terms of pressure, but may also in principle be solved either with a direct finite difference discretization or alternatively with Homotopy Analysis Method (HAM) approaches as investigated earlier by Ramnath (2013). This equation is mainly used for silicon oil operated high pressures from 50 MPa to 1 GPa but the flow model is in principle also valid for gas operated pressure balances. Under the same modelling assumptions, an equivalent one dimensional differential equation for the unknown pressure profile corresponds to the well known Reynolds equation of the form

$$\frac{\mathrm{d}}{\mathrm{d}z}\left[\frac{\rho(p_z)h_p^3}{12\eta(p_z)}\frac{\mathrm{d}p_z}{\mathrm{d}z}\right] = 0 \ : \ \text{general differential equation model} \tag{37}$$

The above one dimensional differential equation that is suitable for a mathematical analysis is usually solved iteratively by assuming a linear pressure variation in the interface gap from $p(0) = p_1$ at the inlet to $p(L) = p_2$ at the outlet until convergence is achieved. Modelling assumptions of the classical Reynolds equation for studying the pressure distribution in fluid film lubricated bearings, is discussed in more detail by Khonsari (2013) who reports that the equation combines the conservation equations for mass and momentum by assuming isothermal conditions by assuming:

1. a Newtonian fluid
2. insignificant inertia and body forces (i.e. an assumption of a small Reynold's number $Re = \frac{Uh}{\nu} \ll 1$, $\nu = \frac{\mu}{\rho}$ where U is the bulk velocity, h is the half-height of the channel, and ν is the kinematic viscosity in terms of the dynamic viscosity)
3. a negligible pressure gradient across the film thickness
4. a laminar flow consistent with a small Reynolds number
5. negligible curvature effects.

These modelling assumptions are sufficiently general in order to study the influence of real gas effects in a fluid mechanics simulation. An integral equation formulation or alternately the thin-film Reynolds lubrication equation is also valid when oil is used as working fluid and is considered to be a special limiting case of the Navier–Stokes equations as reported by Sabuga et al. (2006).

2.4 Extensions of the Reynolds Equation to Two Dimensions

The Reynolds equation is usually considered to model the fluid flow between two horizontally inclined semi-infinite surfaces where the bottom surface is considered as a flat plane that contains an xy coordinate system at $z = 0$, the top surface has a geometry where the height from the bottom surface to the top surface is considered to continuously vary with a known height function $z = h(x, y)$, and the bulk flow velocity U enters and moves in the x direction. The geometry configuration of a Reynolds equation and how it may be adapted to model the variation within a pressure balance interface gap is shown in Fig. 2 where (a) shows a planar flow configuration and (b) shows the approximate equivalent flow configuration for two circular surfaces that correspond to a cylinder inner wall and a solid piston outer wall. It may be noted that when the Reynolds equation is adapted from a rectangular Cartesian $Oxyz$ coordinate system to a cylindrical $Or\phi z$ coordinate system, a change of variables from $x \to z$ and $y \to \phi$ may be used for convenience by setting the interface gap variation to $h(z, \phi)$ where the 'height' between a 'lower surface' and an 'upper surface' is more easily physically understood and visualized in order to apply the prevailing flow geometry and boundary conditions. If the boundary surfaces have zero velocity then a special form of the Reynolds equation may then be written as a two dimensional quasi Poisson equation of the form

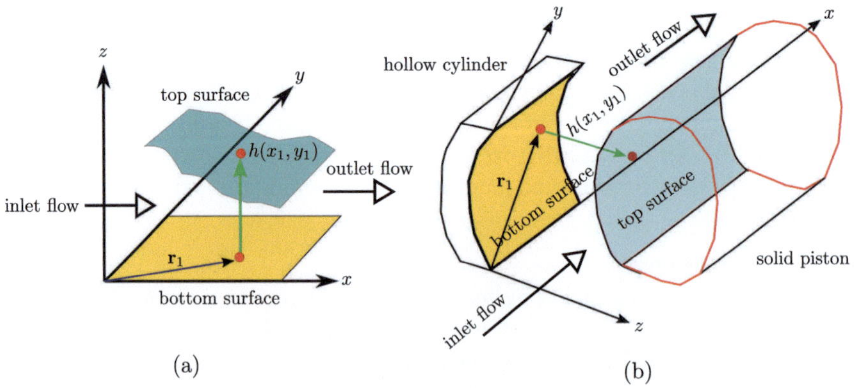

Fig. 2 Schematic comparing geometry configurations of the Reynolds equation

$$\frac{\partial}{\partial x}\left(\frac{\rho h^3}{12\mu}\frac{\partial p}{\partial x}\right)+\frac{\partial}{\partial y}\left(\frac{\rho h^3}{12\mu}\frac{\partial p}{\partial y}\right)=0 \tag{38}$$

The solution to the above two dimensional partial differential equation is the pressure field in-between the bottom and top surfaces (or equivalently a 'bottom' cylinder inner wall surface and a 'top' piston outer wall), where the pressure across the two surfaces is assumed constant from the modelling assumption $\frac{\partial p}{\partial z}=0$, or equivalently that $\frac{\partial p}{\partial r}=0$ in cylindrical coordinates. The bottom and top surfaces to define the PDE domain Ω for the flow problem are built up from $r(z,\phi)$ measurement points for a choice of available radial values in the angular range $\phi_{\min}\leqslant\phi_{\max}$. The utility of the above Reynolds equation as a special case over the Navier–Stokes equation which exhibits a pressure-velocity coupling, is that the above Reynolds equation is considerably mathematically simpler as it only contains a single unknown scalar field in terms of the pressure $p(x,y)$ which may be formulated and solved in terms of a generalized Poisson PDE.

A physical interpretation of the Reynolds equation modelling approximation is that the pressure in contact with the 'top' piston outer surface is equal to the pressure in contact with the 'bottom' cylinder inner surface. Once the pressure field is solved then the pressure on the piston along the engagement length for a value of ϕ that is held constant may be extracted as $p(z)$ for substitution into the Dadson model. The underlying quasi Poisson equation for known radial values of the piston $r(z,\phi)$ and the cylinder $R(z,\phi)$ data may then be conveniently directly solved with a finite difference approach as the domain $\Omega=\{(z,\phi):0\leqslant z\leqslant L,\phi_{\min}\leqslant\phi\leqslant\phi_{\max}\}$ is by definition a rectangular region.

Of the five modelling assumptions by Khonsari, it may be observed that the key deviations in an actual pressure balance which may potentially violate the validity of the equation is the assumption of isothermal modelling which "uncouples" the mass and momentum conservation equations from the energy conservation coefficient, and the negligible curvature effect assumption. An additional limiting assumption of the Reynolds equation when applied to model a pressure balance is that typically the density and viscosity are more strongly dependent on the temperature than on the pressure for many working fluids such as nitrogen and oil. Although a pressure/temperature dependence for the viscosity may be known, in the absence of an energy conservation equation the Reynolds equation is not mathematically closed and the linkage to a known temperature is not possible. If the Reynolds equations is combined with an energy conservation equation, then the system is more easily solved directly with the Navier–Stokes equations.

3 Numerical Simulations

3.1 Construction of Non-ideal Equation of State

In order to construct a coupled pressure-density-viscosity thermophysical model, the starting point is the Peng–Robinson equation of state so that the pressure is a function of the density and temperature so that $p(\rho, T)$ is an explicit equation. By then considering the viscosity model of Latto-Sanders for the viscosity as a function of the density and temperature it follows that the viscosity is then of the form $\eta(\rho, T)$. The main technical challenge is that in order to solve the integral equation it is necessary to work out the values of the density and viscosity knowledge of the instantaneous pressure, however due to the extreme algebraic complexity of the respective equations of states it is not in general possible to determine the corresponding inverse functions. A graphical plot showing the error in the pressure from the real gas Peng–Robinson equation and the ideal dilute density equation is shown in Fig. 3, whilst a graphical plot showing the error in the viscosity from the real gas Latto-Saunders equation and the Sutherland ideal gas equation is shown in Fig. 4. Referring to these graphs it may be observed that the accuracy errors are not negligible and have to be considered in an actual engineering analysis.

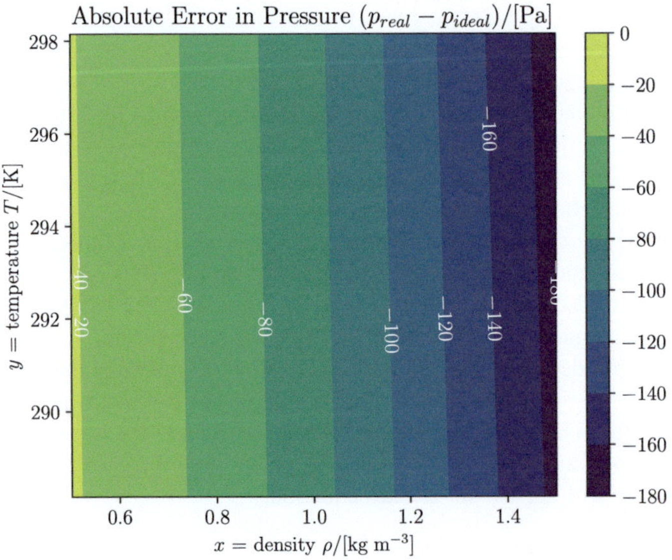

Fig. 3 Graph showing the absolute error in pressure between the Peng–Robinson real gas and ideal gas equations of states models

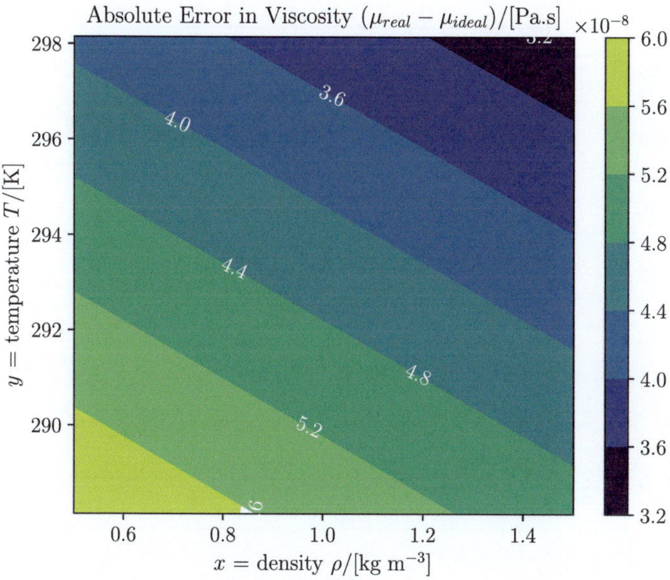

Fig. 4 Graph showing the absolute error in viscosity between the Latto-Saunders real gas and Sutherland ideal gas equation of states models

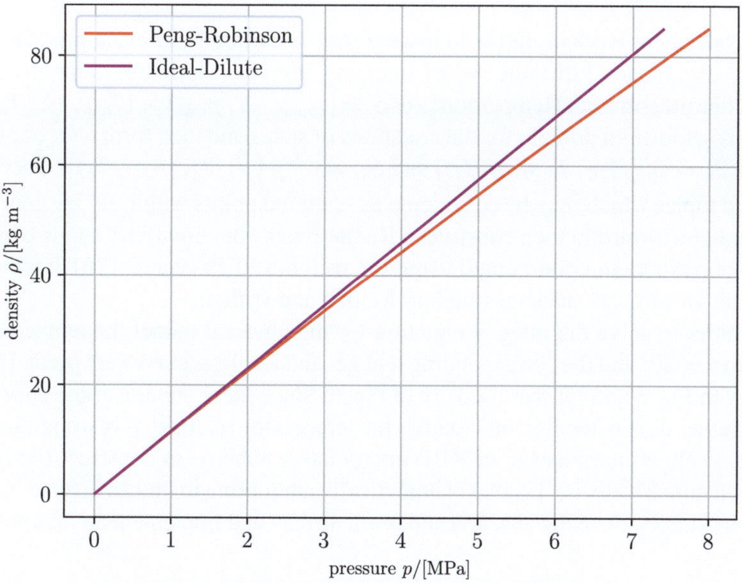

Fig. 5 Plot of deviations for the density between a real gas and an ideal gas

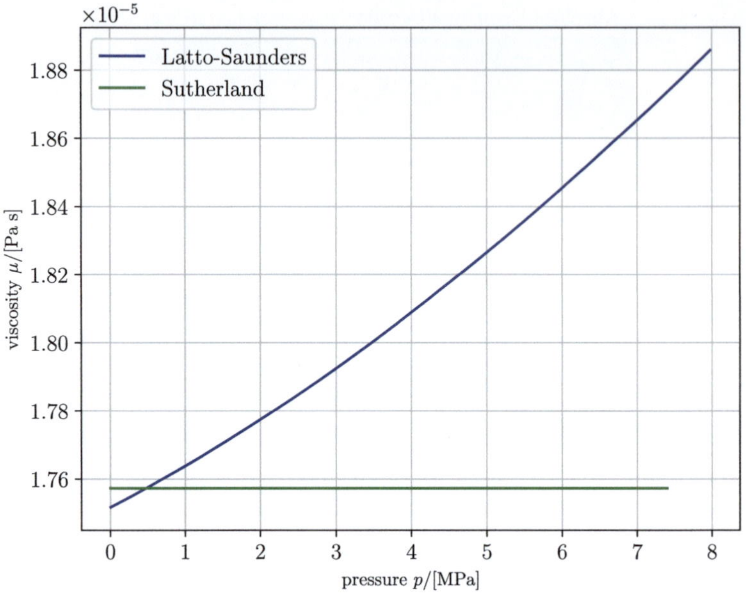

Fig. 6 Plot of deviations for the viscosity between a real gas and an ideal gas

A convenient workaround is to instead first work out a Cartesian product of the vectors of a range of discrete values spanning from the minimum to the maximum respective pressures and temperatures as $S = [p_{\min}, \ldots, p_{\max}] \times [T_{\min}, \ldots, T_{\max}]$ so that this set forms a domain for the equations of states and then form a set of ordered tuples $\mathcal{S}_\eta = \cup_{i=1}^{N} [p_i, T_i, \eta(p_i, T_i)]$ and $\mathcal{S}_\rho = \cup_{j=1}^{M} \left[p_j, T_j, \rho(p_j, T_j)\right]$. Once these ordered tuples which may be considered as scattered points within $\mathbb{R}^3$ are generated, it is straightforward to then construct a Radial Basis Function (RBF) approximation of the corresponding point cloud using the methods of Fasshauer (2010) which are available in software packages such as Matlab and Python.

In order to solve the integral equation for the physical model the temperature is fixed to $t = 20°$ and the corresponding real gas and ideal gas curves are plotted for the density in Fig. 5 and for the viscosity in Fig. 6. Since these are univariate curves and not a higher dimensional point cloud as the temperature is fixed, it is straightforward to use a spline interpolation or RBF interpolation scheme to construct the inverse function numerically i.e. instead of analytically attempting to find an inverse function one may simply swop the abscissa and ordinate data and implement an interpolation.

3.2 Solution of Reynolds Integral Equation

Utilizing the data curves of the auxiliary thermodynamic properties from the previous sections, the underlying integral equation may now be solved using representative geometry for the underlying fluid mechanics problem as shown in Fig. 7 through iteration by assuming a linear profile as a starting solution and then iterating until convergence.

When the nonlinear Reynolds integral is solved in this manner it then produces the ideal gas results shown in Fig. 8, the real gas results shown in Fig. 9, and the deviations between the fluid mechanics simulation using the ideal gas equations and the real gas equations as shown in Fig. 10.

By carefully examining these results it may be observed that the mathematical analysis has proven that significant deviations may exist when comparing fluid mechanics simulations with ideal gas and real gas equations of states. Referring to the previous results in Fig. 5 where it was observed that the deviation in ideal/real gases deviated significantly at elevated pressures and in Fig. 4 where it was demonstrated that by neglecting the influence of pressure on the viscosity by only considering just

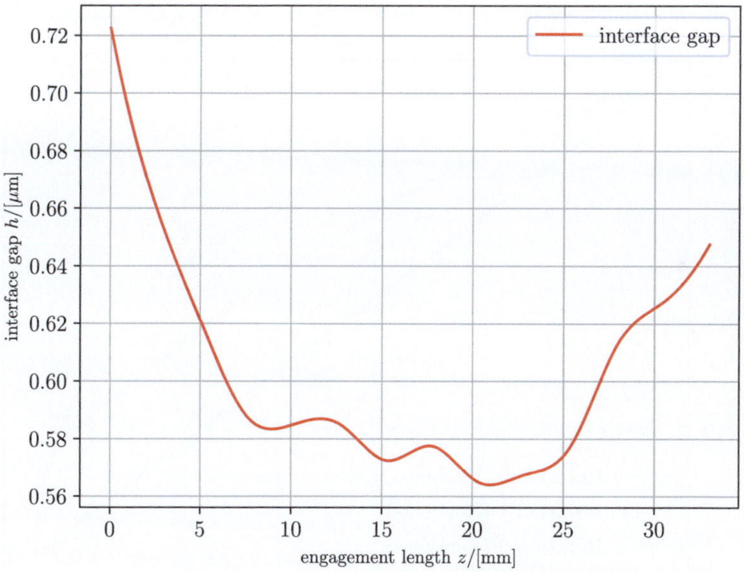

Fig. 7 Representative real world geometry for a channel/crevice flow

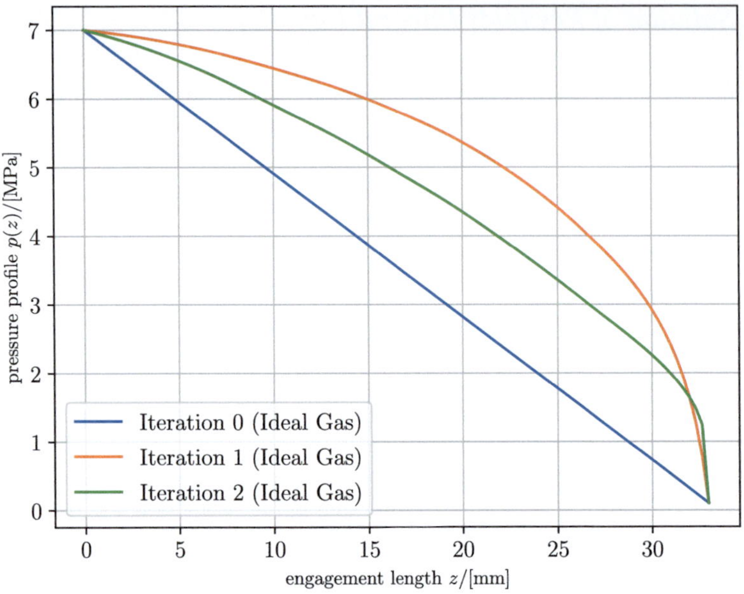

Fig. 8 Final fluid simulation results for channel/crevice flow with ideal gas equations

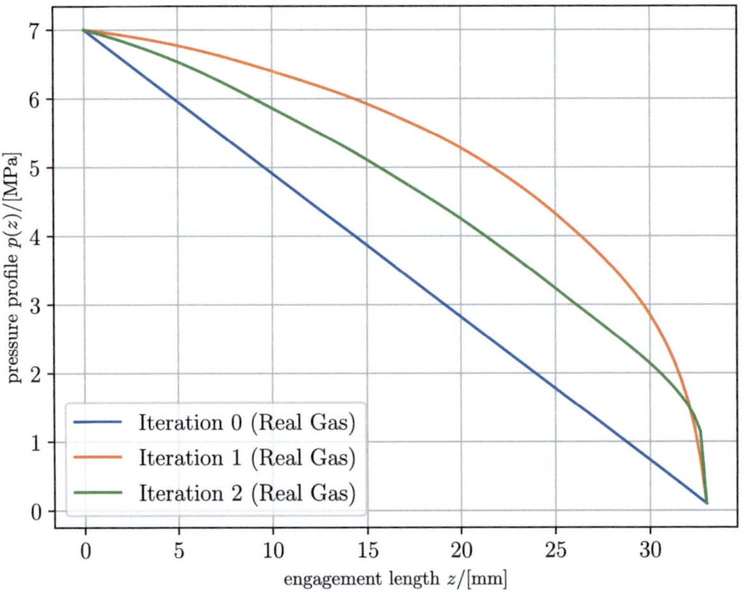

Fig. 9 Final fluid simulation results for channel/crevice flow with real gas equations

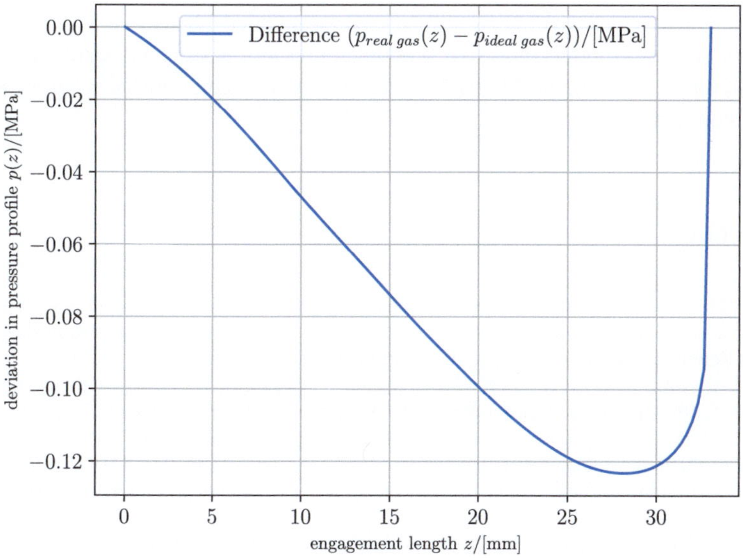

Fig. 10 Error analysis for fluid simulation results comparing ideal gas and real gas equations

the temperature dependence, it may then be observed and concluded that in certain situations these non-ideal real gas effects have significant accuracy implications and consequent detrimental validation/verification concerns in high accuracy/precision work. The particular study that was investigated in this paper yielded errors of up to ± 0.12 MPa over a 7 MPa range i.e. errors of $100 \times \frac{0.12}{7} \approx \pm 1.7\%$ occurred where the particular engineering equipment was intended to target a pressure accuracy of $\pm 0.1\%$, thus the simplification of using ideal equations of states whilst quicker and more convenient was demonstrated to have unacceptable accuracy or safety implications.

4 Discussion and Conclusion

In this study a mathematical analysis was conducted to investigate various approaches to modelling and incorporating real gas effects in practical engineering analysis research work. It was determined that the typical modelling assumptions in many CFD studies implicitly rely on a combination of simplifying assumptions that are not always physically and mathematically justified. A representative complex engineering problem was analysed from mathematical first principles by reformulating the original problem as a nonlinear Reynolds integral equation with variable non-ideal density and viscosity equations of states for an isothermal flow, and it was determined that the overall cumulative effect may have detrimental accuracy and validity concerns.

Based on the results from this study it is recommended that a full real gas non-ideal equation of states models inclusive of coupled pressure-density-viscosity models are utilized for engineering problems that may exhibit a combination of accuracy and/or safety concerns.

References

Akhlaghi H, Roohi E, Stefanov SA (2023) Comprehensive review on micro- and nano-scale gas flow effects: slip-jump phenomena, Knudsen paradox, thermally-driven flows, and Knudsen pumps. Phys Rep 997:1–60. https://doi.org/10.1016/j.physrep.2022.10.004

Atmaca AU, Erek A, Ekren O (2020) One dimensional analysis of the convergent divergent motive nozzle for the two phase ejector: effect of the operating and design parameters. Appl Therm Eng 181(115866). https://doi.org/10.1016/j.applthermaleng.2020.115866

Babu V (2021) Fundamentals of gas dynamics, 2nd ed. Elsevier. ISBN 978-3-030-60818-7

Bird GA (1994) Molecular gas dynamics and the direct simulation of gas flows. Oxford University Press, Oxford, England

Blazek J (2001) Computational fluid dynamics: principles and applications, 2nd ed. Elsevier

Bobylev AV, Meleshko SV (2020) Group analysis of the generalized Burnett equations. J Nonlinear Math Phys 27(3):494–508. https://doi.org/10.1080/14029251.2020.1757238

Bock T, Bernien M, Buchmann C, Rubin T, Jousten K (2021) Investigation of the effects of valve closing in a static expansion system. Vacuum 184:109918. https://doi.org/10.1016/j.vacuum.2020.109918

Chapman S, Cowling TG (1970) The mathematical theory of non-uniform gases: an account of the kinetic theory of viscosity, thermal conduction and diffusion in gases. Cambridge Mathematical Library, 3rd ed. ISBN 052140844X

Cheng Y, Chen S, Wang C, Xi Z, Sun W, Liu D (2021) Gas expansion process in the dynamic vacuum calibration. J Shanghai Jiaotong Univ (Science) 26(2):193–200. https://doi.org/10.1007/s12204-021-2279-3

Dai Y, Zou J, Zhu C, Wu JT, Hu D (2012) Real gas effect on unsteady flow of natural gases in shock tubes. Adv Mater Res 361–363:603–603. https://doi.org/10.4028/www.scientific.net/amr.361-363.603

Dongari N, Sharma A, Durst F (2009) Pressure-driven diffusive gas flows in micro-channels: from the Knudsen to the continuum regimes. Microfluid Nanofluid 6:679–692. https://doi.org/10.1007/s10404-008-0344-y

Fasshauer G (2010) Meshfree approximation methods with MATLAB, 1st ed. World Scientific. ISBN 981-270-634-8

Fefferman CL (2025) Existence and smoothness of the Navier–Stokes equation. Clay Mathematics Institute. https://www.claymath.org/wp-content/uploads/2022/06/navierstokes.pdf. Cited 9 February 2025

Forssen C, Silander I, Zakrisson J, Zelan M, Axner O (2022) An optical pascal in Sweden. J Opt 24:033002. https://doi.org/10.1088/2040-8986/ac4ea2

Guo ZL, Shi BC, Zheng CG (2007) An extended Navier–Stokes formulation for gas flows in the Knudsen layer near a wall. Eur Phys Lett 80:24001. https://doi.org/10.1209/0295-5075/80/24001

Jousten K, Hendricks J, Barker D, Douglas K, Eckel S, Egan P, Fedchak J, Flugge J, Gaiser C, Olson D, Ricker J, Rubin T, Sabuga W, Scherschligt J, Schodel R, Sterr U, Stone J, Strouse G (2017) Perspectives for a new realization of the pascal by optical methods. Metrologia 54(6):S146–S161. https://doi.org/10.1088/1681-7575/aa8a4d

Khonsari MM (2013) Reynolds equation. In: Encylopedia of tribology. Springer, Boston, USA, pp 2769–2772. https://doi.org/10.1007/978-0-387-92897-5 146

Kopriva DA (2009) Implementing spectral methods for partial differential equations: algorithms for scientists and engineers. Springer. ISBN 978-90-481-2260-8

Latto B, Saunders MW (1972) Viscosity of nitrogen gas at low temperatures up to high pressures: a new appraisal. Can J Chem Eng 50:765–770. https://doi.org/10.1002/cjce.5450500615

Lemmon EW, Jacobsen RT, Penoncello SG, Friend DG (2000) Thermodynamic properties of air and mixtures of nitrogen, argon, and oxygen from 60 to 2000 K at pressures to 2000 MPa. J Phys Chem Ref Data 29(331):331–385. https://doi.org/10.1063/1.1285884

Lemmon EW, Jacobsen RT (2014) Viscosity and thermal conductivity equations for nitrogen, oxygen, argon and air. Int J Thermophys 25(1):21–69. https://doi.org/10.1023/B:IJOT.0000022327.04529.f3

Lions P-L (1996) Mathematical topics in fluid mechanics: volume 1: incompressible models (Oxford lecture series in mathematics and its applications). Oxford University Press, Oxford, England

Lions P-L (1998) Mathematical topics in fluid mechanics: volume 2: compressible models (Oxford lecture series in mathematics and its applications). Oxford University Press, Oxford, England

NIST (2023) Quantum pascal: fixed length optical cavity (floc) pressure standards. https://www.nist.gov/programs-projects/quantum-pascal-fixed-length-optical-cavity-flocpressure-standards. Cited 9 February 2025

Oliva AA, Steady MSC (2023) Quasi-one-dimensional, internal compressible flow with area change, heat adition and friction. J Fluid Mech 957(A15):A15–25. https://doi.org/10.1017/jfm.2023.44

Peng DY, Robinson DB (1976) A new two-constant equation of state. Ind Eng Chem Fundam 15(1):59–64. https://doi.org/10.1021/i160057a011

Ramnath V (2012) Computational solution of the extended Navier–Stokes PDE's: incorporating nonlinear fluid-solid boundary conditions for microfluidic simulations. In: 4th international metrology conference CAFMET, pp 78–88. ISBN 978-1-62276-063-3

Ramnath V (2013) Solution of a nonlinear Volterra integral equation for concentric annulus channel pressure profiles with mixed viscous/molecular flow regimes. In: International conference on advances in engineering sciences and applied mathematics (ICAESAM2013), Cape Town, South Africa: International Institute of Engineers (IIE), pp 68–73

Sabuga W, Molinar G, Buonanno G, Esward T, Legras JC (2006) Finite element method used for calculation of the distortion coefficient and associated uncertainty of a PTB 1 GPa Pressure Balance—EUROMET Project 463. Metrologia 43(3):311–325. https://doi.org/10.1088/0026-1394/43/3/015

Sabuga W, Sharipov F, Priruenrom T (2011) Determination of the effective area of piston cylinder assemblies using a rarefied gas flow model. In: Jousten K (ed) Proceedings of the 5th CCM international conference on pressure and vacuum metrology IMEKO, Berlin, Germany, pp 260–262. ISSN: 0030-834X

Saffari H, Zahedi A (2013) A new alpha-function for the Peng–Robinson equation of state: application to natural gas. Chin J Chem Eng 21(10):1155–1161. https://doi.org/10.1016/s1004-9541(13)60581-9

Sforza PM (2016) Theory of aerospace propulsion, 2nd ed. Elsevier. ISBN 978-0-12-809326-9

Shapiro AH (1953) The dynamics and thermodynamics of compressible fluid flow, vol 1, 1st ed. Ronald Press Company. ISBN 0471066915

Sharipov F (2016) Rarefied gas dynamics: fundamentals for research and practice. Wiley-VCH, Weinheim, Germany

Sharipov F, Yang Y, Ricker JE, Hendricks JH (2016) Primary pressure standard based on piston-cylinder assemblies. Calculation of effective cross sectional area based on rarefied gas dynamics. Metrologia 53(5):1177. https://doi.org/10.1088/0026-1394/53/5/1177

Song L, Li P-W, Gu Y, Fan C-M (2020) Generalized finite difference method for solving stationary 2D and 3D Stokes equations with a mixed boundary condition. Comput Math Appl 80(6):1726–1743. https://doi.org/10.1016/j.camwa.2020.08.004

Sun X, Fang Y, Zhao W, Xiang S (2022) New alpha functions for the Peng–Robinson cubic equation of state. ACS Omega 24:5332–5339. https://doi.org/10.1021/acsomega.1.c06519

Tannehill JC, Anderson DA, Pletcher RH (1997) Computational fluid mechanics and heat transfer, 2nd ed. Taylor and Francis
Thakur VN, Yadav S, Kumar A (2023) Quantum pascal realization from refractometry. In: Aswal DK, Yadav S, Takatsuji T, Rachakonda P, Kumar H (eds) Handbook of metrology and applications. Springer, Singapore, pp 1–38. doi:https://doi.org/10.1007/978-981-19-1550-518-1
Versteeg HK, Malalasekera W (1995) An introduction to computational fluid dynamics: the finite volume method. Pearson
White FM (1991) Viscous fluid flow. McGraw-Hill
Zienkiewicz OC, Taylor RL, Nithiarasu P (2005) The finite element method for fluid dynamics, 6th ed. Elsevier Butterworth Heinemann. ISBN 0-7506-6322-7

A Novel Multi-objective Programming Model for Sustainable Agriculture Planning Based on the Water-Energy-Food Nexus Framework

Moatafa Mardani Najafabadi and Elahe Ahani

Abstract As the world's population grows and diets become more diverse, the demand for food has increased and the use of non-renewable resources to produce it has skyrocketed. In this context, the new Water-Energy-Food Nexus approach has been developed to achieve sustainability by balancing the different objectives, benefits, and needs of communities and the environment. In this context, the new Water-Energy-Food Nexus approach has been developed to achieve sustainability by balancing the different objectives, benefits, and needs of communities and the environment. This approach is based on quantifying the relationships between water, food, and energy through qualitative and quantitative modeling and advancing research for integrated resource management. A key mechanism of this approach is to prevent the implementation of inappropriate single-sectoral policies in crop production. Therefore, mathematical programming models play an important role in setting multiple objectives and achieving comprehensive results. This chapter aims to present a multi-objective mathematical programming model using the WEF nexus approach to achieve sustainable agricultural development. This model can identify the complex and interrelated relationships among available resources, leading to more comprehensive programs, decisions, and policies in water, food, and energy security. Ultimately, it guides the country in overcoming barriers and challenges to long-term sustainability. In addition to discussing the general characteristics and types of mathematical programming models, this chapter elaborates on the development and optimization of these models based on the systemic approach, providing a

M. Mardani Najafabadi (✉)
Department of Agricultural Economics, Faculty of Agricultural Engineering and Rural Development, Agricultural Sciences and Natural Resources University of Khuzestan, Mollasani, Iran
e-mail: m.mardani@asnrukh.ac.ir

E. Ahani
Lecturer, Department of Accounting and Department of Agricultural Economics, Bozorgmehr University of Qaenat and Torbat Heydariyeh, Razavi Khorasan, Iran
e-mail: e_ahani@buqaen.ac.ir

M. Ram et al. (eds.), *Advances in Mathematics for Engineering Sciences*, Mathematical Engineering, https://doi.org/10.1007/978-3-031-95693-5_5

practical example. Finally, the economic and environmental benefits of this method for sustainable agricultural development are evaluated.

Keywords Optimization · Agricultural sector · Mathematical programming model

1 Introduction

According to United Nations projections, the population of developing countries, including Iran, is expected to increase by about 50% by 2050. This population growth will increase the demand for food and put additional pressure on limited water and energy resources at both regional and national levels (UNEP 2021). Therefore, in most parts of the world, especially in the Asian region, proper planning and optimal use of limited resources to meet the social and economic needs of society, both now and in the future, in line with sustainable development and environmental protection, remains a major challenge (Madani et al. 2016).

Producing food for the world's population through agriculture consumes, on average, 90% of freshwater resources and 30% of the energy used in food production and supply chains (Hoff 2011). As a result, the demand for food, energy, and water is projected to increase by 30%, 40%, and 30%, respectively, by 2030 as the world's population grows and economies develop. Water, food, and energy are fundamental human needs, and their intricate and boundless interrelationships highlight their interdependencies and impacts during the production process (Wang et al. 2023).

Essentially, food production depends on water and energy resources, energy production depends on water, and the process of extracting and transporting water resources requires energy. In addition, clean energy or biogas can be generated from the food system. This clearly illustrates the interdependence and mutual impact of each resource, which are inextricably linked. Therefore, a balance must be maintained between the extraction and use of production resources and the level of agricultural production (Emamzadeh et al. 2016). Proper planning and effective management of the consumption of non-renewable resources are crucial to meet the social and economic needs of communities now and in the future, in line with sustainable development policies and environmental protection (Madani et al. 2016; Ahani et al. 2024). The depletion of groundwater resources in Iran poses a significant threat to various sectors. Without significant action to optimize the use of these limited water resources, the country will experience land subsidence. This problem is particularly acute in the arid and semi-arid regions of Iran, putting urban infrastructure and agricultural land at risk (Li et al. 2017).

The need to secure without causing damage led the United Nations to establish the Sustainable Development Goals (SDGs) in 2015. These goals aim to achieve sustainable continuity and unwavering perseverance across generations of societies to ensure the availability of these essential resources for future generations (Samadi-Foroushani et al. 2022). Recognizing the close linkages and mutual impacts of the water-food-energy systems, a new concept known as the nexus approach has been introduced.

In 1992, the Earth Charter was adopted at the Rio Conference. The Charter provides a framework for a just, sustainable, and peaceful global society in the twenty-first century. During the World Economic Forum in Germany, to address issues of resource scarcity. This approach aims to provide a transparent, intelligent, and logical framework for exploring the various dimensions of the linkages between these three areas, providing an integrated analysis to better understand the interactions of the dynamic interplay between humankind and the natural world (Hoff 2011).

Poor coordination and weak interactions between different sectors have led to unstable and unbalanced resource management. This has disrupted food security and will hinder the path to sustainable development in the long run (Emamzadeh et al. 2016; Ringler et al. 2013). Understanding these interrelationships allows for the adoption of more comprehensive programs, decisions, and policies in water, food, and energy security, and ultimately guides national development plans to overcome obstacles and challenges to sustainable development on a much broader scale (Zhang et al. 2022). Resource use and avoiding the negative impacts of single-sector development policies are among the key benefits of this approach.

Therefore, focusing on a single-sector policy creates various problems for other resources and poses a threat to the future of the country (Ahani et al. 2024; Esteve et al. 2015). As discussed, with the increase in population, the need for food production and consequently the extraction of water resources has increased. With the increased demand for water and food, there is also an increased demand for non-renewable energy to achieve socio-economic development, which ultimately leads to increased greenhouse gas emissions (Li et al. 2017). Therefore, sustainable development programs, government policies, and strategies of regulatory and operational institutions strive to optimize resource consumption while minimizing environmental challenges (Dumont et al. 2013).

1.1 Background

Researchers believe that Iran will face significant water, energy, and food crises in the future. To address this, the establishment of a nexus network to provide modern planning and policy knowledge in these sectors is essential. This network will facilitate the exchange of ideas and collaboration, especially among policy-makers and stakeholders, through the development of technical knowledge (Lee et al. 2020). Because modeling is a scientific method for understanding phenomena, understanding external realities depends on successful modeling.

The more advanced the models, the easier it is to understand the external world. Of course, understanding phenomena is crucial for controlling and changing policy (Lee et al. 2020). By applying a comprehensive nexus approach, it is possible to improve the efficiency of resource use and reduce environmental risks and degradation (Ahani et al. 2024; Nkiaka et al. 2023). In this context, a planning methodology for sustainable resource management has been proposed. This model can handle management decisions that consider not only the maximization of farmers' net profits but also the control of greenhouse gas and environmental emissions. The sections Challenges and Solutions to Determine Sustainable Agricultural Planning, Agricultural Planning and Reduction of Environmental Impacts Based on the WEF Nexus, Optimization of Irrigation Water Consumption Based on the WEF Nexus, and Mathematical Optimization Based on the WEF Nexus review the WEF Nexus and its relationship to sustainability based on mathematical planning models in various types of literature.

1.1.1 Sustainable Agriculture Planning

The essential questions in Fig. 1a need to be answered correctly to determine optimal crop area planning. In this part of the figure, there are two groups of questions about the main challenges of this topic; the first group looks for general questions about the amount of demand, available resources, and potential for agricultural production. These questions lead to opportunity studies by organizations involved in agricultural planning. The second group of questions deals in more detail with the subject of crop area planning. This group asks questions such as what, how much, how long, and where to produce. Correct answers to these questions lead to optimal crop area planning. Of course, to reach these answers, it is crucial to consider the first groups of questions as limitations in determining the optimal cropping pattern (Mardani Najafabadi and Shirzadi Laskookalayeh. 2023).

Different scientific sources have provided definitions of crop area planning, sometimes covering some of the objectives of the cropping pattern or introducing other issues such as crop rotation or crop composition (He and Cai 2016).

According to Mardani Najafabadi et al. (2019), crop area planning determines the cropping system that includes:

1. based on climatic conditions,
2. optimal utilization of resources and production factors commensurate with regional potentials and economic advantage,
3. Consider the principles of agricultural production sustainability,
4. Observe environmental considerations,
5. Provide in line with the country's major policies, and
6. Ensure food security is a top priority.

Agriculture researchers and experts have been and are seeking to address these issues through various studies.

The main challenges and solutions for determining sustainable agricultural planning in the eight categories in Fig. 1b are shown. These categories are presented based on the focus on specific objectives in optimal cropping pattern studies. To achieve a high level of sustainability in agriculture, it's important to research and implement a variety of methods (Fig. 1c). In today's global agricultural landscape, sustainable agriculture is highly valued and widely recognized. It adheres to scientific

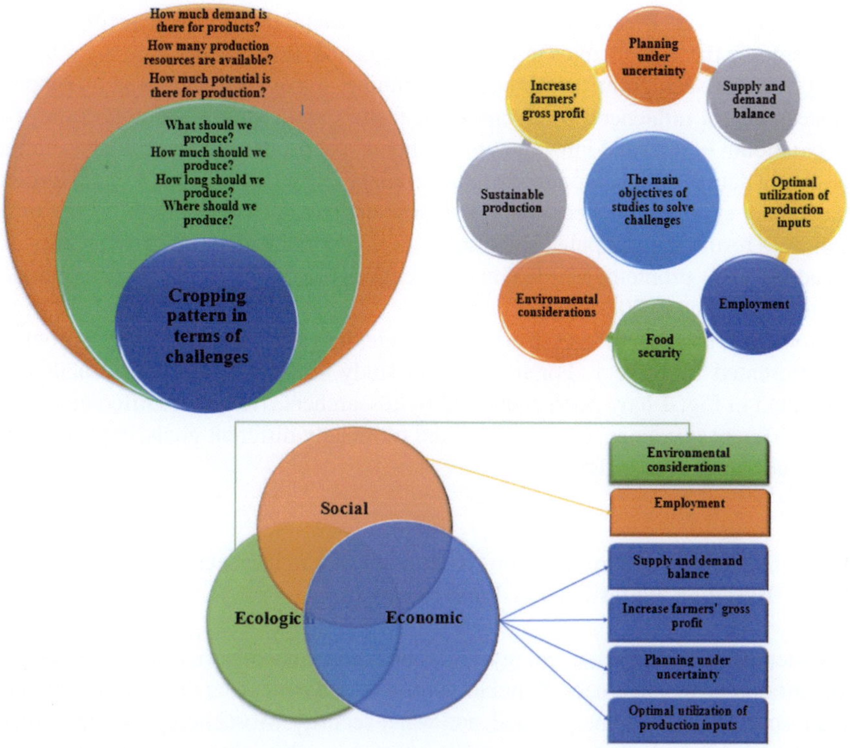

Fig. 1 Challenges and solutions to determine Sustainable Agriculture Planning

and ecological principles with the primary goal of achieving a balanced and stable production system (Sands 1995).

Ecological sustainability refers to the maintenance of the system's ability to fully adapt to the environment and its changes. Sustainability in the economic dimension is an important concept in the literature of sustainable development. One of the most important goals is to improve the gross profit of farmers (Medellín-Azuara et al. 2010; Qureshi et al. 2007).

This objective is critical in that it easily influences others. For example, it is not possible to propose the cultivation of less profitable and environmentally friendly crops that cannot provide a farmer with a minimum subsistence. Supply and demand balances, optimal use of production inputs, and attention to uncertainty can also be considered in the economic dimension of sustainable production.

The core pillars of sustainable development include people, land, and profit, which can be divided into three categories (Emamzadeh et al. 2016): economic, social, and environmental. Every country, including Iran, faces obstacles in achieving sustainable development goals (Chaudhuri and Banerjee 2010). The main challenges in Iran include dependence on oil revenues, lack of public awareness of these goals, and deforestation. Despite these problems, it is essential to make concerted efforts to overcome these challenges in Iran (Brambilla and Testa 2014).

As can be seen from the above, determining a cropping pattern is a complex process that is influenced by a variety of factors and requires the collection of a large amount of data and information. Mathematical programming problems, unlike assuming data certainty, are subject to uncertainty under current conditions (Mardani Najafabadi et al. 2023). Accounting for these uncertainties does not affect the solution. As a result, the computed solution becomes impractical in real-world scenarios. To address this problem, Mardani and colleagues used a robust optimization model (Badih Barzin et al. 2024).

Some researchers' efforts to address the challenges of the crop pattern problem are presented in Table 1. For instance, in study number 1, all of the challenges presented in Fig. 1 have been addressed by researchers. However, Different studies have different questions, perspectives, and ultimately different goals, depending on the issues facing the region.

1.1.2 Agricultural Planning and Reduction of Environmental Effects Based on Water-Energy-Food Nexus

The increasing use of water, energy, and food resources poses significant challenges for sustainable management. Population growth and increasing demand for services and resources threaten energy, food, and water security. Most energy consumption is based on fossil fuels and non-renewable sources, resulting in greenhouse gas emissions (Kalbali et al. 2021). Scientists widely agree that these emissions contribute significantly to global warming and climate change. For example, in the United States, 5% of total greenhouse gas emissions are associated with energy use in the water sector, a trend that is also observed in many other countries (Lalawmpuii and Rai 2023). In addition, thermal power plants and agricultural technologies contribute significantly to greenhouse gas emissions, with agriculture alone accounting for approximately 30% of total emissions (FAO 2014). Thus, the growing demand for energy and food, coupled with the water demand, is leading to increased greenhouse gas emissions. Effective resource management for sustainable development must focus on controlling these emissions to mitigate the effects of climate change (Batlle-Bayer et al. 2020; Esteve et al. 2015). Managing the nexus between water, energy, and food while controlling emissions is critical to reducing environmental impacts. Addressing the negative consequences of greenhouse gas emissions is also essential (Dumont et al. 2013). Given the critical role of crops in food security, agricultural land will become increasingly important for future food production, making its proper use critical (Li et al. 2019).

The study by Ahani et al. (2024) showed that although the crops analyzed have environmental constraints (such as water consumption, energy use, and greenhouse gas emissions) for cultivation in different catchment areas, they provide economic benefits to farmers in terms of calorie production and increased profits. Therefore, the area planted for crops should be determined based on the analysis of carbon

Table 1 Articles reviewed are classified based on categories and common keywords

Title	Study area	Optimal utilization of inputs	Gross profit increase	Production sustainability	Environmental considerations	Employment	Applying uncertainty condition	Supply and demand balance	Food security
Mathematical programming model (MMP) for optimization of regional cropping patterns decisions: A case study (Mardani Najafabadi et al. 2019)	Isfahan Province (Iran)	✓	✓	✓	✓	✓	✓	✓	✓
Agricultural Optimal Cropping Pattern Determination Based on Fuzzy System (Neamatollahi et al. 2017)	Mashhad Plain (Iran)	✓	✓	✓	✓	✗	✓	✗	✗
Optimization of the cropping pattern in Egypt (Osama et al. 2017)	Old lands of Egypt	✓	✓	✗	✗	✗	✗	✗	✗
An inexact robust two-stage mixed-integer linear programming approach for crop area planning under uncertainty (Zhang et al. 2018a, b)	Heihe River Basin, northwest China	✓	✓	✗	✓	✗	✓	✗	✗

(continued)

Table 1 (continued)

Title	Study area	Optimal utilization of inputs	Gross profit increase	Production sustainability	Environmental considerations	Employment	Applying uncertainty condition	Supply and demand balance	Food security
Linking a farmer crop selection model (FCS) with an agronomic model (EPIC) to simulate cropping pattern in Northeast China (He and Cai 2016)	Northeast China	✓	✓	✗	✗	✗	✗	✗	✗
Climate change and crop choice in Zambia: A mathematical programming approach (Wineman and Crawford 2017)	Zambia	✓	✓	✗	✗	✗	✗	✗	✓
An interval multistage joint-probabilistic chance-constrained programming model with left-hand-side randomness for crop area planning under uncertainty (Zhang et al. 2017)	Heihe River Basin, northwest China	✓	✓	✗	✓	✗	✓	✗	✗
Fuzzy multi-objective linear programming applying to crop area planning (Zeng et al. 2010)	Gansu province of northwest China	✓	✓	✓	✓	✗	✓	✗	✗

(continued)

Table 1 (continued)

Title	Study area	Optimal utilization of inputs	Gross profit increase	Production sustainability	Environmental considerations	Employment	Applying uncertainty condition	Supply and demand balance	Food security
An interval nonlinear multi-objective programming model with fuzzy-interval credibility constraint for crop monthly water allocation (Zhang et al. 2018a)	Yingke Irrigation District, China	✓	✓	✗	✓	✗	✓	✗	✗
Mixed integer linear programming models for optimal crop selection (Filippi et al. 2017)	Veneto region, Italy	✓	✓	✗	✗	✗	✓	✗	✗

emissions, or the crops should be grown in a more environmentally friendly manner. Overall, the use of the nexus approach through mathematical optimization models is crucial not only for economic development but also for optimal resource management. To achieve sustainability in the agricultural sector, production resources such as water and land must be allocated as efficiently as possible. In this context, based on the developed multi-objective programming, the cultivated area is 6% according to the WEF nexus approach of Ahani et al. (2024). This allows farmers to reduce resource consumption, increase profitability and welfare by 39.42%, and reduce greenhouse gas emissions by 21.13%.

1.1.3 Optimizing Irrigation Water Use Based on System Modeling

The water-food-energy nexus perspective is essential for a thorough assessment of water resource management policies. As demand increases, competition for renewable and non-renewable resources in agriculture, industry, health, and the environment will have significant impacts on livelihoods and the environment (Ding et al. 2023). The interaction between water, food, and energy components in the production cycle is critical. For example, the expansion of bioenergy crop production for energy purposes may increase energy supply and create employment opportunities, but it also increases competition for water and soil resources and reduces the production of other agricultural products, thereby affecting food security (Nasrollahi et al. 2021; Wolfert et al. 2017). It is projected that an average of 60% more food production will be required by 2050 to meet the nutritional needs of the world's population. In addition, total water withdrawals for irrigation are expected to increase by 10% by 2050 (FAO 2014). Therefore, this comprehensive strategy helps clarify the complex and dynamic linkages and relationships between these sectors and supports sustainable decision-making for the allocation of limited resources.

Increasing crop production with minimal energy and water use, especially in regions experiencing water scarcity due to climate change, is critical for global food and energy security (Emamzadeh et al. 2016). Traditional water management techniques in agriculture aim to increase crop productivity through various agricultural and engineering methods. These include improving water-efficient irrigation technologies, adopting methods to reduce soil evaporation, and increasing the drought tolerance of crops through selective breeding and genetic modification (Zhang et al. 2022; Kalbali et al. 2021). As a result, resource management is mainly focused on water distribution and improving water resource efficiency throughout the production process. Meanwhile, quantitative research and analysis on reducing water consumption in agriculture and energy production using the nexus model provides a new strategy to reduce pressure on water resources. According to Ahani et al. (2024), water availability for agriculture in the upland areas of the northeastern river basins of Iran is reduced by 44% based on the systemic approach. In addition, electricity consumption was optimized by 50% in this study.

Multiobjective methods are commonly used in the scientific literature on cropping patterns. Most studies show that using different planning models to achieve an optimal cropping pattern not only improves current conditions but also ensures efficient use of resources and inputs (Han et al. 2022). These studies often include objectives related to program yield and water use, taking into account regional constraints. The primary objective of this study is to implement a comprehensive and systematic approach to resource management and determine cropping patterns with a focus on reducing water consumption, maximizing profits, and achieving sustainable agriculture.

1.1.4 Mathematical Optimisation Based on Water-Energy-Food Nexus

In recent decades, an overemphasis on the economic well-being of farmers has led to the neglect of detrimental environmental and social impacts. As a result, there is a noticeable gap in the use of effective models that can simultaneously address economic, social, and environmental dimensions (Forni et al. 2016; Ringler et al. 2013).

Mathematical planning models are adept at assessing the impacts of various decisions on urban, industrial, agricultural, and environmental uses of agricultural production resources. These models have been widely used (Romero et al. 2004). When managers face multiple specific objectives, decision-making requires not only tools but also a wide range of information. Managing a system based on such information and multiple objectives over time and across different locations is complex and requires a method to guide agricultural managers in making logical decisions based on available data and different objectives. In the optimization literature, this type of decision model is considered appropriate (Wang et al. 2023; Shang et al. 2018). In such circumstances, it is categorized under "methods that involve mutual goals, trade-offs, and conflicting" What all these methods have in common is the simultaneous consideration of multiple goals (Romero et al. 2004). Therefore, mathematical planning approaches offer it has a crucial role in the decision and selection of the appropriate type of crop cultivation and have been widely used in various studies.

Mardani Najafabadi et al. (2019) utilize both single-objective and multi-objective planning models to increase productivity by maximizing total yield, minimizing water use, reducing risk, and reducing agricultural input consumption with a focus on environmental considerations.

In general, within the WEF nexus approach, various multi-objective optimization models have been used to increase hydropower production and conserve water supply sources through hydro-economic models (Dumont et al. 2013; Lee et al. 2020; Yue and Guo 2021; Lalawmpuii and Rai 2023; Ahani et al. 2024), stochastic optimization to reduce the financial burden of farmers and preserve the environment (Forni et al. 2016; Wang et al. 2023; Zhang et al 2022; Esteve et al. 2015; Li et al. 2019), simulations to reduce water and energy consumption and improve crop performance (Mardani Najafabadi et al. 2019; Nie et al. 2019; Shang et al. 2018; Yue et al. 2020; Ringler et al. 2013), and mixed-integer nonlinear multi-objective planning models aimed at reducing water and energy use to increase economic yield (Mardani Najafabadi et al. 2019; Ahani et al. 2024; Emamzadeh et al. 2016).

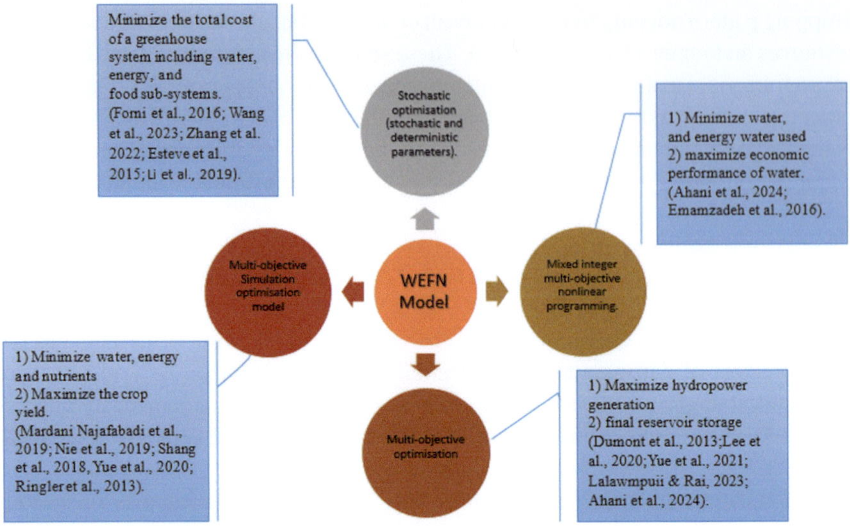

Fig. 2 Summary of different optimization studies

Figure 2 illustrates examples of studies, such as reducing the use of non-renewable resources and maximizing the yield of a program, that uses them systematically. These models include stochastic optimization (with both stochastic and deterministic parameters), multiobjective simulation optimization models, multiobjective optimization, and multiobjective nonlinear optimization. These models have been aligned with economic and environmental objectives based on various studies.

The primary objective of this chapter is to present a novel optimization planning model for sustainable agricultural planning using the WEF Nexus framework. The secondary objectives are to (1) maximize farmers' profits and (2) reduce greenhouse gas emissions. By applying the nexus approach, the linkages and integration between the economic, environmental, and agricultural sectors are clearly illustrated. The development of this model provides valuable insights for developing strategies that reduce environmental impacts and increase farmers' incomes, thereby supporting the growth of the agricultural sector.

2 Materials and Methods

Mathematical planning is an important tool in research and is widely used to analyze resource management issues. These models are particularly effective in agricultural economics due to their unique ability to link economic components with biophysical and ecological aspects (Mardani Najafabadi et al. 2019). Hazell and Norton (1986) noted that while other models may have similar capabilities to varying degrees, mathematical planning models excel in agricultural policy analysis due to their ability to model the complex and interdependent nature of agricultural activities (Kalbali et al. 2021). By integrating relationships between model components and existing constraints based on the water-food-energy nexus, mathematical planning provides more comprehensive and realistic results. However, individual approaches often lack sufficient and comprehensive information. Therefore, studies using the WEFN approach to control environmental impacts and manage input consumption under different climatic conditions to achieve sustainable development goals have been reviewed (Emamzadeh et al. 2016; Forni et al. 2016). The primary goal in multiobjective planning is to design an optimal solution, typically using algorithms and min–max models. Given the heterogeneity of many desired objectives, they must be harmonized (Mardani Najafabadi et al. 2019).

2.1 Conceptual Structure of the Sustainable Agriculture Water-Food-Energy Nexus (SAWEFN)

Following the introduction of the nexus approach, multiobjective optimization has been used as a mathematical tool to address resource challenges. Because the independent goal of securing each nexus resource can negatively impact other resources, researchers have used multi-objective optimization to address these conflicts in environmental resource management. The growing environmental crisis and increasing global population, which have increased the demand for resources, led to the development of the Systemic approach (Li et al. 2017). This approach emphasizes the internal structure, relationships, and interactions of these three resources as an integrated system, which requires the development or localization of assessment tools or methods to identify synergies based on the problem statement and geographical factors. In addition, sustainable management of nexus resources is challenging for managers, decision-makers, researchers, and others because it often involves multiple and conflicting objectives, such as economic development, environmental protection, and social improvement, adding to the complexity and uncertainty of the problem (Cui et al. 2023).

The use of multi-objective optimization techniques can be critical to finding solutions that meet the goals of the problem. These models can identify synergies and trade-offs among resources. However, managing and making decisions about nexus resources should involve quantifying these relationships and making decisions based on the resulting data. Group decision-making methods, such as social choice, can be beneficial to decision-makers. By understanding the relationships among water, food, and energy resources, they can create scenarios and management plans that minimize damage to the sustainability of the system (Mardani Najafabadi et al. 2019).

In the following sections, since all equations and relationships between decision variables in the objective functions and constraints are linear, the optimization model has been solved using linear programming methods (Simplex algorithm).

By applying the new mathematical programming approach, a systemic optimization management model has been developed in various economic and environmental sectors. The core structure of the mathematical programming model, which uses the latest multi-objective programming (MOP) techniques for the water-food-energy nexus, is shown in Fig. 3. As shown, the primary objectives of agricultural activities are to maximize profit and minimize greenhouse gas emissions, taking into account constraints related to water, land, fertilizers, pesticides, and energy, which are detailed below.

In the water-food-energy cycle, each phase of energy development plays a critical role, including fossil fuel extraction, production and processing, electricity generation, and waste treatment. Energy is also used for mechanization, land preparation, fertilizer production, packaging, and food processing. As the world continues to develop and move toward greater industrialization, energy consumption in its various

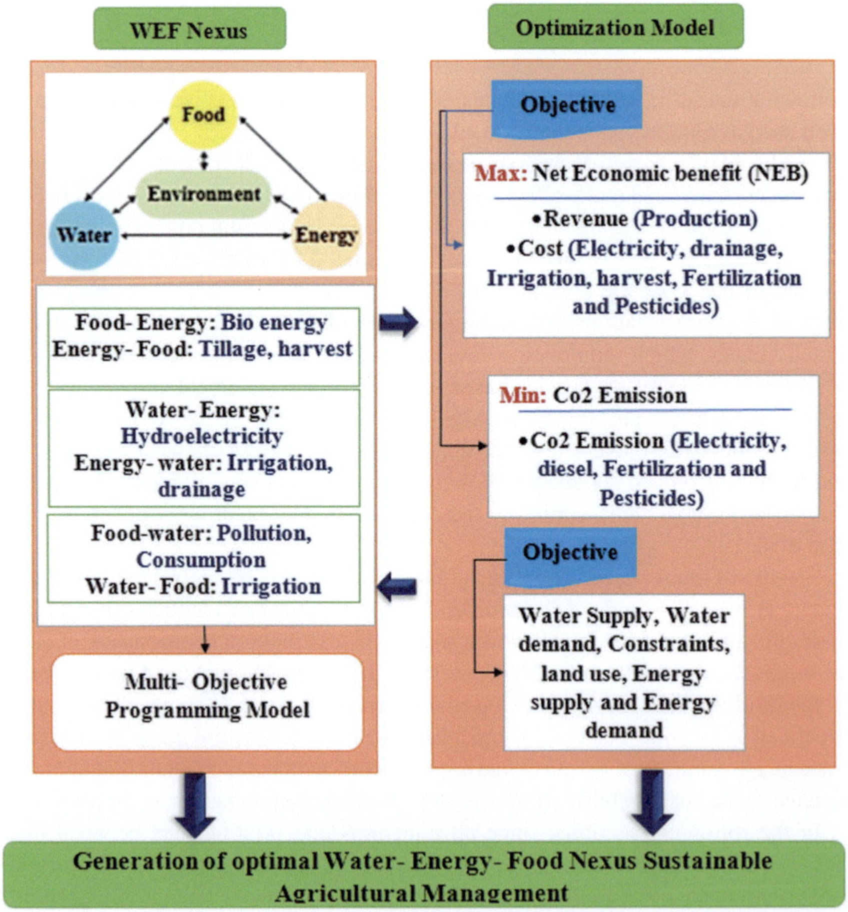

Fig. 3 Structural Algorithm of the WEF System

forms has increased by an average of about 30% (Emamzadeh et al. 2016). In addition, the agricultural sector uses more non-renewable resources compared to other sectors, resulting in greenhouse gas emissions. Therefore, substituting biofuels and clean energy in this sector can have a significant positive impact.

2.2 *The Structural Model of the Water-Food-Energy Nexus*

Mathematical optimization is often used as a decision-support tool to improve system efficiency and help decision-makers achieve the best possible outcomes. It provides a mathematical framework to model real-world problems and allows for the selection of the optimal choice among various alternatives. In addition, optimization tools play a critical role in ensuring resource security and maintaining system sustainability (Romero et al. 2004). Therefore, given the objectives of the nexus approach, mathematical optimization is considered an appropriate method for assessing the nexus.

This section also includes environmental emissions from energy activities, covering emissions to air, water, and soil, air quality, and greenhouse gas emissions, as an integral part of the energy-environment model. The second section is dedicated to the development of water supply and demand, also in the short and long term. This section consists of four main components: the water demand model, the hydrology model, the supply model, and water resources and quality. The demand component calculates water demand for four primary sectors: agriculture, industry, residential and commercial, and other sectors, on both a short-term (monthly) and long-term basis. The supply and water resources component identifies major water sources, including surface water, oceans, perennial streams, dams, and groundwater. In addition, this section models water transmission lines to consumption points and return lines to rivers, as well as water treatment and sewerage systems.

2.2.1 Multi-function Model

Considering that systemic models assess different dimensions, the proposed model includes different and diverse objectives (Ahani et al. 2024; Mardani Najafabadi et al. 2019). In general, the water-food-energy nexus aims to: (1) identify trade-offs and synergies among water, food, and energy resources; (2) assess the social and environmental impacts of each resource within the system; and (3) guide development through cross-sectoral policies to achieve economic growth and improve social and environmental conditions.

Therefore, this study outlines several objectives for the proposed cropping model. Achieving the primary objectives of the cropping pattern requires consideration of different and sometimes conflicting objectives, such as maximizing net farmer profit and minimizing environmental damage. As a result, the metrics for these objectives differ significantly. For example, the goal of maximizing profit, measured in monetary terms, contrasts with minimizing greenhouse gas emissions. Therefore, the application of MOP (Multi-Objective Programming) methods requires homogenization techniques. The method proposed by Jones and Barnes in 2000 was adopted to harmonize multiobjective planning. The algebraic forms of the goals of maximizing net profit and reducing greenhouse gas emissions are detailed in Eqs. 1 and 2.

$$
\begin{aligned}
& Max: Obj1Profit\_V = \sum_{i=1}^{I}\Bigg[\Bigg(\Pr ice\_p_k \cdot \sum_{k=1}^{K} Yield\_P_{ik} \cdot Area\_V_{ik}\Bigg) \\
& -\Bigg(Cost\ surf\ water\_{P_{ik}} \cdot \Bigg(\sum_{k=1}^{K} \frac{Net\ surf\ water\_V_{ik}}{Surf\ wate\ coffici\_P_i}\Bigg)\Bigg) \\
& + Cost\ groun\ wate\_{P_{ik}} \cdot \Bigg(\sum_{k=1}^{K} \frac{Net\ grou\ water\_V_{ik}}{Ground\ wate\ coffici\_P_i}\Bigg) \\
& + \Bigg(Elect\ Cost\ wate\ trans\_{P_i}\Bigg(\sum_{k=1}^{K} Elect\ surf\ wate\ \_V_{ik} + Elect\ Ground\ wate\ \_V_{ik}\Bigg)\Bigg) \\
& + Elect\ Costs food\ proc\ \_{P_i}\Bigg(\sum_{k=1}^{K} Elect food\ proc\ \_{P_i}\Bigg) + \sum_{k=1}^{K} Fertili\ cost\_P_{ik} \\
& + Costs\ pesti\_{p_{ik}} + Cost\ agri\ machin\_{P_{ik}}) \\
& + \Bigg(Cost\_pik + Cost\ Flim\_P_k \cdot \Bigg(\sum_{k=1}^{K} Area\_V_{ik}\Bigg)\Bigg)\Bigg] \qquad (1)
\end{aligned}
$$

where its variables and parameters are defined as follows:

$Obj1Profit\_V$: maximize the gross profit of farmers (million IRR /ha).

$Price\_p_k$: The selling price of the product (Rial/Kg).

$Yield\_P_{ik}$: Production yield of each region (Ton/ha).

$Cost\ surf\ water\_{P_{ik}}$: Consumed used to move surface water to crops (million IRR/ m^3).

$Cost\ groun\ Wate\_P_{ik}$: Expenditures for the extraction of groundwater for consumption purposes (million IRR/ m^3).

$Net\ surf\ water\_V_{ik}$: the allocation of surface water for the production of food (m^3).

$Net\ group\ water\_V_{ik}$: The amount of groundwater that is allocated for food production (m^3).

$Ratio\ Surf\ Wate\_P_i$: Ratio utilization ratio of agricultural surface waters in each region.

$Ratio\ Groun\ Wate\_P_i$: Agricultural groundwater utilization ratio of each region.

*Elect Cost wate trans*_P_i: Electricity consumption Expenditures for water transport (million IRR/ kWh).

*Elect surf wate*_V_{ik}: The amount of electricity consumed in the process of irrigation of the crops (kWh).

*Elect Ground wate*_V_{ik}: Amount of electricity used to power agricultural pumps (kWh).

*Elect costs food proc*_P_i. Regional food processing consumption costs (million IRR/ kWh).

*Elect food proc*_$_{P_i}$: Processing power and auxiliary power consumption (kWh).

*Fertili cost*_P_{ik}: Cost of the fertilizer per hectare (million IRR/ha).

*Costs pesti*_p_{ik}: Costs of pesticides per ha.

(million IRR/ha).

*Cost agri machin*_P_{ik}: Equipment for use in agricultural machinery and equipment (million IRR/ha).

*Cost*_p_{ik}: Expenditure on agriculture (million IRR /ha).

*Cost Flim*_P_k: Agricultural Film is suitable for roof and wall covering of greenhouses for effective protection and insulation (million IRR/ha).

*Area*_V_{ik}: Cultivated area (ha).

For sustainable development, economic development should not take place at the expense of the environment (Li et al. 2019). Therefore, the effect of food production and processing on the environment when harmonically allocating water and energy resources should be considered alongside economic development in an agricultural system. CO_2 emissions are generated during food production and processing which will adversely affect the environment. Therefore, minimization of CO_2 emissions as much as possible is another objective function of the SAWEFN model, which can be expressed as Eq. 2.

$$\begin{aligned} Obj2\ Green\ house\_V = \sum_{i=1}^{I} (&carbon\ diesel\_p.Disel\ fuel\ consum\_p_i \sum_{k=1}^{K} Area\_V_{ik} \\ &+ carbon\ flim\_p.Consum\ Flim\_P_i \sum_{k=1}^{K} Area\_V_{ik} \\ &+ carbon\ pesti\_p.pesticides\ consum\_p_{ik} \sum_{k=1}^{K} Area\_V_{ik} \\ &+ Carbon\ emiss\ coffici\_p.Eectri\ consum\_p_{ik} \sum_{k=1}^{K} Area\_V_{ik} \end{aligned}$$

$$+ Carbon\,food\,proce\_p. \sum_{k=1}^{K} Elect\,food\,process\_V_{ik}$$

$$+ Carbon\,fertili\_p, Ferili\,consum\_p_{ik} \sum_{k=1}^{K} Area\_V_{ik}) \quad (2)$$

where its variables and parameters are defined as follows:

Obj2 Greenhouse_V: Minimize greenhouse gas emissions (Million tons).

Carbon diesel _P: the amount of carbon dioxide (CO_2) emitted per unit of diesel consumed (kg/CO_2 liters).

Diesel fuel consum_p$_i$: Diesel fuel consumption per ha per area (liters/ha).

Carbon Film_p: The amount of carbon dioxide (CO_2) emitted per unit of film used (kg/CO_2 kg).

Consum Film_P$_i$: The amount of agricultural film used (kg/ha).

Carbon pesti_p: The amount of carbon dioxide (CO_2) emitted per unit of pesticide used (kg/ CO_2 kg).

Electric consum_p$_{ik}$*:* Amount of electricity consumed per ha per area (kWh/ ha).

Carbon food proce_p: The amount of carbon dioxide (CO_2) emitted by the electricity used to process the food (kg/ CO_2 kWh).

Carbon fertili_p: The amount of carbon dioxide (CO_2) that is emitted per unit of fertilizer used (kg/ CO_2 kg).

pesticides consum$_{\_p_{ik}}$: Amount of pesticides consumption per ha (kg/ha).

Carbonemisscoeffici_p: The amount of carbon dioxide (CO_2) emitted by the electricity used to power the irrigation system (kg/CO_2 kWh).

Fertili consum_p$_{ik}$*:* Fertilizer consumption per ha per area (kg/ha).

2.2.2 Constraints

The model's constraints include land, water, and energy constraints, as well as constraints specific to the constrained MOP (Multi-Objective Programming) methodology. For example, a defined program yield level and a specified amount of water use were specified. First, the optimization for each objective is determined to create the membership functions for that objective. In general, the constraints are described below.

1. Energy demand for water collection and delivery constraint

The amount of electricity used to allocate and deliver surface water and pump groundwater cannot exceed the amount allowed for each region. Equation 3 is used for the calculations.

$$
\begin{aligned}
& Elect\ surf\ wate\_P_i \left[\left(\frac{\sum_{k=1}^{k} \text{Net surf } water\_V_{ik}}{Ratio\ Surf\ Wate\_P_i} \right) \right. \\
& \left. + \text{Elect Ground wate}\_P_i \left[\left(\frac{\sum_{k=1}^{k} \text{Net grou } water\_V_{ik}}{Ratio\ Groun\ Wate\_P_i} \right) \right] \right] \\
& \le Energy\ allocate\ water\_V_{ik}
\end{aligned}
\tag{3}
$$

where $Energy\ allocate\ water\_V_{ik}$: The amount of electricity required to extract the water (kWh).

2. The electrical power required for the auxiliary activities

Electricity use in food processing and auxiliary industries must remain within the allocated energy limits established for that sector. Equation 4 is used for the calculations.

$$
\sum_{k=1}^{k} (Elect\ Food\ produ\_P_{ik}.Yield\_P_{ik}).Area\_V_{ik} \le Energy\ allocate\ Food\_V_i \forall_i
\tag{4}
$$

where $Energy\ allocate\ Food\_V_i$: The electricity required for water extraction (kWh).

3. Energy availability constraint

The energy used for agricultural water pumps to distribute water to different areas of farms and orchards must remain within the total allowable energy limits set for the agricultural sector. Equation 5 is used for calculations.

$$
\begin{aligned}
& Energy\ allocate\ water\_V_i + Energy\ allocate\ Food\_V_i \\
& \le Availa\ energy\ \_P_i \\
& - Energycons\ indus\ \&\ Home\_P_i \qquad \forall_i
\end{aligned}
\tag{5}
$$

where, $Availa\ energy\_P_i$: Available amount of energy for zone i (kWh). $Energy\ cons\ indus\ \&\ Home\_P_i$: The use of agricultural water pumps offers significant benefits to farmers. These devices provide direct and consistent water delivery to fields and orchards, effectively meeting the irrigation needs of crops. As a result, crop yields are increased and overall agricultural conditions are improved (kWh).

4. Surface water supply constraint

While precipitation in the watershed is the only natural inflow to any surface water system, the total amount of water should not exceed the storage capacity of lakes, wetlands, and artificial reservoirs, the permeability of the soils beneath these reservoirs, and the runoff characteristics of the land in the watershed. Exceeding these limits would increase water loss. Equation 6 represents this limitation.

$$\frac{\sum_{k=1}^{k} \text{Net surf } water\_V_{ik}}{Surf\ wate\ coeffici\_P_i} \leq wate\ supply\ proje\_P_i.Ratio\ Surf\ Wate\_P_i \forall_i \quad (6)$$

where, $wate\ supply\ proje\_P_i$: Supply of surface water (m^3) $Surf\ wate\ coeffici\_P_i$: Surface water utilization coefficient.

5. Groundwater supply constraint

Groundwater resources, which are recharged directly or indirectly by surface water and precipitation, require sustainable management, which includes limiting abstraction. Consequently, this limitation emphasizes that water abstraction should not exceed the rate of recharge to the aquifer. Equation 7 is used for the calculations.

$$\frac{\sum_{k=1}^{k} \text{Net grou } water\_V_{ik}}{Ground\ wate\ coeffici\_P_i} \leq Groundwater\ supply\_P_i.RatioGrounWate\_P_i \forall_i \quad (7)$$

where, $Groundwat\ coeffici\_P_i$: Groundwater utilization coefficient

$Groundwater\ supply\_P_i$: Groundwater supply from well pumping for area (m^3).

6. Water demand for food production constraint

A plant's water requirements are determined by its evapotranspiration rate. It is important to remember that in different climates and weather conditions, the amount of water applied should be adapted to the growth conditions of the plant to meet the net water requirements of the plants. Equation 8 is used in the calculations.

$$\begin{aligned} &\text{Net } surf\ water\_V_{ik} + Net\ grou\ water\_V_{ik} + (Effect\ preci\ \_P_{ik}.Area\_V_{ik}) \geq \\ &Wate\ require\_P_{ik}.Area\_V_{ik} \quad \forall_i \end{aligned} \quad (8)$$

where, $Effect\ preci\_P_{ik}$: Effective rainfall (m3/ha)

$Wate\ require\_P_{ik}$: Crop Watering Needs (cubic meters/ ha).

7. Land use constraint

The conversion of uncultivated land into farmland significantly increases the area available for crop production. This process transforms barren and unproductive land into fertile and productive land, improving the livelihoods and economic conditions of local people. Equation 9 is used for the calculations.

$$Area\_P_{ik}^{min} \leq Area\_V_{ik} \leq Area\_P_{ik}^{max}, \forall_I, k \quad (9)$$

where, $Area\_P_{ik}^{min}$: Min cultivated area (ha)

$Area\_P_{ik}^{max}$: Max cultivated area (ha).

8. Crop produced constraint

The production capacity of products that have historically faced challenges and fallen short of societal demand should be reassessed. This review should ensure that production levels are sufficient to meet the needs of the population. Equations 10 and 11 are used for the calculations.

$$\sum_{i=1}^{I} Area\_V_{ik}.Yield\_P_{ik} \leq Deman\_P_{ik}^{Max}, \forall_k, i \quad (10)$$

$$\sum_{i=1}^{I} Area\_V_{ik}.Yield\_P_{ik} \geq Demand\_P_{ik}^{Min}, \forall_k, i \quad (11)$$

where, $Deman\_P_{ik}^{Max}$: Maximum product demand (ton/ha)

$Demand\_P_{ik}^{Min}$: *m*in demand of each region for the product (ton/ha).

Equations 10 and 11 highlight food security as a critical priority for a nation's economic and social progress. Ignoring a country's food infrastructure can lead to severe hunger crises and widespread disease among its people. This problem is especially evident in many developing countries.

9. Non- negative constraint.

The Eq. 12 are used to determine the non-negative variables in the model.

$$\text{Net surf } water\_V_{ik} \geq 0, \forall_i, k \quad (12)$$

$$\text{Net grou } water\_V_{ik} \geq 0, \forall_i, k$$

$$Area\_V_{ik} \geq 0, \forall_i, k$$

$$Elect\ surf\ water\_V_{ik} \geq 0, \forall_i, k$$

$$\text{Elect Ground water}\_V_{ik} \geq 0, \forall_i, k$$

$$Elect\, food\ process\_V_{ik} \geq 0, \forall_i, k$$

$$Energy\ allocate\ water\_V_{ik} \geq 0, \forall_i, k$$

$$Energy\,allocate\,Food\_V_{ik} \geq 0, \forall_i, k$$

2.3 The Multi-objective Fuzzy Non-linear Programming

Fuzzy logic is used to reconcile the above objectives. The proposed model is constructed based on the framework of the maximum composite ideal distance of these objectives from their respective critical thresholds. The general form of this model, inspired by the research of Jones and Barnes (2000), is detailed in Eq. 13.

$$Max : \lambda = \left[\sum_{j=1}^{M} w_j \lambda_j^p\right]^{\frac{1}{p}} \tag{13}$$

Subject to:

$$\lambda_j\left(Z_j^{Max} - Z_j^{Min}\right) + Z_j(x^*) \leq Z_j^{Max} \text{ when } Z_j^{Min} \text{ is best}$$

$$Z_j(x^*) - \lambda_j\left(Z_j^{Max} - Z_j^{Min}\right) \geq Z_j^{Max} \text{ when } Z_j^{Max} \text{ is best}$$

$$\sum_i \sum_j A_{hit} x_{it} \leq b_h$$

$$x_{it} \geq 0$$

$$0 \leq \lambda \leq 1$$

$i = 1, 2, \ldots, n$ $t = 1, 2, 3$ j = 1,2, …, 4 $h = 1, 2, \ldots, q$

In the given relations, x(it) represents the agricultural area designated for the crop management method, and j is the distance to the ideal limit of the (j) objective. These are the decision variables of the model to be optimized. Equation (13) serves as the model's objective function, where λ is the weighted composite ideal sum of the objectives in this study, which is to be maximized. W_j is the weight of the $(\sum_{j=1}^{M} W_j = 1, W_j) \geq 0)$ (j) objective, and (P) is the balancing factor among the objectives. This factor indicates the degree of substitution between the goals. When (P) equals 1, the degree of substitution is the highest, and when (P) equals 3, it is the lowest. Equation (13) define the membership function of the goals in the study model. Here, Z_j^{min} and Z_j^{max} Are the minimum and maximum values of the

(j) objective, respectively, $Z_j^{(x*)}$ Is the optimal value of the objective, and $x*$ is the optimal solution. The difference between Z_j^{max} and Z_j^{min} Is called the tolerance limit.

Equation (14) applies to objectives where the goal is to minimize the relevant target variable (such as water use and chemical fertilizers), while Eq. (15) applies to objectives where the goal is to maximize (such as program yield and social benefits). It is assumed that the critical value for each objective. Z_j^{max} for Eq. 14 and Z_j^{min} Eq. 15) corresponds to the current conditions of farmers' cropping patterns. Therefore, the optimal value derived from the model $Z_j^{(x*)}$. . Should not fall below this critical value (for maximization objectives) or exceed it (for minimization objectives). Consequently, the ideal value for each objective Z_j^{min} for Eq. 15 and Z_j^{max} Eq. 14) must be determined through a single-objective optimization process, which will be explained in detail later.

The equations from 13 to 15 are subject to the technical constraints given in Eq. 17. Here, $A_{\lambda it}$ represents the technical coefficient for the (i) crop under the (h) constraint (production resources) and the (t) management practice, while (b_λ) represents the availability of the (h) constraint (production resources). The model is limited by land and water constraints, as well as specific constraints from the multi-objective programming approach, such as defined program yield levels and specific water use amounts.

Once the pattern for optimizing activities to achieve the specified goals is established, the constraints and decision criteria for each option must be identified. This process is typically subjective, not fully defined or documented, and relies on the preferences and knowledge of the decision-maker. To incorporate the decision maker's knowledge and preferences, weights are assigned to the competing objectives based on their importance.

$n = [1, N]$ Assuming that represents the number of operators and $j = [1, M]$ denotes the criterion (objective) in question, and Z_j^N Is the value that criterion (j) takes for an operator (n), the method for determining the membership function of an index (j) for an operator (n) under conditions where the minimum value of the index is considered ideal (such as water consumption, risk probability, and chemical fertilizer use) is defined by Eq. (14).

$$\lambda_j(n) = \begin{cases} 1 & \text{if } Z_j^n \le Z_j^{\min} \\ \frac{Z_j^{\max} - Z_j^n}{Z_j^{\max} - z_j^{\min}} & \text{if } Z_j^{\min} \le Z_j^n \le Z_j^{\max} \\ 0 & \text{if } Z_j^n \ge Z_j^{\max} \end{cases} \tag{14}$$

In this scenario, the function $\lambda_j^{(n)}$ Evaluate how well the (n) operator satisfies the (j) criterion. Similarly, if the ideal boundary is defined by the maximum value of the objectives (such as program yield and employment), the membership function $\lambda_j^{(n)}$ Is determined according to Eq. (15).

$$\lambda_j(n) = \begin{cases} 1 & \text{if } Z_j^n \geq Z_j^{\min} \\ \frac{Z_j^n - Z_j^{\min}}{Z_j^{\max} - z_j^{\min}} & \text{if } Z_j^{\min} \leq Z_j^n \leq Z_j^{\max} \\ 0 & \text{if } Z_j^n \leq Z_j^{\min} \end{cases} \tag{15}$$

These functions are increasing functions of the degree of achievement of the model, ranging from zero to one. These membership functions were used to rank the objectives considered by the operators. The geometric mean weight for the operators' objective membership functions was calculated based on Eq. (16).

$$\lambda(i) = \sum_{i=1}^{M} w_j \lambda_j(i) \tag{16}$$

The objective weights must fall within the specified maximum and minimum limits. Consequently, the weights are determined according to Eq. (17).

$$w_j = ln(\frac{1}{\lambda_j}) / \sum_{j=1}^{M} \sum \frac{1}{\lambda_j} ln \tag{17}$$

In this scenario, wj represents an inverse function of the average level of the goals to the (j) goal. In addition, the logarithmic function suggests that the priority of each optimal pattern is a nonlinear function of the goals used.

Given the nonlinear nature of the composite ideal distance function in this study, a nonlinear programming approach is used to solve it. Consequently, by fuzzifying the study objectives and striving to achieve an overall ideal by maximizing the composite ideal distance function, the decision model is structured as a multi-objective fuzzy nonlinear programming model. This model facilitates the simultaneous reconciliation of multiple objectives within the constraints of available resources.

3 A Real-World Example

The developed model was validated in real-world applications by studying the Kashafrud sub-basin, which is part of the Qaraqum Basin in northeastern Iran. This river, one of the largest in the region and part of the Qaraqum basin has a length of 290 km. The average elevation of the basin is 1197 m above sea level, with an annual rainfall of 263 mm and an annual river flow of 28 million cubic meters allocated to various sectors. The total area of the basin is 16,500 km, of which 5,000 square kilometers are plains and the rest are highlands. The varying elevations and proximity to the central desert significantly influence the region's climate, resulting in a variety of weather conditions. According to the Kashafrud Irrigation and Drainage Network, surface water from the Kashafrud River is diverted through the Doosti Diversion Dam to irrigation canals and main streams in the plain. The network is divided into three sections: upstream, midstream, and downstream, with the upper section being the beginning of the network. The main crops grown in this basin and analyzed in this study include wheat, barley, sugar beets, potatoes, onions, tomatoes, cucumbers, forage corn, and alfalfa. Notably, this basin ranks among the top three in the production of barley, cucumbers, tomatoes, onions, and vegetables.

Table 2 shows that wheat and barley dominate the cultivated area in the study region with 22,254 and 21,261 respectively. Potato has the smallest area with 323 ha, while wheat has the largest area with 22,254 ha. The total area of the selected crops in the study region is 64,006. Alfalfa has the lowest consumption of fertilizers and pesticides among the crops studied, with 253 tons and 0,957 kg, respectively.

Figure 4 illustrates the changes in cultivated areas within the agricultural regions of the Kashafrud Basin. The model shows that the cultivated areas for cucumbers, onions, wheat, sugar beets, and tomatoes decrease by 1%, 28%, 2% and 6%, respectively. Tomatoes have the largest decrease in area, while cucumbers have the smallest decrease. Overall, the total cultivated area in the basin decreases by 6%, from 64,006 under current conditions to 60,189 in the multi-objective model.

Figure 5 illustrates the changes in farmers' net profit and CO_2 emissions using the integrated approach. In the optimal model, the total farmers' profit is 3981.930 billion rials, which is an increase of 39.42% from the current 2856.123 billion rials. In addition, CO_2 emissions in the optimal model decreased by 21.13% from the current 205.96 million tons to 162.44 million tons.

The proposed implementation of the resource nexus not only provides environmental benefits and efficient resource management but also supports farmers and ensures the sustainability of agricultural practices. Choosing a cropping pattern without integrating the water-food-energy system is primarily driven by economic objectives. While it provides beneficial outcomes within a single sector, it does not

Table 2 Average consumption of crop inputs in the watershed (*Source* Anonymous 2020)

Pesticides (Kg/ha)	Fertilizer consumption (tons/ha)	Cultivation (ha)	Yield (Ton/ ha)	Crops
1.391	392	22,254	4.2	Wheat
0.962	364	21,261	3.6	Barley
4.38	572	2385	35	SugarBeet
4.18	606	5385	50	Onion
3.64	706	323	35	Potato
2.97	641	1149	40	Tomato
0.957	253	920	11	Alfalfa
2.13	543	5869	55	Maize
4.18	456	4460	25	Cucumber
24.79	4533	64,006	258.8	Total

Source Anonymous (2020)

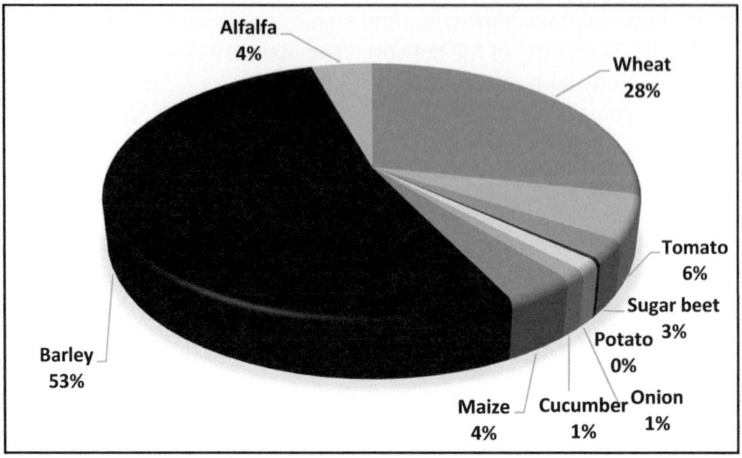

Fig. 4 The cultivation areas of crops Basin. *Source* Research results

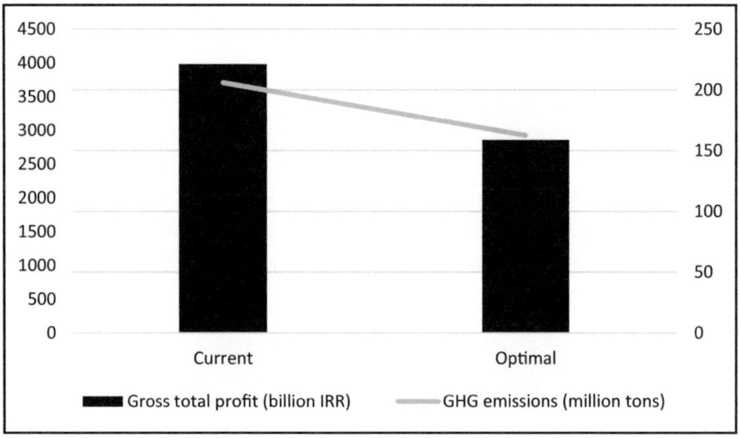

Fig. 5 The rate of change in farmers' net profit and greenhouse gas emissions. *Source* Research results

provide a comprehensive assessment of the effectiveness of other policies. Therefore, using this approach in agriculture addresses economic goals while thoroughly examining environmental impacts and the linkages between water, food, and energy resources.

Table 3 shows the main economic variables such as costs, product yields, and input consumption for the nexus approach using the optimal model at the basin level. The net profit of farmers in the basin increased by 39.42% to 3981.93 billion rials compared to the current 2856.123 billion rials. Consumption cost decreased by 40.19% to 6232.42 billion rials in the multi-objective model from the current 10,419.67 billion rials. Fertilizer consumption based on the climate change scenario decreased by 23.84 tons, which is 9.64% less than the current level. Pesticide use also decreased by 18.76% in the optimal model. In addition, the cultivated area decreased by 5.96% to 60,189 from the current 64,006.

Water consumption decreased by 23.36% to 566.798 million cubic meters in the multi-objective model compared to 739.554 million cubic meters in the current scenario. Finally, product yield decreased by 9.18%, from 920.18 tons in current conditions to 605.22 tons in the optimal model. Overall, the main objective Systemic approach in this study is the optimal use of resources and the achievement of sustainable development, the results show an increase in farmers' profits and a reduction in resource and input consumption in the multi-objective model, which makes this approach justifiable and necessary for economic and environmental issues.

Table 3 Key economic variables of the optimal model. (*Source* Research results)

Variables		Total
Gross total profit (billion IRR)	Current	2856.123
	Optimal	3981.93
	Percentage change	39.42
Total production cost (billion IRR)	Current	10,419.67
	Optimal	6232.423
	Percentage change	−40.19
Fertilizer consumption (Thousand tons)	Current	26.38
	Optimal	23.84
	Percentage change	−9.64
Pesticides consumption (Thousand tons)	Current	99.98
	Optimal	81.22
	Percentage change	−18.76
Irrigation water in each area (million cubic meters)	Current	739.554
	Optimal	566.798
	Percentage change	−23.36
Total production (thousand tons)	Current	920.18
	Optimal	605.22
	Percentage change	−34.23
Crop area (has)	Current	64,006
	Optimal	60,188.87
	Percentage change	−5.96

4 Conclusion

This chapter presents a mathematical programming model based on the water-food-energy nexus approach for crop production. A major challenge in the agricultural sector is the single-sector perspective, which cannot conclusively determine the effectiveness of other policies when only one sector is considered. The results of the study using the proposed water-food-energy nexus approach show a 9.64% reduction in the negative environmental impacts of fertilizers and chemical pesticides and a 40.19% reduction in the costs associated with the use of renewable and non-renewable resources. The economic benefits of this approach increase by 39.42%, and in the long term it positively impacts the economic sector by 21.13% and water resource management by 23.36%. The main outcome is the achievement of sustainability by taking into account not only environmental aspects but also economic and social dimensions. In essence, development is sustainable when it simultaneously addresses societal needs, economic prosperity, and environmental protection. Research on single-sector policy implementation has shown that focusing on one area of the resource nexus in isolation highlights the one-dimensional impact of each policy in agriculture, but does not conclusively determine the effectiveness of other policies. Therefore, the adoption and application of the optimal water-food-energy nexus model from this study enables farmers to maintain economic income while considering environmental factors and optimal sustainable use of resources (water-food-energy) in policy selection and implementation. This technique unifies approaches to assess the impacts of decisions within energy and water systems on the environment, food production, energy security, climate, and ecosystems, and thereby determines the optimal mechanism for resource allocation. Overall, the single-dimensional approach has underscored the importance of the nexus approach due to policy inconsistency and low resource use efficiency. As shown in this chapter, the nexus approach can improve comprehensive WEF resource security, ensure sustainable and equitable access to resources, and help farmers choose optimal cropping patterns. Therefore, to avoid the implementation of inappropriate single-dimensional policies, the nexus approach is recommended, which covers all dimensions and factors affecting sustainable development in a systematic and comprehensive framework using various mathematical programming methods and multi-objective models.

References

Ahani E, Ziaee S, Mohammadi H, Mardani Najafabadi M, Mirzaei A (2024) An investigation into the effects of climate change on water-energy-food nexus: a new mathematical programming approach. Int J Environ Res 18(2). https://doi.org/10.1007/s41742-024-00567-9

Anonymous (2020) Agricultural jihad organization Khorasan Razavi

Batlle-Bayer L, Aldaco R, Bala A, Fullana-i-Palmer P (2020) Toward sustainable dietary patterns under a water–energy–food nexus life cycle thinking approach. Curr Opinion Environ Sci Health 13. https://doi.org/10.1016/j.coesh.2019.11.001

Badih Barzin H, Hoseini SM, Hashemitabar M, Mardani Najafabadi M (2024) The allocation of crop production resources in the southeast of Iran: the application of the water-energy-food nexus approach. Frontiers Environ Sci, 12. https://doi.org/10.3389/fenvs.2024.1349095

Brambilla G, Testa C (2014) Food safety/food security aspects related to the environmental release of pharmaceuticals. Chemosphere 115:81–87

Cui S, Zhang J, Wang X, Wu M, Cao X (2023). Fuzzy composite risk assessment of water-energy-food-carbon nexus in the dispark pumped irrigation system. J Hydrology 624:129879. https://doi.org/10.1016/j.jhydrol.2023.129879

Chaudhuri S, Banerjee D (2010) FDI in agricultural land, welfare, and unemployment in a developing economy. Res Econ 64(4):229–239

Ding T, Fang L, Chen J, Ji J, Fang Z (2023) Exploring the relationship between water-energy-food nexus sustainability and multiple ecosystem services at the urban agglomeration scale. Sustain Prod Consum 35(12):184–200. https://doi.org/10.1016/j.spc.2022.10.028

Dumont A, Salmoral G, Llamas MR (2013) The water footprint of a river basin with a special focus on groundwater: The case of Guadalquivir basin (Spain). Water Resour Industry 1–2:60–76. https://doi.org/10.1016/j.wri.2013.04.001

Emamzadeh SM, Forghani MA, Karnema A, Darbandi S (2016) Determining an optimum pattern of mixed planting from organic and non-organic crops about economic and environmental indicators: a case study of cucumber in Kerman, Iran. Inform Process Agric 3(4):207–214. https://doi.org/10.1016/j.in.2016.08.001

Esteve P, Varela-Ortega C, Blanco-Gutiérrez I, Downing TE (2015) A hydro-economic model for the assessment of climate change impacts and adaptation in irrigated agriculture. Ecolo Econ 120:49–58. https://doi.org/10.1016/j.ecolecon.2015.09.017

FAO (2014) The water-energy-food nexus: a new approach in support of food security and sustainable agriculture

Filippi C, Mansini R, Stevanato E (2017) Mixed integer linear programming models for optimal crop selection. Comput Oper Res 81:26–39

Forni LG, Medellín-Azuara J, Tansey M, Young C, Purkey D, Howitt R (2016) Integrating complex economic and hydrologic planning models: An application for drought under climate change analysis. Water Resour Econ 16:15–27. https://doi.org/10.1016/j.wre.2016.10.002

Hazell PB, Norton R (1986) Mathematical programming for economic analysis in agriculture. Colliei MacMillan, London, UK

Han X, Hua E, Engel BA, Guan J, Yin J, Wu N, Sun S, Wang Y (2022) Understanding implications of climate change and socio-economic development for the water-energy-food nexus: a meta-regression analysis. Agric Water Manag 269:107693. https://doi.org/10.1016/j.agwat.2022.107693

Hoff H (2011) Understanding the NEXUS. Background paper for the Bonn, nexus conference: The water, energy, and food security nexus solutions for the green economy. The Water, Energy, and Food Security Nexus. publications/understanding-the-nexus/

He Y-B, Cai W-M (2016) Linking a farmer crop selection model (FCS) with an agronomic model (EPIC) to simulate cropping patterns in Northeast China. J Integr Agric 15(10):2417–2425

Jones D, Barnes EM (2000) Fuzzy composite programming to combine remote sensing and crop models for decision support in precision crop management. Agric Syst 65(3):137–158. https://doi.org/10.1016/S0308-521X(00)00026-3

Kalbali E, Ziaee S, Najafabadi MM, Zakerinia M (2021) Approaches to adapting to impacts of climate change in northern Iran: the application of a Hydrogy-Economics model. J Cleaner Prod 280:124067. https://doi.org/10.1016/j.jclepro.2020.124067

Lalawmpuii, Rai PK (2023) Role of water-energy-food nexus in environmental management and climate action. Energy Nexus 11:100230. https://doi.org/10.1016/j.nexus.2023.100230

Lee S-H, Choi J-Y, Hur S-O, Taniguchi M, Masuhara N, Kim KS, Hyun S, Choi E, Sung J, Yoo S-H (2020) Food-centric interlinkages in agricultural food-energy-water nexus under climate change and irrigation management. Resour Conserv Recycling 163:105099. https://doi.org/10.1016/j.resconrec.2020.105099

Li M, Fu Q, Singh VP, Ji Y, Liu D, Zhang C, Li T (2019) An optimal modeling approach for managing agricultural water-energy-food nexus under uncertainty. Sci Total Environ 651:1416–1434. https://doi.org/10.1016/j.scitotenv.2018.09.291

Li M, Fu Q, Singh VP, Ma M, Liu X (2017) An intuitionistic fuzzy multi-objective non-linear programming model for sustainable irrigation water allocation under the combination of dry and wet conditions. J Hydrol 555:80–94. https://doi.org/10.1016/j.jhydrol.2017.09.055

Madani K, AghaKouchak A, Mirchi A (2016) Iran's socio-economic drought: challenges of a water-Bankrupt nation. Iran Stud 49(6):997–1016

Mardani Najafabadi M, Ziaee S, Nikouei A, Ahmadpour Borazjani M (2019) Mathematical programming model (MMP) for optimization of regional cropping patterns decisions: a case study. Agric Syst 173:218–232. https://doi.org/10.1016/j.agsy.2019.02.006

Mardani Najafabadi M, Shirzadi Laskookalayeh S (2023) Solving cropping pattern optimization problems using robust positive mathematical programming. In: Kulkarni AJ, Gandomi AH (Eds), Handbook of formal optimization (pp 1–26). Springer Nature Singapore. https://doi.org/10.1007/978-981-19-8851-6_52-1

Mardani Najafabadi M, Magazzino C, Valente D, Mirzaei A, Petrosillo I (2023) A new interval meta-goal programming for sustainable planning of agricultural water-land use nexus. Ecol Model 484:110471. https://doi.org/10.1016/j.ecolmodel.2023.110471

Medellín-Azuara J, Harou JJ, Howitt RE (2010) Estimating the economic value of agricultural water under changing conditions and the effects of spatial aggregation. Sci Total Environ 408(23):5639–5648

Nasrollahi HSR, Reza S, Pourali Omid AM (2021) Unraveling the water-energy-food-environment nexus for climate change adaptation in Iran: Urmia lake basin case-study. Water 13(9):1282. https://doi.org/10.3390/w13091282

Neamatollahi E, Vafabakhshi J, Jahansuz MR, Sharifzadeh F (2017) Agricultural optimal cropping pattern determination based on fuzzy system. Fuzzy Inform Eng 9(4):479–491

Nie Y, Avraamidou S, Xiao X, Pistikopoulos EN, Li J, Zeng Y, Song F, Yu J, Zhu M (2019) A food-energy-water nexus approach for land use optimization. Sci Total Environ 659:7–19. https://doi.org/10.1016/j.scitotenv.2018.12.242

Nkiaka E, Bryant RG, Manda S, Okumah M (2023) A quantitative understanding of the state and determinants of water-energy-food security in Africa. Environ Sci Policy 140:250–260. https://doi.org/10.1016/j.envsci.2022.12.015

Osama S, Elkholy M, Kansoh RM (2017) Optimization of the cropping pattern in Egypt. Alex Eng J 56(4):557–566

Qureshi ME, Qureshi SE, Bajracharya K, Kirby M (2007) Integrated biophysical and economic modeling framework to assess impacts of alternative groundwater management options. Water Resour Manage 22(3):321–341

Ringler C, Bhaduri A, Lawford R (2013) The nexus across water, energy, land and food (WELF): potential for improved resource use efficiency?. Curr Opin Environ Sustain 5(6):617–624. https://doi.org/10.1016/j.cosust.2013.11.002

Romero R, Imberger J, Ewing T, Antenucci J (2004) A real-time decision support system for reservoir management. Agric Water Manag 57:207–221

Samadi-Foroushani M, Keyhanpour MJ, Musavi-Jahromi SH, Ebrahimi H (2022) Integrated water resources management based on water governance and water-energy-food nexus through system dynamics and social network analyzing approaches. Water Resour Manag 36(15):6093–6113. https://doi.org/10.1007/s11269-022-03343-6

Sands P (1995) International law in the field of sustainable development. British Yearbook Int Law 65(1):303–381

Shang Y, Hei P, Lu S, Shang L, Li X, Wei Y, Jia D, Jiang D, Ye Y, Gong J, Lei X, Hao M, Qiu Y, Liu J, Wang H (2018) China's energy-water nexus: assessing water conservation synergies of the total coal consumption cap strategy until 2050. Appl Energy 210:643–660. https://doi.org/10.1016/j.apenergy.2016.11.008

UNEP (2021) Making peace with nature: a scientific blueprint to tackle the climate, biodiversity and pollution emergencies. In: Nairobi: United Nations Environment Assembly. Obtained from. resources/making-peace-nature

Wang X, Fielding KS, Dean AJ (2023) Nature is mine/ours: Measuring individual and collective psychological ownership of nature. J Environ Psychol 85:101919. https://doi.org/10.1016/j.jenvp.2022.101919

Wineman A, Crawford EW (2017) Climate change and crop choice in Zambia: a mathematical programming approach. NJAS- Wageningen J Life Sci 81:19–31

Wolfert S, Ge L, Verdouw C, Bogaardt M-J (2017) Big data in smart farming—a review. Agric Syst 153:69–80. https://doi.org/10.1016/j.agsy.2017.01.023

Yue Q, Guo P (2021) Managing agricultural water-energy-food-environment nexus considering water footprint and carbon footprint under uncertainty. Agric Water Manag 252:106899. https://doi.org/10.1016/j.agwat.2021.106899

Yue Q, Zhang F, Zhang C, Zhu H, Tang Y, Guo P (2020) A full fuzzy-interval credibility-constrained nonlinear programming approach for irrigation water allocation under uncertainty. Agric Water Manag 230:105961. https://doi.org/10.1016/j.agwat.2019.105961

Zeng X, Kang S, Li F, Zhang L, Guo P (2010) Fuzzy multi-objective linear programming applying to crop area planning. Agric Water Manag 98(1):134–142

Zhang Y, Cui J, Liu X, Liu H, Liu Y, Jiang X, Li Z, Zhang M (2022) Application of water-energy-food nexus approach for optimal tillage and irrigation management in intensive wheat-maize double cropping system. J Clean Prod 381:135181. https://doi.org/10.1016/J.JCLEPRO.2022.135181

Zhang C, Engel BA, Guo P, Zhang F, Guo S, Liu X, Wang Y (2018a) An inexact robust two-stage mixed-integer linear programming approach for crop area planning under uncertainty. J Clean Prod 204:489–500

Zhang C, Li M, Guo P (2017) An interval multistage joint-probabilistic chance-constrained programming model with left-hand-side randomness for crop area planning under uncertainty. J Clean Prod 167:1276–1289

Zhang F, Zhang C, Yan Z, Guo S, Wang Y, Guo P (2018b) An interval nonlinear multiobjective programming model with fuzzy-interval credibility constraint for crop monthly water allocation. Agric Water Manag 209:123–133

Proposed a Food Insecurity Index: An Economic and Mathematical Modeling Approach

Maryam Rezaeifar and Mostafa Mardani Najafabadi

Abstract Over the past twenty years, food security and public health have become central to the overarching policies of governments worldwide. Within this framework, poverty reduction and food security are goals of sustainable development. In this regard, improving the income of the studied community is essential to improve various poverty indicators. In agriculture, this improvement can be achieved through effective resource management and selection of appropriate cropping patterns. This chapter begins by examining studies on food insecurity, followed by an overview of the indicators used to measure it and the introduction of the Foster-Greer-Thorbecke (FGT) poverty index. A mathematical programming model is then presented, aiming to optimize the FGT index by increasing income through the implementation of an optimal cropping pattern. An empirical study is conducted to evaluate food insecurity in Iranian provinces, showcasing the application of the index. Additionally, a case study was selected in several rural areas of Khuzestan province in Iran to test the capability of the proposed mathematical programming model. The findings indicate that food insecurity is more prevalent in urban areas compared to rural areas in Iran. Also, the proposed mathematical programming model can significantly help to reduce the poverty index by determining a new optimal cropping pattern.

Keywords Food security · FGT Index · Cropping pattern · Optimization · Poverty

---

M. Rezaeifar
PhD in Agricultural Economics, Department of Agricultural Economics, Tarbiat Modares University, Tehran, Iran

M. Mardani Najafabadi (✉)
Department of Agricultural Economics, Faculty of Agricultural Engineering and Rural Development, Agricultural Sciences and Natural Resources University of Khuzestan, Mollasani, Iran
e-mail: m.mardani@asnrukh.ac.ir

M. Ram et al. (eds.), *Advances in Mathematics for Engineering Sciences*, Mathematical Engineering, https://doi.org/10.1007/978-3-031-95693-5_6

1 Introduction

1.1 General

The concept of household food security gained prominence in the 1980s, largely through the efforts of international organizations such as FAO (2001). As defined by the World Food Conference, food security is achieved when all people, at all times, have physical and economic access to sufficient, safe, and nutritious food that meets their dietary needs and food preferences for an active and healthy life (Stephens et al. 2018). Food security is a critical component of development and poverty reduction, and is a primary goal of many international and national public organizations. Moreover, food security is a complex and multidimensional concept with various implications (Abafita and Kim 2014). To achieve food security, access to sufficient food is necessary for household well-being as well as for other development activities (Endale et al. 2014).

Food security is a critical element of both individual and societal health and has been recognized as a key goal of sustainable development. Following the Millennium Development Goals, the 2030 Agenda for Sustainable Development was adopted by all United Nations member states in 2015, with the primary goal of eradicating poverty in all its forms worldwide. At the Rio+20 Summit, it was decided that 2015 would mark the end of the MDGs, to be followed by the Sustainable Development Goals (SDGs) and a new development agenda. After numerous meetings and discussions, the culmination of the efforts and collaboration of stakeholders around the world is the 2030 Agenda for Sustainable Development, which will guide the actions and activities of all countries for the next 15 years (Sustainable Development Goals 2016). The Sustainable Development Goals are interlinked, so that achieving one goal strengthens the others, while failing to achieve one goal can have a negative impact on the others (Code 2020). The second goal of the United Nations Sustainable Development Goals is to end hunger, ensure food security, and improve nutrition for all by 2030.

In contrast, food insecurity is defined as inadequate or uncertain access to food due to insufficient resources (Loopstra and Tarasuk 2012). According to the American Institute of Nutrition, food insecurity is characterized by limited or uncertain access to adequate, safe, and nutritious food or limited or uncertain ability to obtain acceptable food in a socially acceptable manner (Cook and Frank 2008). Food insecurity has emerged as a critical global issue (Shushu et al. 2023), with its impacts being particularly severe in developing nations (Smith and Meade 2019). Despite substantial financial investments by governments and international organizations to combat hunger and food insecurity, these challenges persist as major concerns in these regions (Abebaw et al. 2010). Food insecurity poses a major threat to global public health due to its detrimental effects on the psychological, physical, and social health and well-being of those who experience it (Fotakis et al. 2024). As a result, food insecurity has far-reaching implications for health and well-being (Pollard and Booth 2019), affecting weight (Rogers et al. 2015; Weedn et al. 2014), general health

(Thomas et al. 2019), mental health (Weinreb et al. 2002), and quality of life (Casey et al. 2005).

Efforts to combat food insecurity involve various strategies, including improving agricultural practices, enhancing food distribution systems, and implementing policies that support food access for vulnerable populations (Endris et al. 2017). However, the persistent high rates of malnutrition in developing countries highlight the need for more effective and targeted interventions (Erinle et al. 2021; Unicef 2020). Table 1 shows the global prevalence of undernutrition from 2000 to 2021. These data help to understand the trends and regional disparities in undernutrition, indicating which areas need more attention and resources.

According to the World Bank, 12.9% of the world's population, or 850 million people, were hungry in 2001. By 2021, that percentage had fallen to 9.2%. However, during the years 2019, 2020, and 2021, global malnutrition has seen an increase. Since 2019, the number of individuals suffering from severe hunger has surged from 135 million to 345 million (Animasaun et al. 2023). Reports from FAO, IFAD, UNICEF, WFP, and WHO indicate that approximately 768 million people, or 9.8% of the world's population, are malnourished, with 760 million of these individuals residing in developing countries (FAO; IFAD; UNICEF; WFP; WHO 2022). Additionally, poverty and food insecurity move in tandem, with both remaining major challenges to achieving economic development in countries worldwide (Mulugeta et al. 2018). Therefore, the increase in malnutrition in recent years has led to a rise in poverty.

In recent years, food security has emerged as a major concern and a top priority for developing countries (Abafita and Kim 2014). In this regard, the statistics of malnutrition in Iran are presented in Table 2. According to the World Bank, malnutrition in Iran increased from 5% in 2001 to 6.1% in 2021. The highest malnutrition rates in Iran were observed in 2011 (7.6%), 2012 (7.3%), 2017 (7.2%), and 2018 (7.2%) (Table 2).

Table 1 Prevalence of undernourishment in World (% of population)

Year	2001	2005	2010	2015	2016	2017	2018	2019	2020	2021
Prevalence of undernourishment (% of population)	12.9	12	8.8	7.8	7.8	7.7	7.7	8.2	8.7	9.2

Source World Bank (2024)
In the World Bank database, this data is available to anyone

Table 2 Prevalence of undernourishment in Iran (% of population)

Year	2001	2005	2010	2011	2012	2015	2016	2017	2018	2019	2020	2021
Prevalence of undernourishment (% of population)	5	5.4	7.1	7.6	7.3	6.3	6.8	7.2	7.2	7	6.2	6.1

Source World Bank (2024)
In the World Bank database, this data is available to anyone

Over the last twenty years, Iran has joined other nations in prioritizing the fight against hunger and malnutrition, particularly after committing to the Sustainable Development Goals at the Millennium Summit. This commitment has led to a significant focus on evaluating and ensuring food security. Despite these efforts, malnutrition rates have unfortunately risen during this period (World Bank 2024).

In this regard, Research highlights that food insecurity affects 37.6% of households in Arsanjan (Najafi and Shushtarian 2004), 44% in Shiraz (Ramesh et al. 2009), 58% in rural households in Khalajestan, Qom province (Purtaheri et al. 2011), 73.8% in Marvdasht (Akbarpour et al. 2017), and 24% in rural households across all Iranian provinces (Akbari et al. 2020). Additionally, data from 2019 indicates that urban areas experience higher levels of food insecurity compared to rural areas (Rezaeifar et al. 2022). As a result, the findings of these studies highlight the significant importance of food insecurity. On the other hand, in the rural economy, agriculture is considered one of the most important components that can play a significant role in reducing poverty and ensuring sustainable rural livelihoods (Darban Astaneh et al. 2018). Since the likelihood of poverty is higher in rural households whose heads are engaged in agriculture, forestry, and fishing (Yousefi et al. 2016), and due to the importance of the agricultural sector in the economies of developing countries, it is essential to emphasize the resources available to farmers and the factors influencing decision-making in resource allocation to sustain agriculture as a source of income and especially for the production and supply of necessary food (Hosseini-Yekani and Kashiri Kolaei 2017). In addition, with the increase in agricultural production, it is essential to Increasing productivity and optimize the use of scarce resources. Optimal use of these resources, in addition to meeting societal demand as a macro objective, can increase the income of beneficiaries for whom agricultural activity is not only an economic activity but also a way of life. Therefore, it is essential to adopt various methods that address the scarcity of agricultural resources, especially irrigation water (Mardani Najafabadi et al. 2019).

One of the effective strategies for increasing productivity in the agricultural sector is the optimal use of production resources and, consequently, determining the optimal cropping pattern (Naderi Mahdei 2015). The lack of an appropriate cropping pattern and poor land management can negatively affect agricultural production and rural households' livelihoods (Bazrafshan et al. 2017; Devkota et al. 2011), which ultimately leads to food insecurity among rural households, especially smallholder farmers (Bazrafshan et al. 2017; Saadi et al. 2014). The optimal cropping pattern is a program developed with the aim of optimally managing the spatial composition of the crop and Given the vast border areas and the climatic diversity of different regions, achieving an appropriate cropping pattern that maximizes the use of production factors and inputs is an unavoidable necessity (Mardani et al. 2016).

The strong correlation between poverty and income in various poverty intensity indices can link the issues of poverty and optimal use of agricultural production factors (optimal cropping pattern) (Namara et al. 2010). The reduction of poverty has always been one of the most important goals of human society and is a clear indicator of economic, social and cultural underdevelopment, which endangers political stability, social cohesion and mental health of different segments of society (Fotros and Ghodsi 2010; Naderi Mahdei 2015). This issue, as the first goal of the Millennium Development Goals (MDGs), is the greatest global challenge. Despite the 2014 MDG Report stating that global extreme poverty would be halved by 2015, many people still live in extreme poverty.

Different definitions of poverty have been proposed over time. In the 1960s and 1970s, Townsend suggested that poverty occurs when an individual, family, or group lacks the resources necessary for a diverse diet, participation in activities, and access to standard living conditions and amenities (Arshadi and Karimi 2013). According to Arzrom Chiller (2005), poverty is the deprivation of assets and opportunities that everyone deserves. Rural dwellers and households dependent on agriculture are more vulnerable to poverty due to their high dependence on natural resources, vulnerability to natural shocks such as drought, and low income. Therefore, increasing farmers' income to meet basic needs and conserving agricultural resources, given their dependence on these resources, can alleviate poverty. Therefore, the development of cropping patterns aimed at maximizing income (Baniasadi and Zare Mehrjerdi 2010; Alkire and Foster 2011) can reduce poverty in agricultural and rural areas and, as mentioned above, increase the productivity of production factors.

The most comprehensive definition of an optimal cropping pattern can be seen as the establishment of an agricultural system with sustainable economic advantages based on the country's macro policies, farmers' indigenous knowledge, and optimal use of local potentials (Naderi Mahdei 2015). This model can be used to determine the maximum income from the consumption of a certain amount of inputs, or at least the cost of producing a certain combination of products, which can reduce poverty in agricultural and rural areas (Shirvanian and Ismaili 2009). The World Bank's 1999 Poverty Reduction Indicators show that countries are determined to tackle poverty. Researchers such as Alkire and Foster (2011), Almas et al. (2013), and Arsyad et al. (2020) have emphasized the importance of studying poverty (Mohamadiyan and Negahdari 2018). In various studies in the country, including those by Eftekhari and Karami (2014), Khodadad Kashi and Heydari (2009), and Shirvanian and Ismaili (2009), poverty has been measured using an income-based approach. The results of studies by Adams and H (2003), Khoshnood (2019), and Mortazavi et al. (2011) in some developing countries indicate that poverty decreases as economic growth increases. Thus, given that food security is a key factor in reducing poverty, the following section explores the status of food insecurity across various countries globally.

1.2 The Food Insecurity Situation in Different Countries Around the World

The study of food insecurity is crucial for health and socio-economic stability, prompting numerous global assessments. Researchers have focused on measuring food insecurity's prevalence, gap, and severity, such as in Addis Ababa, Ethiopia, where 58.16% of households are food insecure. The study found significant gaps and severity in food insecurity, highlighting the need for targeted interventions. Continuous monitoring and multifaceted approaches, including policy reforms and social safety nets, are essential to address urban food insecurity effectively (Gebre 2012). Mannaf and Uddine estimated food insecurity in rural areas of Bogra in (2012). Out of 60 households, 20 (33.33%) were food insecure, unable to meet the recommended daily intake of 2122 kilocalories per capita, while 40 (66.67%) were food secure. A 2014 study by Álvares and Amaral found that 16.5% of Portuguese citizens face food insecurity, affecting one in six people. This issue, driven by economic instability, unemployment, and low income, highlights significant public health concerns. In Ethiopia, Welderufael's (2014) study estimated that 48% of the population experienced food insecurity. The research also indicates that food insecurity is more prevalent in rural households compared to urban ones. Cheema and Abbas's (2016) study in Pakistan estimated that the prevalence of food insecurity was 28.63%, the food insecurity gap was 5.28%, and the severity of food insecurity was 1.46% during 2010–2011. In Iran, data from 2010 to 2020 showed an increasing trend in food insecurity, with urban households facing higher rates compared to rural households, according to Rezaeifar et al. (2023). Another study conducted by Telila and Sima (2024) revealed that 52.1% of Ethiopians experienced moderate or severe food insecurity in 2019, with severe food insecurity increasing to 15.8%.

The study by Welderufael et al. (2014) highlights that food insecurity is significantly more prevalent among households in rural areas compared to those in urban areas, but Research by Rezaeifar et al. (2022, 2023), Pakravan et al. (2015) and Migotto et al. (2007) indicates that urban households experience higher levels of food insecurity than their rural counterparts. A review of these studies highlights the critical importance of food insecurity. The following is a review of the indicators used to measure food insecurity and a review of the optimization of cropping patterns and the calculation of poverty indicators. An economic and mathematical modeling approach (FGT index) is then proposed to measure food insecurity. In the next phase, a mathematical programming model will be introduced to optimize the FGT poverty index. Finally, using this proposed index, food insecurity for the year 2022 will be estimated for the provinces of Iran and the FGT poverty index will be optimized using the proposed mathematical programming model.

1.3 An Overview of Methods for Measuring Food Insecurity

There are five common methods for assessing food insecurity: (1) the Food and Agriculture Organization (FAO) method for estimating per capita calorie availability at the national level, (2) household income and expenditure surveys, (3) individual dietary intake assessment, (4) anthropometry, and (5) experience-based food insecurity scales. Of these, four are indirect or derived measures of food insecurity (FAO method, household expenditure surveys, dietary assessment, and anthropometry). The only method that directly measures food insecurity is the one based on experience-based food insecurity scales. The FAO method estimates per capita calorie availability at the national level using energy intake data from household income and expenditure surveys. Almost all countries produce these data to estimate the daily calorie intake of individuals. The Household Income and Expenditure Survey method involves interviewing respondents about their families. Respondents provide information on the amount of money spent on food and other necessities. Different reference periods, such as weeks or months prior to the survey, are used. This method requires data on the amount of food purchased (or spent) and the costs associated with different foods consumed both at home and away from home. It estimates the average daily calorie intake per family member. However, this method accepts a high degree of measurement error and is costly to collect. The individual dietary intake method directly measures not only food availability but also food consumption. This method considers both diet quality (macro- and micronutrients) and caloric intake at the individual level. It allows the study of the determinants and consequences of food insecurity at the individual level. Anthropometric indicators commonly used in national surveys are based on the weight and height (or length) of infants, young children, adolescents and adults. An advantage of this method is that weight and height measurements are repeatable between individuals. In addition, the cost of conducting these measurements is relatively low, making it a very popular method in national surveys worldwide. On the other hand, significant progress has been made in the basic measurement of household food insecurity using experience-based scales. Experience-based scales are the only basic method that directly measures food insecurity based on the experiences of those affected. This method includes not only the physical dimensions but also the psychosocial aspects of food insecurity. In addition, data collection, processing, and analysis are simple and relatively inexpensive (Pérez-Escamilla and Segall-Corrêa 2008).

1.4 A Review of Cropping Pattern Optimization and Poverty Index Calculation

The issues of optimizing cropping patterns and calculating poverty indices have been extensively studied separately by many researchers. However, considering these issues together has been neglected due to the complexity of modeling them simultaneously. However, when considering studies on this topic, three groups of existing research can be reviewed. The first group consists of studies that have focused only on providing optimal cropping patterns and appropriate use of production factors. Many of these studies have used linear programming models (Mousavi et al. 2011; Osama et al. 2017; Sethi et al. 2006), Goal Programming (Van Hop 2007; Zhang 2016), and multi-objective programming (Mardani Najafabadi et al. 2019; Ren et al. 2017; Zhang et al. 2018; Zhang 2016). The set of constraints in the proposed model has been derived from studies conducted in this area in recent years. The second group includes studies that have focused exclusively on measuring poverty indices, in particular the Foster-Greer-Thorbecke (1984) or FGT poverty index. Numerous studies, including those by Alkire and Foster (2011), Alkire and Santos (2013), Anríquez and Stamoulis (2007), and Fotros and Ghodsi (2010), have measured poverty indices using an income-based approach.

The third group consists of studies that examine the impact of optimizing cropping patterns on poverty indices. Research has been conducted to develop appropriate cropping patterns, increase income from these patterns, and consequently improve poverty indices. Namara et al. (2010) in an article titled "Agricultural water management and its relationship with the category of poverty" investigated the relationship between water resources management using the determination of cropping pattern and the sustainability of poverty. These researchers provided a list of promising strategies for sustainable poverty reduction through agricultural resource management. BeniAsadi and Zare Mehrjerdi (2010) calculated the FGT poverty index for households in the Arzuiyeh region of Baft County and determined the optimal cropping pattern for the area using a mathematical programming method. The results of the study showed that using the optimal cropping pattern leads to poverty reduction. Mousavi et al. (2011) studied the effect of cropping pattern on water consumption, income, and rural poverty in Kazerun County and showed that changing the cropping pattern reduced groundwater extraction and increased net profit. In addition, the use of the optimal cropping pattern resulted in 17.5% of villagers being lifted above the poverty line.

Research on optimal cropping patterns and poverty has focused either on determining optimal cropping patterns and appropriate resource use or on calculating poverty indices, particularly the FGT poverty index. A few studies have examined the impact of optimal cropping patterns on poverty indices. The fundamental difference between the present study and the previous ones is that this study minimizes poverty, represented by the FGT index, as the main objective of the optimal cropping pattern, considering various constraints. In contrast, the objective function in other studies has been to maximize program returns or minimize costs. It is worth noting that the FGT poverty index was chosen because it is measured using only quantitative values. In mathematical programming and cropping pattern determination, all parameters used must be expressed quantitatively, so the most commonly used purely quantitative index was chosen for this study. Therefore, considering these points, the main objective of this chapter is to analyze food insecurity in rural and urban areas of Iranian provinces. It also aims to introduce a mathematical programming model to optimize the poverty index, ensure the optimal use of agricultural production factors, and determine the optimal cropping pattern.

2 Measuring Food Insecurity

To ensure household food security, multiple tools must be developed to measure and evaluate food indicators using food measurement indices. In this regard, a food insecurity index with an economic and mathematical modeling approach has been proposed. Subsequently, an empirical study has been conducted by researchers to examine food insecurity in the provinces of Iran.

2.1 FGT Index

The FGT (Foster-Greer-Thorbecke) index, initially created to measure poverty, has been widely applied in various contexts. In 1999, Hoddinott adapted this index to assess food insecurity. The FGT index for food insecurity utilizes three key indicators: the headcount ratio, the food insecurity gap, and the severity of food insecurity (Hoddinott 2001). This index providing a comprehensive measure of food insecurity levels (Eq. 1). These indicators help in understanding not just the prevalence of food insecurity, but also the depth and intensity of the issue, making the FGT index a valuable tool for policymakers and researchers.

$$P_\alpha = \left(\frac{1}{\mathrm{N}}\right) \sum_{i=1}^{q} [\frac{Z - Y_i}{Z})]^{\alpha} \tag{1}$$

In Eq. 1, P, N, Yi, Z, q, and α are the Index of food insecurity, total number of respondents, the calories or protein available to each person, Calories or protein a person needs for food security, the number of food insecure individuals, and the weight given to the severity of food insecurity, respectively.

Three indicators of food insecurity prevalence, food insecurity gap, and food insecurity severity are calculated based on the value (α). Food insecurity prevalence includes households whose per capita energy consumption is below the predetermined energy level of 2100 kilocalories. This indicator also describes households whose per capita protein consumption is below the predetermined protein level of 60 grams. When α is zero, the food insecure people in the community are obtained based on Eq. 2.

$$P_0 = \left(\frac{q}{\mathrm{N}}\right) \tag{2}$$

The second index, the food insecurity gap, measures the depth of food insecurity, assuming all households below the required calorie level are equally food insecure, with (α) set to one, calculated as Eq. 3.

$$P_1 = \left(\frac{1}{N}\right)\sum_{i=1}^{q}[\frac{Z - Y_i}{Z})] \tag{3}$$

The third index, the severity of food insecurity, gives more weight to households with greater food insecurity gaps, typically setting (α) to 2, making this measure closely related to the food insecurity gap, calculated as Eq. 4.

$$P_2 = \left(\frac{1}{N}\right)\sum_{i=1}^{q}[\frac{Z - Y_i}{Z})]^2 \tag{4}$$

These indices provide a comprehensive understanding of food insecurity, from its prevalence to its depth and severity, aiding policymakers in designing targeted interventions (Mitiku et al. 2012).

2.2 *Design of a Mathematical Programming Model for the Optimization of the Poverty Index of the FGT*

In this section, four main phases with different steps were considered. Figure 1 shows the conceptual framework and the stages of the research. In the first phase, the necessary data for this study were collected through interviews with experts for each product and experts involved in collecting information on the production costs of the products under study. The second phase aims to develop an initial optimal cropping pattern using the information from the previous phase. It should be noted that several revisions were made to the model in this phase. In the third phase, the FGT index is calculated for the study areas under current conditions. Finally, using the information from the previous three phases, the proposed model for optimizing the FGT poverty index is presented.

This section first presents the mathematical programming model. Before explaining the model and to avoid confusion with the symbols used in the study, the sets, parameters and variables of the model are presented in Table 3. The indices (c), (r) and (m) represent the product, the region and the month under study, respectively.

2.2.1 Objective Function

The goal of this section is to minimize the FGT index. This index was proposed by three economists: Foster, Greer, and Thorbecke, and has gained significant application in recent years. The index is calculated using Eq. 5 (Mousavi et al. 2011).

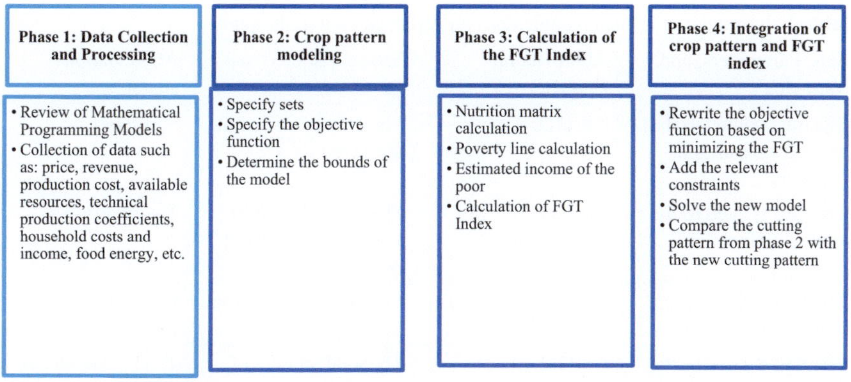

Fig. 1 The research conceptual framework

$$\mathrm{FGT\_V} = \sum_{\mathrm{i}} \left(\frac{\mathrm{PL} - \mathrm{Y_i}}{\mathrm{PL}} \right)^{\alpha} \tag{5}$$

Given Eq. 5, it is clear that the proposed model is a nonlinear programming model. Per capita income is the only parameter that can improve (reduce) the FGT index in the short run. In other words, if the change in income (ΔY) is positive, the value of this index will decrease. The minimization of the FGT poverty index for the studied areas was carried out according to Eq. 6.

$$\text{Min: } \mathrm{FGT\_V} = \sum_{\mathrm{r}} \mathrm{FGT\_V_r} \tag{6}$$

2.2.2 Set of Constraints

The first step is to calculate the FGT index for each region. This is done using Eq. 7.

$$\mathrm{FGT\_V_r} = (\mathrm{PL} - (\mathrm{Y} + \Delta \mathrm{Y_r}/\mathrm{PL}))^{\alpha} \tag{7}$$

where (ΔY) determines the amount of reduction in the FGT poverty index for each region, given the ability to generate income by optimizing the cropping pattern. Now, the relationship (ΔY) in each region must be aligned with the income generated by the cropping pattern. This is done using Eq. 8.

Table 3 Introduction of sets, parameters, and variables used in the study

Sets, Parameters, and Variables	Abbreviation character	Sets, parameters and variables	Abbreviation character
The Sets related to the studied agricultural products	$c \in \{1,2,3,\ldots\}$	The amount of fertilizer used type f	$Fertilizer_{c,r,f}$
The Sets related to the studied villages	$r \in \{1,2,3,\ldots\}$	The total amount of available fertilizer of type f	$Total\ Fertilizer_{r,f}$
The Sets related to the months of the year	$m \in \{1,2,3,\ldots,12\}$	The amount of poison used type z	$Pesticide_{c,r,z}$
The Sets related to the type of chemical fertilizer	$f \in \{1,2,3\}$	The total amount of available poison of type z	$Total\ Pesticide_{r,z}$
The Sets related to the type of pesticide	$z \in \{1,2,3\}$	The number of labor required to cultivate one hectare	$Labor_{c,r}$
The deciles of society	$i \in \{1,2,3,\ldots,10\}$	Total workforce available	$Total\ Labor_r$
Household food items	$j \in \{1,2,3,\ldots\}$	Working hours of machinery to cultivate one hectare	$Machine_{c,r}$
Poverty line	PL	Total machine working hours available	$Total\ Machine_r$
Income of poor people	Y	The price of agricultural products	$Price_{c,r}$
The total expenditure of the decile	$Expen_i$	Current cultivated area	$CurrentPattern_{c.r}$
The amount of energy received by the decile i	$Calo_i$	The minimum amount of land that can be allocated	$MinLand_{c.r}$
Expenditure incurred decile i for each calorie unit	$Excal_i$	The maximum amount of land that can be allocated	$MaxLand_{c.r}$
Performance rate	$Yield_{c,r}$	The amount of land (cultivated area)	$Land\_V_{c,r}$
The cost of producing agricultural products, except the cost of water	$Cost_{c,r}$	Poverty index	FGT_V
Land occupancy rate	$Landsch_{c,r,m}$	Nutritional performance matrix	$Mcalorie_{i,j}$
Total arable land available	$Total\ Land_r$	Amount of food intake j in decile i for each person	$Food_{i,j}$

(continued)

Table 3 (continued)

Sets, Parameters, and Variables	Abbreviation character	Sets, parameters and variables	Abbreviation character
The net water requirement of the plant	$\text{Net Water}_{c,r,m}$	Excess energy received The amount of calories from food consumption j	Extra_ Calo_i
The total amount of water available	Total Water_r		

$$\Delta Y = \left[\left(\sum_c \text{Price}_{c,r} \text{Yield}_{c,r} \text{Land \_V}_{c,r} - \sum_c \text{Cost}_{c,r} \text{Land\_V}_{c,r} \right. \right. \\ - \sum_c \text{Priac}_{c,r} \text{Yield}_{c,r} \text{Current Pattern}_{c,r} \qquad \forall_r \\ \left. - \sum_c \text{Cost}_{c,r} \text{Current Pattern}_{c,r}) / \text{Pop}_r \right]^{\alpha} \tag{8}$$

An important constraint that needs to be considered is to prevent income from exceeding the per capita poverty line due to the limitation of the FGT index, as shown in Eq. 9.

$$Y + \Delta Y_r \leq PL \quad \forall_r \tag{9}$$

The first set of physical constraints relates to the available agricultural land, which is applied according to Eq. 10.

$$\sum_c \text{Land\_V}_{cr} \text{LandSch}_c, r, m \ \leq \ \text{TotalLand}_r \ \ \forall_r \tag{10}$$

In Eq. 10, the land use coefficient is taken into account to avoid model errors in land allocation for crops with the same growing season.

In constraint 11, the irrigation efficiency of different regions is considered based on reports from the Water and Power Organization. Net water demand is also extracted from the NetWat software.

$$\sum_c \sum_m ((\text{NetWater}_{c,r,m}) / \text{Eff}_r) \text{Land\_V}_{c,r,m} \leq \text{TotalWater}_r \quad \forall_r \tag{11}$$

Constraints for other resources such as chemical fertilizers, pesticides, machinery and labor are applied according to Eqs. 12 to 15.

$$\sum_c \text{Fertilizer}_c, r, \text{fLand}_{V_{c},r} \ \leq \ \text{TotalFertilizer}_r, f \ \ \forall_{r \cdot f} \tag{12}$$

$$\sum_{c} \text{Pesticide}_{c,r,z}\text{Land\_V}_{c,r} \leq \text{TotalPesticide}_{r,z} \quad \forall_{r\cdot z} \tag{13}$$

$$\sum_{c} \text{Machine}_{c,r}\text{Land\_V}_{c,r,m} \leq \text{Total Machine}_{r} \quad \forall_{r} \tag{14}$$

$$\sum_{c} \text{Labor}_{c,r}\text{Land\_V}_{c,r,m} \leq \text{TotalLabor}_{r} \quad \forall_{r} \tag{15}$$

In the optimal cropping pattern, the current profit for each region must be maintained. Equation 16 enforces this constraint.

$$\begin{aligned}&\sum_{c} \text{Price}_{c,r}\,\text{Yeild}_{c,r}\text{Land\_V}_{c,r} - \sum_{c} \text{Cost}_{c,r}\text{Land\_V}_{c,r} \geq \qquad \forall_{r}\\ &\sum_{c} \text{Price}_{c,r}\text{Yeild}_{c,r}\text{CurrentPatern}_{c,r} - \sum_{c} \text{Cost}_{c,r}\text{CurrentPatern}_{c,r}\end{aligned} \tag{16}$$

The minimum and maximum cultivated area for agricultural products are applied according to Eqs. 17 and 18, respectively.

$$\text{Land}_{V_{c,r}} \geq \text{MinLand}_{c,r} \quad \forall_{r} \tag{17}$$

$$\text{Land}_{V_{c,r}} \leq \text{MaxLand}_{c,r} \quad \forall_{r} \tag{18}$$

2.3 Estimating the Poverty Line Based on Body Calorie Needs

The daily calorie requirement method was used to calculate the poverty line (PL). This method was used by Rowntree (2000) and was also used in Iran by Khodadad-Kashi and Heydari (2009). The reason for choosing this method is that there is a bidirectional relationship between poverty and health, and nutrition is considered an important factor in determining the nature of this relationship. During the growth period, significant changes occur in the physical, mental, intellectual, and psychological state of an individual, and during this period, the need for food increases sharply. However, poor groups are unable to meet these needs, and due to the problems they have carried over from previous periods of deprivation, they suffer irreparable damage to their health and mind, which ultimately leads to disease, low efficiency, and the perpetuation of poverty.

To apply this method, the first step was to estimate the Nutritional Performance Matrix. This matrix is the product of two matrices: the quantities of food consumed during the year and the amount of calories obtained from each kilogram of food, according to the relationship (19).

$$\mathrm{Mcalorie}_{\mathrm{i,j}} = \mathrm{Food}_{\mathrm{i,j}} \times \mathrm{CalorieCoeff}_{\mathrm{j}} \tag{19}$$

If the ith decile is the decile that receives the minimum required energy, the amount of energy received by that decile is calculated using Eq. 20.

$$Calorie_i = \sum_j Mcalorie_{i,j} \tag{20}$$

In the next step, using relation (21), the cost in rials of providing each unit of calories is determined by dividing the total expenditure of the ith decile by the amount of energy received by this decile.

$$\mathrm{Expcal}_{\mathrm{i}} = \frac{\mathrm{Expen}_{\mathrm{i}}}{\mathrm{Calo}_{\mathrm{i}}} \tag{21}$$

In the next step, using relations (22) and (23), the excess energy received by the ith decile over the minimum recommended energy is calculated, and the resulting number is multiplied by the cost of each unit of energy.

$$\mathrm{Extra\_Calo}_{\mathrm{i}} = \mathrm{Expen}_{\mathrm{i}} - 2300 \tag{22}$$

$$\mathrm{Extra\_Expcal}_{\mathrm{i}} = \mathrm{Extra\_Calo}_{\mathrm{i}} \times \mathrm{Expcal}_{\mathrm{i}} \tag{23}$$

In the final step, by subtracting the extra expenditure of the ith decile ($Extra\_Expcal_i$) from the total expenditure of the ith decile ($Expen_i$), the overall poverty line is calculated using relation (24).

$$\mathrm{PL} = \mathrm{Expen}_{\mathrm{i}} - \mathrm{Exha} - \mathrm{Expcal}_{\mathrm{i}} \tag{24}$$

After identifying the decile with the minimum energy consumption, an optimal measure of the income of poor individuals (Y) is obtained by averaging the income of this decile and the preceding deciles (with lower income). The necessary data for this section was collected through interviews with experts for each product, experts related to the collection of production cost information of the products under study. In addition, information related to the dietary matrix, household Expenditure and income were collected from the Statistical Center of Iran for the year 2018 and analyzed using GAMS and Excel software.

3 Empirical Example 1: Measuring Food Insecurity in Urban and Rural Households of Iranian Provinces

In this study, raw data from the Household Expenditure and Income Survey of urban and rural households conducted by the Statistical Center of Iran were utilized in the year 2022. The relevant indices were calculated for a substantial sample size, including 19,511 households in urban areas (64,659 individuals), and 18,321 households in rural areas (61,670 individuals). Based on the expenditure and income data from the provinces of Iran, the sample size is detailed in Table 4.

The basic information needed to analyze food insecurity in different provinces was compiled using food consumption data. The Statistical Center's questionnaire classifies all food items into 18 distinct groups. This categorization enables a detailed analysis of food consumption patterns across different regions. By examining these 18 food groups. These foods were then converted to nutrients to calculate the nutrients consumed by the household. The calorie and protein tables for the various food groups were used to determine the nutritional value of the foods that were consumed. Per capita values were obtained by dividing total calories and protein by household size. The study used the minimum daily requirements for energy (2100 kilocalories) and protein (60 grams) as standards. Households were categorized by province using codes from the Expenditure and Income Questionnaire of the Statistical Center of Iran. By identifying the households that did not receive the standard amounts, food insecurity were calculated using the FGT index. The results are presented in Table 5.

Based on the results of Table 5, in terms of calorie availability, urban areas in Kohgiluyeh and Boyer Ahmad, Qazvin, Hormozgan, West Azerbaijan, Qom, Yazd, Hamedan, Gilan, Alborz, Tehran, Golestan, and Isfahan have higher levels of food insecurity than other provinces. Meanwhile, urban areas in Semnan, Ilam, Kermanshah, North Khorasan, South Khorasan, and Kurdistan have the lowest levels. In rural households, the prevalence of food insecurity in Hormozgan and West Azarbaijan is higher than other provinces with a food insecurity gap of 28.16% and 22.08% and severity of 13.10% and 9.84%. The lowest levels of food insecurity in rural areas are found in Kermanshah and Ilam, with gaps of 2.04% and 3.17%, and severity levels of 0.47% and 0.72%, respectively. Overall, rural areas in Hormozgan, West Azerbaijan, Gilan, Tehran, Kerman, Golestan, Isfahan, Alborz, Qazvin, and Sistan and Baluchestan have the highest levels of food insecurity, while Kermanshah, Ilam, Semnan, and Lorestan have the lowest levels of access to calories.

In conclusion, a comparison of the 2022 index for urban and rural areas reveals that 48.06% of urban areas and 44.34% of rural areas in Iran experience food insecurity in terms of calorie access. The food insecurity gap and severity in urban areas are 12.77% and 5%, respectively, while in rural areas, they are 12.19% and 4.95%. This indicates that urban areas face higher levels of food insecurity and greater gaps and severity compared to rural areas. In addition, similar calculations have been made for the assessment of household protein consumption in different provinces. Indicators of protein deficiency were calculated based on daily food consumption,

Table 4 Sample size

	Urban areas	Rural areas
	Number of Households	Number of Households
Markazi	587	841
Gilan	837	402
Mazandaran	431	548
East Azerbaijan	587	674
West Azerbaijan	518	580
Kermanshah	526	624
Khuzestan	615	701
Fars	662	800
Kerman	581	698
Razavi Khorasan	722	814
Isfahan	636	641
Sistan and Baluchestan	732	660
Kurdistan	463	316
Hamedan	759	587
Chaharmahal and Bakhtiari	657	487
Lorestan	458	569
Ilam	435	502
Kohgiluyeh and Boyer-Ahmad	536	582
Bushehr	565	524
Zanjan	607	512
Semnan	478	514
Yazd	615	645
Hormozgan	865	661
Tehran	1433	467
Ardabil	505	463
Qom	511	381
Qazvin	420	542
Golestan	1006	672
North Khorasan	706	690
south Khorasan	650	697
Alborz	408	527
Iran	19511	18321

Source Statistics Center of Iran

Table 5 Food insecurity indicators related to calories access 2022 (percent)

Province name	Urban			Rural		
	P_0*	P_1**	P_2***	P_0*	P_1**	P_2***
Markazi	45.83	13.64	5.44	38.17	11.16	4.66
Gilan	57.71	14.99	5.44	61.44	17.47	6.97
Mazandaran	46.87	9.61	2.94	41.97	8.51	2.75
East Azerbaijan	42.59	11.06	4.20	47.03	12.94	5.21
West Azerbaijan	62.16	21.77	10.34	63.97	22.08	9.84
Kermanshah	23.76	3.89	1.01	14.74	2.04	0.47
Khuzestan	45.69	9.09	2.92	43.22	7.90	2.29
Fars	48.49	12.34	4.47	37.13	11.22	5.11
Kerman	43.72	8.98	2.70	58.02	13.95	4.46
Razavi Khorasan	44.04	10.04	3.58	43.12	11.44	4.23
Isfahan	52.83	13.61	5.16	54.76	14.34	5.44
Sistan and Baluchestan	40.57	10.70	4.71	50.30	16.45	8.26
Kurdistan	36.93	9.23	3.38	44.30	13.87	6.25
Hamedan	60.21	21.02	9.74	46.85	15.35	6.98
Chaharmahal and Bakhtiari	38.51	8.51	2.78	31.83	6.87	2.26
Lorestan	43.89	8.67	2.62	31.11	6.76	2.16
Ilam	22.30	3.31	0.66	19.92	3.17	0.72
Kohgiluyeh and Boyer-Ahmad	70.34	19.14	7.76	48.97	14.71	6.70
Bushehr	45.49	11.40	4.35	39.69	10.67	4.37
Zanjan	39.87	9.54	3.55	33.59	8.41	3.05
Semnan	16.32	2.66	0.68	21.79	4.56	1.36
Yazd	60.81	18.14	7.64	48.99	15.89	7.42
Hormozgan	66.24	22.73	10.47	77.76	28.16	13.10
Tehran	55.62	13.50	4.93	58.67	12.97	4.25
Ardabil	44.36	9.51	3.25	39.31	7.99	2.51
Qom	61.25	17.52	7.17	46.98	10.54	3.24
Qazvin	70.24	19.13	6.86	53.51	13.40	4.51
Golestan	53.28	15.21	6.14	56.70	16.82	7.32
North Khorasan	32.01	7.52	2.52	34.06	8.38	3.03
south Khorasan	32.92	8.42	3.15	32.71	8.15	3.22
Alborz	57.35	21.57	10.33	54.27	19.46	9.41
Iran	48.06	12.77	5.00	44.35	12.19	4.95

Source Research Findings (* Prevalence of Food Insecurity, ** Food Insecurity Gap, *** Intensity of Food Insecurity) with a minimum required intake of 60 grams per day. These indicators are presented in Table 6.

Based on the results in Table 6, it is observed that the urban areas of Semnan, Kermanshah, Ilam, and South Khorasan provinces have the least food insecurity in terms of required protein consumption. In contrast, the urban areas of West Azerbaijan, Qazvin, Hormozgan, Kohgiluyeh and Boyer-Ahmad, Hamedan, Yazd, Kerman, Alborz, and Qom provinces have the highest food insecurity in terms of required protein consumption. The highest level of food insecurity is in West Azerbaijan Province, with a food insecurity gap of 23.18% and a severity of 10.86%. The lowest level of food insecurity is in Semnan Province, with a food insecurity gap of 2.26% and a severity of 0.55%. In rural areas, the provinces of Hormozgan, West Azerbaijan, Kerman, Alborz, and Isfahan are the most food insecure, while the rural areas of Kermanshah, Ilam, and Semnan are the least food insecure in terms of required protein consumption. This indicates significant regional disparities in food security, with some provinces facing much higher levels of food insecurity than others. Also, the results of Tables 5 and 6 show that, while calorie intake is a critical issue, protein deficiency is also a significant concern.

4 Empirical Example 2: Development of a Mathematical Programming Model for the Optimization of the Poverty Index of the FGT in the Rural Areas of the Behbahan District

Khuzestan is a province in southwestern Iran, with its southern part located on the coast of the Persian Gulf. The area of Khuzestan Province is 64,057 square kilometers, and with a population of 4,710,509 (2016), it is considered the fifth most populous province in Iran. Studies show that the poverty line has increased in both urban and rural areas of Khuzestan Province. In rural areas, the poverty line increased from 83,116 rials in 1997 to 387,696 rials in 2006, an increase of 6.36% (Statistical Center of Iran 2016a). Nearly 53% of Khuzestan's population lives in rural areas, and the average share of employment in the agricultural sector was 2.7% in 2016. Behbahan County, with an area of 3,195 square kilometers, is located in the southwest of the province and had a population of 180,593 according to the 2016 census. Due to its favorable climatic conditions, the presence of Marun and Kheirabad rivers, and fertile soil, there are suitable opportunities for the expansion of the agricultural sector in this county (Statistical Center of Iran 2016b).

The areas of Houmeh, Sardasht, Langirat, Dodangeh and Tashan in this county were selected as the study areas. According to the statistics of Khuzestan Agricultural Jihad Organization, about 86% of the total cultivated area of crops in this county is concentrated in these five areas, and the remaining cultivated area is completely scattered. The main advantage of selecting these five areas is the presence of agricultural service centers in each of them, which can easily provide the necessary and

Table 6 Food insecurity indicators related to protein access 2022 (percentage)

Province name	Urban			Rural		
	P_0*	P_1**	P_2***	P_0*	P_1**	P_2***
Markazi	44.97	12.25	4.64	42.09	11.13	4.39
Gilan	44.32	10.25	3.55	47.51	12.31	4.62
Mazandaran	28.77	5.00	1.31	24.27	4.02	1.18
East Azerbaijan	40.37	9.98	3.72	45.70	12.46	5.08
West Azerbaijan	64.29	23.19	10.87	68.28	24.84	11.32
Kermanshah	20.72	3.12	0.80	11.54	1.47	0.32
Khuzestan	39.19	7.93	2.58	34.38	6.33	1.91
Fars	44.71	10.04	3.36	35.25	9.50	3.93
Kerman	53.18	11.13	3.67	59.74	13.89	4.56
Razavi Khorasan	41.14	9.72	3.51	46.31	11.38	3.98
Isfahan	44.97	10.97	4.04	51.01	13.46	5.18
Sistan and Baluchestan	37.02	9.28	4.04	46.36	15.10	7.79
Kurdistan	30.89	7.76	2.88	39.56	11.58	4.73
Hamedan	58.50	18.88	8.37	46.00	14.41	6.23
Chaharmahal and Bakhtiari	30.14	5.83	1.75	21.77	4.36	1.37
Lorestan	33.84	6.02	1.67	28.82	6.35	1.94
Ilam	21.15	2.72	0.51	20.12	3.00	0.74
Kohgiluyeh and Boyer-Ahmad	60.26	14.84	5.70	40.72	12.13	5.61
Bushehr	36.46	9.28	3.67	34.92	8.96	3.56
Zanjan	38.22	8.31	2.80	34.18	8.08	2.86
Semnan	14.64	2.26	0.55	20.62	4.03	1.17
Yazd	56.91	17.37	7.19	48.53	15.16	6.67
Hormozgan	62.31	18.76	7.57	74.13	23.54	9.91
Tehran	41.52	9.44	3.29	45.61	8.65	2.67
Ardabil	29.31	5.26	1.73	23.97	4.26	1.28
Qom	52.64	14.41	5.77	39.11	7.48	2.12
Qazvin	63.81	14.72	4.48	49.63	10.57	3.06
Golestan	46.22	11.77	4.39	49.85	13.87	5.92
North Khorasan	33.00	9.48	3.90	36.38	10.48	4.39
south Khorasan	28.62	6.75	2.51	26.11	5.93	2.30
Alborz	52.94	16.38	6.49	51.23	15.12	6.36
Iran	42.39	10.63	4.01	40.69	10.68	4.22

Source Research Findings (* Prevalence of Food Insecurity, ** Food Insecurity Gap, *** Intensity of Food Insecurity) validated data by statistical experts for developing the cropping pattern. Therefore, to evaluate the effectiveness of the proposed model, the rural regions of Sardasht, Langirat, Houmeh, Dodangeh and Tashan in Behbahan County, Khuzestan Province were selected. The poverty line was determined using the daily calorie requirement method derived from the Nutritional Performance Matrix. Data for this analysis were collected from agricultural service centers in the county and the Statistical Center of Iran for the agricultural year 2017–2018, and were processed using Excel and GAMS software.

Table 7 shows the results of the estimation of the dietary matrix and the determination of the poverty line in the study area. As observed, the third decile is the first decile with an energy intake exceeding 2300 calories. The cost of each unit of energy consumed by this decile is 12,291 thousand rials, which is calculated by dividing the total expenditure by the energy intake (Eq. 21). The excess energy intake of the third decile was calculated using Eqs. 22 and 23, resulting in 5 thousand rials. Equation 24 was used to calculate the poverty line. The calculated poverty line for the third decile was 11,790 thousand rials. Comparing this amount with the poverty line calculated by Khodadad Kashi and Heidari (2009) using a similar method for the year 2007, which was 318 thousand rials, shows an increase of 3607%. By averaging the income of this decile and the lower income deciles, the income of poor individuals was determined to be 9,630 thousand rials per person.

Table 7 Results of calculating the nutritional matrix across different income deciles (unit: calories)

Main food groups	First decile	Second decile	Third decile	Fourth decile	Fifth decile	Sixth decile	Seventh decile	Eighth decile	Ninth decile	Tenth decile
Bread and cereals	468	896	1080	1236	1393	1657	1890	2127	2391	3911
Meat	93	130	153	196	229	298	399	441	651	1984
Fish and shrimp	7	17	16	22	27	32	44	53	61	25
Milk and products	64	111	137	117	157	230	199	292	304	428
Oils	207	229	272	318	358	447	429	491	486	777
Fruits and nuts	260	357	410	511	625	773	994	1103	1097	1833
Vegetable	36	55	60	80	91	110	119	125	146	167
Sugar	74	114	143	220	221	284	352	400	444	586
Seasoning spice	56	107	122	173	210	258	243	251	261	447
Tea and coffee	0.01	0.05	0.07	0.09	0.1	0.12	0.12	0.14	0.17	0.2
Sum	1269	2019	2397	2877	3314	4093	4673	5287	5845	10,161

Source Research Findings

Now let's examine the results of the proposed model for optimizing the FGT poverty index. Table 8 shows the cultivated area of agricultural products in Behbahan County, categorized by the studied regions under both the current and optimized cropping patterns. According to the data in this table, the cultivated area of wheat in the five studied regions has decreased by 23% in the optimized pattern compared to the current pattern, with most of this decrease occurring in the Houmeh area. In this area, the cultivated area of wheat has decreased from 2800 hectares in the current model to 1016 hectares in the optimized model (about 64%). This reduction can be attributed to the lower price of wheat compared to other products and the preference of farmers to grow more profitable crops such as watermelon, dates, and sugar beets, with the area of these crops increasing by about 41%, 41%, and 48%, respectively, in the optimized model. The main cultivation of sesame in this county occurs in the Langirat and Sardasht regions, where the cultivated area has increased by 22% and 53%, respectively, in the optimized model. This increase is due to the suitability of the soil in these regions for sesame cultivation and the significant profit that farmers receive from this crop. One of the most consumed products in this county is Broad bean, with the Dodangeh region having a larger share of its cultivation compared to other regions. In the optimized pattern, the area under Broad beanhas increased by 51.7% compared to the current pattern. Canola is another crop that is cultivated in all five regions studied, with an overall increase of 42.1% in the optimized pattern, as it is used for both fodder and oilseed. Overall, the area of this crop in the optimized pattern decreases by 7% compared to the current pattern, mainly due to the decrease in the area of major crops such as wheat, barley, and tomatoes, which are the most important crops in the optimized pattern.

The results of comparing the constraints on water consumption, fertilizers and pesticides, labor, machine hours, and profit in the current and optimized conditions are presented in Table 9. According to the data in this table, the consumption of all inputs has decreased in the optimized pattern compared to the current pattern, ranging from 0.51% for water to about 10% for labor. The reduction in water consumption occurred only in the Langirat region, about 3%, while in other regions, water consumption remained constant with the change in cropping pattern. This is because the main crop in the Langirat region is barley, and in the optimized model, its cultivated area has decreased significantly. The total fertilizer consumption in the county has decreased by 9%, with the Dodangeh region contributing significantly to this reduction. This is because the soil in this region is more suitable for agriculture compared to other regions, which is one of the reasons for the reduced fertilizer consumption in the optimized model. The Langirat region has the highest consumption of chemical fertilizers. This is due to the less suitable soil texture compared to other regions studied, which requires more chemical fertilizers and longer machine operating hours to compensate. The Langirat region uses less labor compared to other regions due to the higher use of machinery. An important aspect of the optimized cropping pattern is the significant reduction in labor use, which is a disadvantage of this pattern given the existing unemployment in the region.

Figure 2 shows the profit of cultivating the studied crops in the five studied regions. According to this figure, the highest profit is related to the Houmeh and Sardasht

Table 8 Cultivated area of agricultural products in Behbahan County under current and optimized conditions, categorized by studied regions (unit: hectares)

Products	Model	Area					
		Tashan	Dodangeh	Houmeh	Langirat	Sardasht	Total
Wheat	Current	2100	3500	2800	5000	4200	17,600
	Optimal	1750	2562	1016	4468	3756	13,552
	Percentage changes	−16.7	−26.8	−63.7	−10.6	−10.6	−23
Barley	Current	25	0	0	800	4	829
	Optimal	0	0	0	318	6	324
	Percentage changes	−100	–	–	−60.3	50	−60.1
Canola	Current	40	700	1000	900	535	3175
	Optimal	61	969	1508	1342	633	4513
	Percentage changes	52.5	38.4	50.8	49.1	18.3	42.1
Sesame	Current	0	0	0	1000	25	1025
	Optimal	0	0	0	1221	38.3	1259
	Percentage changes	–	–	–	22	53	22.8
Cucumber	Current	0	0	43	0	0	43
	Optimal	0	0	65.7	0	0	65.7
	Percentage changes	–	–	52.7	–	–	52.7
Tomato	Current	0	0	200	0	150	350
	Optimal	0	0	0	0	117	117
	Percentage changes	–	–	−100	–	−22	−66.7
Rice	Current	160	0	150	0	102	412
	Optimal	231	0	218	0	156	605
	Percentage changes	44.4	–	45.3	–	52.9	46.8
Alfalfa	Current	0	0	2300	0	0	2300
	Optimal	0	0	2281	0	0	2281
	Percentage changes	–	–	−0.8	–	–	−0.8
Citrus fruits	Current	0	0	0	0	3	3
	Optimal	0	0	0	0	4.4	4.4
	Percentage changes	–	–	–	–	46	46
Grape	Current	0	0	20	0	0	20

(continued)

Table 8 (continued)

Products	Model	Area					
		Tashan	Dodangeh	Houmeh	Langirat	Sardasht	Total
	Optimal	0	0	30	0	0	30
	Percentage changes	–	–	50	–	–	50
Beans	Current	0	0	0	0	86	86
	Optimal	0	0	0	0	33.8	−33.8
	Percentage changes	–	–	–	–	−60.7	−60.7
Melon	Current	0	0	250	0	64	314
	Optimal	0	0	374	0	95.1	469.1
	Percentage changes	–	–	49.6	–	48.6	49.4
Watermelon	Current	0	0	1000	0	230	1230
	Optimal	0	0	1399	0	334	1733
	Percentage changes	–	–	39.9	–	45.2	40.8
Forage corn	Current	0	300	100	0	0	400
	Optimal	0	437	150	0	0	587
	Percentage changes	–	45.7	50	–	–	46.7
Sugar beet	Current	0	0	200	0	0	200
	Optimal	0	0	295	0	0	295
	Percentage changes	–	–	47.5	–	–	47.5
Broad bean	Current	0	150	0	0	0	150
	Optimal	0	227	0	0	0	227
	Percentage changes	–	51.3	–	–	–	51.3
Dates	Current	0	100	0	0	0	100
	Optimal	0	141	0	0	0	141
	Percentage changes	–	41	–	–	–	41
Onion	Current	80		0	0	0	80
	Optimal	86.8		0	0	0	86.8
	Percentage changes	8.5		–	–	–	8.5
Total	Current	2380	4750	8063	7700	5399	28,292

(continued)

Table 8 (continued)

Products	Model	Area					
		Tashan	Dodangeh	Houmeh	Langirat	Sardasht	Total
	Optimal	2128	4335	7336	7348	5173	26,320
	Percentage changes	−10.6	−8.7	−9.1	−4.6	−4.2	−7

Source Research Findings

Table 9 Comparison of the constraints studied in general in two current and optimal states in the study areas

Constraint	Model	Area					
		Tashan	Dodangeh	Houmeh	Langirat	Sardasht	Total
Water (thousand cubic meters)	Current	3611	6692	23,439	8942	7174	49,858
	Optimal	3612	6690	23,438	8687	7173	49,600
	Percentage changes	0	0	0	−2.8	0	−0.52
Fertilizer (unit: thousand tons)	Current	801.6	3197.5	4757	2408.3	4399	15,562
	Optimal	689.6	2733.4	4187	2470	4082	14,162
	Percentage changes	−14	−14.5	−12	2.6	−7.2	−9
Machinery (thousand hours)	Current	18,700	35,050	74,445	69,532	51,080	251,275
	Optimal	16,495	30,476	70,511	72,000	48,051	235,065
	Percentage changes	−11.8	−13.1	−5.3	−3.5	−5.9	−6.5
Workforce (thousand person–days)	Current	12,015	23,900	40,335	56,100	44,912	177,262
	Optimal	10,781	21,825	33,754	52,006	42,036	160,402
	Percentage changes	−10.3	−8.7	−16.3	−7.3	−6.4	−9.5
Poison (thousand liters)	Current	5560	20,250	58,417	32,100	30,043	146,370
	Optimal	4682	17,859	56,143	30,008	29,783	138,475
	Percentage changes	−15.8	−11.8	−3.9	−6.5	−0.9	−5.4

Source Research Findings

regions. These regions are larger and have more diverse crops compared to other regions. The smallest profit increase is related to the Tashan region, which has increased by about 3% in the optimized state. The crops grown in this region are limited, with wheat and barley being the main crops. In the optimized pattern, the area of wheat has decreased by about 17%, and the area of barley has reached zero. It should be noted that the amount of available water in this region has decreased significantly in recent years, so the reduction in the area of major crops in the optimized

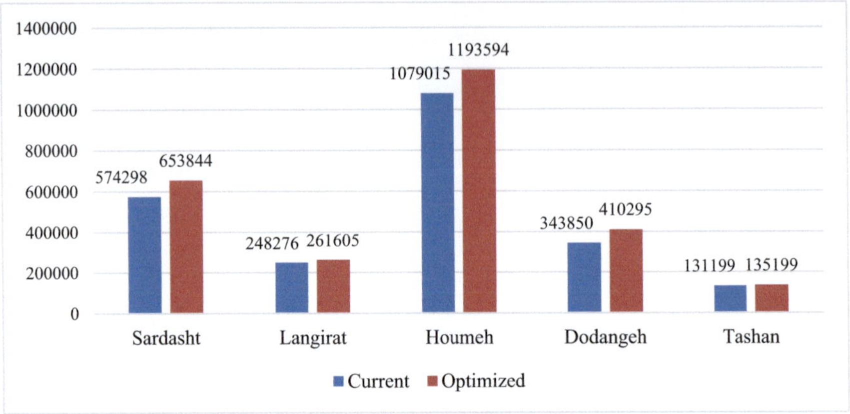

Fig. 2 Profit from growing crops in two current and optimal conditions (unit: million rials)

pattern seems logical. This result may raise the question of whether this result is reasonable given the strategic importance of wheat. The answer lies in Eq. 17, which ensures that no crop is allowed to decrease beyond the acceptable limit set by experts. As shown in the figure, the Dodangeh region, although smaller than other regions, has achieved significant profit in the optimized pattern. The profit from major crops in this region has increased by 19.3% in the optimized state. This is due to the fact that in this region the profitable crop of dates is cultivated, which includes different varieties and is exported not only within the county, province and country, but also abroad. In addition, the high share of these two regions in the increase of cultivated area in the optimized model has also contributed to this result.

A more detailed examination shows that the profit of most of the crops grown in the study regions has increased in the optimized state, with only wheat and tomatoes experiencing profit decreases of 23% and 66%, respectively. It should be noted that both crops experienced a relatively large decrease in area cultivation. Given that in linear programming the relationship between the technical coefficients and the objective function coefficients follows the Leontief production function, this reduction in profit seems logical given the reduction in area cultivation. The highest profit in the optimized state is from watermelon cultivation at 812,140 million rials, while the lowest is from citrus fruits at 1,422 million rials. Despite its high gross water requirement, watermelon has a high gross profit and has generated this profit with a relatively small cultivated area in the region. The model introduced in this study has notably decreased the cultivation areas for crops like wheat and barley, while expanding the areas for more profitable crops such as watermelon to offset the profit reduction. It should be noted that the increase in the area of this water-intensive crop is conditional on the significant reduction in the area of crops such as wheat and barley (Table 8). In other words, the proposed area cultivation for the optimized model are feasible as a whole, and selecting the proposed area cultivation separately is not justifiable or logically interpretable. In general, in most provinces of the country, the excessive increase in the cultivated area of agricultural products in recent years has led to numerous problems (Mardani et al. 2016). In most of the optimized patterns that have been implemented in regions facing severe water scarcity, a reduction in the cultivated area of agricultural products, especially those with large cultivated areas such as wheat and barley, has been suggested. For example, studies by Mardani Najafabadi et al. (2019) in Isfahan, Mosleh et al. (2017) in Shahrekord, and even Hosseini Yekani and kashiri kolaei (2017) in Sari have reached similar conclusions. It should be noted that in each of these studies, the reduction of cultivated area was selected based on regional potentials and economic benefits for specific groups of products. In addition, these studies, as well as the present study, also considered the share of regions in ensuring food security with the constraint cultivated area minimum.

Irrigation water consumption also differs between the optimized cropping pattern and the current cropping pattern. Comparing the optimized cropping pattern with the current cropping pattern, the highest water consumption in the optimized model occurs in April and May, and the lowest in December and January. This is because most of the crops in this county are irrigated in April and May. Regarding the water consumption in the hot months of the year, as shown in Fig. 3, it is noted that the water consumption in the proposed pattern is higher than in the current cropping pattern of the region. For example, the average water consumption in June in the current cropping pattern of the region is 3.02 million cubic meters. However, in the optimized cropping pattern, the water consumption in this month has increased to 3.41 million cubic meters. Water consumption in other hot months of the year is also higher in the optimized pattern, with July, August, and September seeing increases from 3.1, 3.64, and 3.64 million cubic meters in the current pattern to 3.29, 3.99, and 4 million cubic meters, respectively.

Figure 3 compares the water consumption required for various crops under the current and optimized patterns. It is observed that in the optimized pattern, water use for wheat has significantly decreased by about 23%, while water use for crops such as canola, rice, watermelon, forage corn, and dates has increased. Alfalfa water use did not change significantly. The significant reduction in water use for wheat is due to the low purchase price set by the government, and farmers prefer to cultivate crops that provide a reasonable profit for the water consumed. Dates are one of the most important crops grown in this district, and as observed, they provide considerable profit for relatively low water consumption. Other crops that are economically viable despite lower water consumption include onions, broad beans, and sesame.

Finally, by calculating the FGT index using Eq. 5 for both current and optimized conditions, it is observed that the numerical value of this index has decreased across all regions in the optimized state. Averaging the FGT index under both conditions shows that the objective function value has dropped from 3.47 in the current state to 0.03 in the optimized state. This indicates a significant reduction in poverty based on this index when optimal factors of production are used. Studies by BaniAsadi and Zare Mehrjerdi (2010) for households in the Arzuiyeh region of Baft County and by Mousavi et al. (2011) for Kazerun County also confirm the positive impact of the optimized cropping pattern on improving the FGT poverty index (Fig. 4).

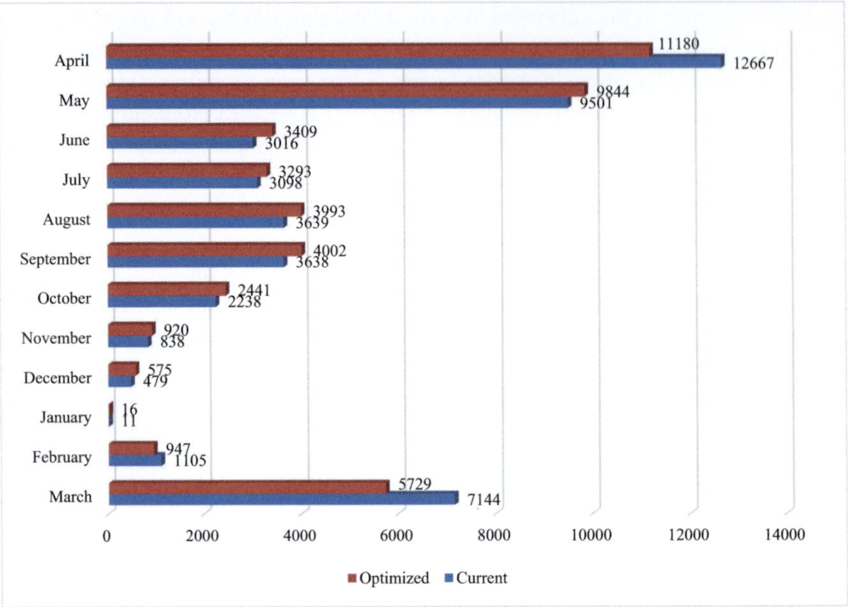

Fig. 3 Comparison of water consumption in different months of the year under the current and optimized cropping patterns (unit: thousand cubic meters)

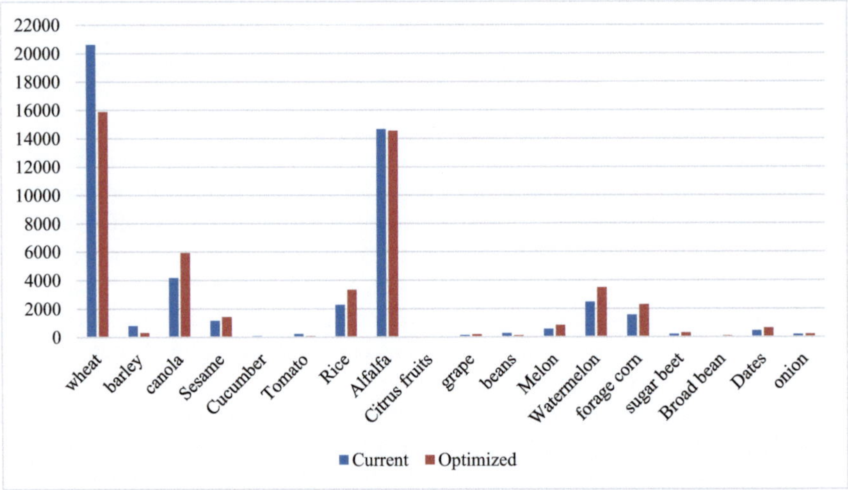

Fig. 4 Comparison of water use for the crops studied under the current and optimized cropping patterns (unit: thousand cubic meters)

5 Conclusion

Food security is critical to both individual and societal health and is recognized as a sustainable development goal. Studying food insecurity in different communities not only highlights household conditions, but also helps governments, and global organizations reduce its impact on people's quality of life. Therefore, the assessment of household food insecurity is critical. Given its importance in maintaining national security and promoting a healthy society, assessing household food insecurity is essential. This study highlights the significant importance of food insecurity in different countries. Most studies indicate that food insecurity and poverty have increased in recent years. On the other hand, the agricultural sector is a powerful tool for developing countries (including Iran) to achieve economic growth and development. Moreover, the growth of production in this sector has led to the reduction of poverty in these countries. To increase agricultural production, it is inevitable to increase productivity and make optimal use of scarce resources. Therefore, using an optimal cropping pattern while properly utilizing production resources will lead us to increase income, which in turn will reduce poverty. After reviewing the studies, this research provides an overview of the indices used to measure food insecurity. It then explains the FGT index for studying food insecurity. In addition, the study presents a nonlinear programming model to optimize the FGT poverty index. Ultimately, this model leads to the presentation of an optimal cropping pattern for agricultural products by creating a gap between current and optimal income, thereby reducing the FGT poverty index. Finally, an empirical example was conducted to assess food insecurity in rural and urban areas of Iranian provinces using the FGT index for 2022. In addition, another empirical example was conducted using the proposed mathematical programming model to optimize the FGT poverty index.

The FGT index results indicated that in 2022, food insecurity was more prevalent in urban households compared to rural ones. This observation is consistent with the findings of Rezaeifar et al. (2023), Pakravan et al. (2015), Migotto et al. (2007) and Hoddinott and Yohannes (2002). In addition, applying the specified method to determine the optimal cropping pattern in Behbahan County revealed differences between the current and optimal patterns. In particular, the optimal pattern resulted in a 7% reduction in cropped area compared to the existing pattern. Overall, the results indicate that the current use of resources is not optimal and there is potential to increase profits by changing the cropping pattern. Therefore, considering the calculated poverty line and the measurement of the poverty index before and after the cropping pattern adjustment, it was found that this pattern increases farmers' income and reduces poverty.

Based on the results obtained, it is recommended that due to the difference between the current and optimal cropping patterns, the government should promote the optimal cropping pattern through the use of agricultural extension agents. In addition, the Agricultural Jihad Organization and other related bodies involved in agricultural production can support farmers by implementing supportive policies such as providing loans, insuring crops, and guaranteeing the purchase of major agricultural products in the county that has seen an increase in cultivated area in the optimal pattern. This can lead to an increase in farmers' income and consequently reduce poverty. Considering that the cropping pattern can be related to various issues such as production sustainability, environmental issues, logistics and transportation of agricultural products, it is suggested that future studies use multi-objective models capable of addressing all these aspects.

References

Abafita J, Kim KR (2014) Children's schooling in rural Ethiopia: the role of Household food security, parental education and income. Children 5(14)

Abebaw D, Fentie Y, Kassa B (2010) The impact of a food security program on household food consumption in Northwestern Ethiopia: a matching estimator approach. Food Policy 35(4):286–293

Adams J, HR (2003) Economic growth, inequality, and poverty : findings from a new data set retrieved from world bank policy research working paper series

Akbari MR, Pish Bahar E, Dashti Q (2020) Identifying the factors affecting food insecurity in rural Iran: application of generalized sequential logit model. Iranian J Appl Econ Studies 9(35):91–125. (In Persian)

Akbarpour M, Mahdavi Damghani A, Dehim Fard R, Veisi H (2017) The role of milk, fruit and vegetable consumption patterns on food security 16(6):726–733. (In Persian)

Alkire S, Foster J (2011) Understandings and misunderstandings of multidimensional poverty measurement. J Econ Inequality 9(2):289–314

Alkire S, Santos ME (2013) Measuring acute poverty in the developing world : robustness and scope of the Multidimensional Poverty Index. OPHI working paper no. 59. Oxford Poverty & Human Development Initiative (OPHI)

Almas I, Kjelsrud A, Somanathan R (2013) A behaviour—based approach to the estimation of poverty in India. CESifo working paper series, 4122

Álvares L, Amaral TF (2014) Food insecurity and associated factors in the Portuguese population Food and Nutrition Bulletin 35(4):395S402

Animasaun DA, Adedibu PA, Shkryl Y, Emmanuel FO, Tekutyeva L, Balabanova L (2023) Modern plant biotechnology: an antidote against global food insecurity. Agronomy 13(8):2038

Anríquez G, Stamoulis K (2007) Rural development and poverty reduction: is agriculture still the key? (2521–1838). Retrieved from http://ageconsearch.umn.edu/record/289048/files/a-ah885e.pdf

Arshadi A, Karimi A (2013) The study of poverty in rural and urban areas in iran during five-year development plan. J The Marco Strategic Policies 1(1):23–42. Retrieved from https://www.sid.ir/en/journal/ViewPaper.aspx?ID=389279

Arsyad M, Pulubuhu DAT, Kawamura Y, Maria IL, Dirpan A, Unde AA, … Yusuf S (2020) The role of public health services (PHS) in agricultural poverty alleviation. Enfermería Clínica 30:194–197

Arzrom Chiller N (2005) Various dimensions of poverty in Iran. Econ Res Collection 92:21–27

Baniasadi M, Zare Mehrjerdi MR (2010) Studying the effects of optimal cultivation pattern on rural poverty: case study of Orzoo'iyeh district in Baft (Kerman–Iran). Agric Econ Iranian J Agric Econ (Econ Agric J) 4(2):209–226. Retrieved from https://www.sid.ir/en/journal/ViewPaper.aspx?ID=191557.

Bazrafshan JA, Tuolabi Nejad M, Sadeghi K (2017) Analysis of the relationship between sustainable land manage-ment and crop yield and evaluating its effects on food security of households in rural Eastern Miyankoh (Poldokhtar City). J Rural Res 8(2):346–363

Casey PH, Szeto KL, Robbins JM, Stuff JE, Connell C, Gossett JM, Simpson PM (2005) Child health-related quality of life and household food security. Arch Pediatr Adolesc Med 159(1):51–56
Cheema AR, Abbas Z (2016) Determinants of food insecurity in Pakistan: evidence from PSLM 2010–11. Pakistan J Appl Econ 26(2):183–213
Code Q (2020) COVID-19 and Sustainable Development Goals (SDGs). Sustainable Development
Cook JT, Frank DA (2008) Food security, poverty, and human development in the United States. Ann N Y Acad Sci 1136(1):193–209
Darban Astaneh AR, Motiei Langeroudi HS, Ghasemi F (2018) Factors affecting the livelihood of farmers: a case study of Shazand county. J Rural Res 9(2):324–337
Devkota RP, Bajracharya B, Maraseni TN, Cockfield G, Upadhyay BP (2011) The perception of Nepal's Tharu community in regard to climate change and its impacts on their livelihoods. Int J Environ Stud 68(6):937–946
Eftekhari SF, Karami A (2014) Investigating factors affecting poverty in rural areas of Kohgiluyeh and Boyerahmad Province. Agric Econ (Econ Agric J) 8(1):139–156
Endale W, Mengesha ZB, Atinafu A, Adane AA (2014) Food Insecurity in Farta District, Northwest Ethiopia: a community based cross–sectional study. BMC Res Notes 7:1–6
Endris N, Asefa H, Dube L (2017) Prevalence of malnutrition and associated factors among children in rural Ethiopia. Biomed Res Int 2017:1–6
Erinle KO, Ogwu MC, Evivie SE, Zaheer MS, Ogunyemi SO, Adeniran SO (2021) Impacts of COVID-19 on agriculture and food security in developing countries: potential mitigation strategies. South Asia 30:13
FAO (2001) Handbook for defining and setting up a food security information and early warning systein (FSIEWS). Food and Agriculture Organization of the United Nations, New York
FAO; IFAD; UNICEF; WFP; WHO (2022) The state of food security and nutrition in the world 2022; repurposing food and agricultural policies to make healthy diets more Affordable; WHO: Rome, Italy; ISBN: 978-92-5-136499-4
Foster J, Greer J, Thorbecke E (1984) A class of decomposable poverty measures. Econometrica 52(3):761–766
Fotakis EA, Kontele I, Tzoutzou M, Grammatikopoulou MG, Arvanitaki E, Sergentanis TN, … Vassilakou T (2024) Food insecurity in greece and across the globe: a narrative literature review. Foods 13(10):1579
Fotros M, Ghodsi S (2010) Alkire and Foster method for measuring multidimensional poverty. Econ J 11(14):77–92
Gebre GG (2012) Determinants of food insecurity among households in Addis Ababa city, Ethiopia. Interdisciplinary Description Complex Syst INDECS 10(2):159–173
Hoddinott J (2001) Methods for rural development projects: food security in practice (No. 1). International food policy research institute (IFPRI)
Hoddinott J, Yohannes Y (2002) Dietary diversity as a food security indicator. Food and nutrition technical assistance, Academy for Educational Development. Washington, D.C.
Hosseini-Yekani SA, kashiri kolaei F (2017) Investigating the effect of crop price fluctuations on the optimal cropping pattern in Sari. Agric Econ 11(2):75–94
Khodadad Kashi F, Heydari K (2009) The measurement of poverty indices based on Iranian household nutrition. Econ Res Rev 10(3 (34)):205–231
Khoshnood A (2019) Poverty in Iran: a critical analysis. Middle East Policy 26(1):60–74
Loopstra R, Tarasuk V (2012) The relationship between food banks and household food insecurity among low-income Toronto families. Can Public Policy 38(4):497–514
Mannaf M, Uddin MT (2012) Socioeconomic factors influencing food security status of miaze growing households in selected areas of Bogra district. Bangladesh J Agric Econ 35(454–2016–36355):177–187
Mardani Najafabadi M, Ziaee S, Nikouei A, Ahmadpour Borazjani M (2019) Mathematical programming model (MMP) for optimization of regional cropping patterns decisions: a case study. Agric Syst 173:218–232

Mardani M, Nikouei A, Ziaei S, Ahmadpour M (2016) Codifying regional cropping pattern of agricultural and horticultural products in Isfahan Province: multi-objective structural planning approach. Agric Econ Dev 30(3):188–206

Migotto M, Davis B, Carletto C, Beegle K (2007) Measuring food security using respondents' perception of food consumption adequacy. Food Security: Indicators, Measurement, and the impact of trade openness. B. Guha-Khasnobis, SS Archaya and B. Davis, Eds. UNU-WIDER and ICSSR

Mitiku A, Fufa B, Tadese B (2012) Empirical analysis of the determinants of rural household's food security in Southern Ethiopia: the case of Shashemene District. Basic Res J Agric Sci Rev 1(6):132–138

Mohamadiyan S, Negahdari E (2018) Study of poverty indicators in the urban areas of hormozgan province using linear expenditure system (les) during 4th and 5th development plans: Household budjet survey. SOCIAL WELFARE, 18(68 #P00393), 201–223

Mortazavi SA, Seraj Zadeh F, Shokohi M (2011) The relationship between income inequality and economic growth in urban and rural areas of Iran: application of Kuznets curve. J Agric Econ Res 3(1):165–180

Mosleh Z, Salehi MH, Amini Fasakhodi A, Jafari A, Mehnatkesh A, Esfandiarpoor Borujeni I (2017) Sustainable allocation of agricultural lands and water resources using suitability analysis and mathematical multi-objective programming. Geoderma 303:52–59

Mousavi SN, Mobseri M, Akbari S (2011) The effects of optimum cropping pattern water consumption, income and rural poverty: a case study of Kazeroon region. Agric Econ Dev 25(2):219–226

Mulugeta M, Tiruneh G, Alemu ZA (2018) Magnitude and associated factors of household food insecurity in Fedis Woreda East Hararghe zone, Oromia region, Ethiopia. Agric Food Secur 7:1–8

Naderi Mahdei K (2015) Optimal cropping pattern in Bahar county. J Appl Econ Studies Iran 3(12):167–184

Najafi B, Shushtarian A (2004) Targeting subsidies and eliminating food insecurity: a case study of Arsanjan. Bus Res J 8(31):127–151 (In Persian)

Namara RE, Hanjra MA, Castillo GE, Ravnborg HM, Smith L, Van Koppen B (2010) Agricultural water management and poverty linkages. Agric Water Manag 97(4):520–527

Osama S, Elkholy M, Kansoh RM (2017) Optimization of the cropping pattern in Egypt. Alex Eng J 56(4):557–566

Pakravan MR, Hosseini SS, Salami H, Yazdani S (2015) Identifying the factors affecting food security of urban and rural households in Iran. Iranian Agric Econ Dev Res 46(3):395–408 (In Persian)

Pérez-Escamilla R, Segall-Corrêa AM (2008) Food insecurity measurement and indicators. Rev Nutr 21:15s–26s

Pollard CM, Booth S (2019) Food insecurity and hunger in rich countries—it is time for action against inequality. Int J Environ Res Public Health 16(10):1804

Purtaheri M, Talebi H, Rokanuddin Eftekhari A (2011) Analysis of economic and social factors affecting rural poverty reduction with emphasis on food poverty: Khalajistan section of Qom Province. Quart J Teacher Humanities (Spatial Planning and Planning) 2(15):61–80 (In Persian)

Ramesh T, Dorosti Motlagh AR, Abdollahi M (2009) Prevalence of food insecurity in households in Shiraz and the relationship between some socio-economic and demographic factors in 2008. Iranian J Nutrition Sci Food Industry 4(4):53–64 (In Persian)

Ren C, Guo P, Tan Q, Zhang L (2017) A multi-objective fuzzy programming model for optimal use of irrigation water and land resources under uncertainty in Gansu Province, China. J Clean Prod 164:85–94

Rezaeifar M, Khalilian S, Najafi Alamdarlo H (2022) Spatial distribution of food insecurity in urban and rural areas of Iran. Agric Econ 16(1). https://doi.org/10.22034/iaes.2022.540824.1881 (In Persian)

Rezaeifar M, Khalilian S, Najafi Alamdarlo H (2023) Measuring the Impact of COVID-19 on the food insecurity in Iran and the achievement of SDGs. In: Leal Filho W, Abubakar IR, da Silva I, Pretorius R, Tarabieh K (eds) SDGs in Africa and the middle east region . implementing the UN sustainable development goals—regional perspectives. Springer, Cham. https://doi.org/10.1007/978-3-030-91260-4_110-1

Rogers R, Eagle TF, Sheetz A, Woodward A, Leibowitz R, Song M, … Eagle KA (2015) The relationship between childhood obesity, low socioeconomic status, and race/ethnicity: lessons from Massachusetts. Childhood Obesity 11(6):691–695

Rowntree B (2000).Poverty: a study of town life. Macmillan, London, Qouted in Ravallio: Policy Press

Saadi H, Azizi M, Azami M (2014) Extension educational and farmer households food security (case study Ghorveh county). Iranian J Agric Econ Dev Res (IJAEDR) 45(3):483–499

Sethi LN, Panda SN, Nayak MK (2006) Optimal crop planning and water resources allocation in a coastal groundwater basin, Orissa. India. Agric Water Manag 83(3):209–220

Shirvanian A, Ismaili A (2009) The impact of changes in prices on rural poverty (Case study: Fars Province). Iranian J Agric Econ Dev Researches 40(2):13–25

Shushu GNJ, Mmbengwa VM, Swanepoel JW, Manasoe B (2023) Analysis of the indicators of household food insecurity in Frances Baard district municipality of South Africa

Smith MD, Meade B (2019) Who Are the World's food insecure? identifying the risk factors of food insecurity around the world. Amber Waves: The Economics of Food, Farming, Natural Resources, and Rural America, 2019(1490–2020–721)

Statistical Center of Iran (2016a) Indicators of the millennium development goals of the Islamic Republic of Iran. Theran: Country Planning and Budget Organization

Statistical Center of Iran (2016b) Statistical Yearbook of Khuzestan Province. Khozestan, Iran

Statistics Center of Iran (2024). https://amar.org.ir/

Stephens EC, Jones AD, Parsons D (2018) Agricultural systems research and global food security in the 21st century: an overview and roadmap for future opportunities. Agric Syst 163:1–6

Sustainable Development Goals (2016) Javaherian Z, Vahdati, Seyed AF, Rahmati A, Zamani L, Commissioned by the environmental protection agency, office of sustainable development and environmental economics. (In Persian)

Telila HF, Sima EA (2024) Quantifying food insecurity in Ethiopia: Prevalence, drivers, and policy implications. Cogent Soc Sci 10(1). https://doi.org/10.1080/23311886.2024.2318862

Thomas M, Miller DP, Morrissey TW (2019) Food insecurity and child health. Pediatrics 144(4)

UNICEF (2020) Micronutrient initiative. Vitamin and mineral deficiency: a global progress report. Available online: https://www.unicef.org/media/files/vmd.pdf. Accessed 2 December 2020

Van Hop N (2007) Fuzzy stochastic goal programming problems. Eur J Oper Res 176(1):77–86

Weedn AE, Hale JJ, Thompson DM, Darden PM (2014) Trends in obesity prevalence and disparities among low-income children in Oklahoma, 2005–2010. Child Obes 10(4):318–325

Weinreb L, Wehler C, Perloff J, Scott R, Hosmer D, Sagor L, Gundersen C (2002) Hunger: its impact on children's health and mental health. Pediatrics 110(4):e41–e41

Welderufael M (2014) Determinants of households vulnerability to food insecurity in Ethiopia: econometric analysis of rural and urban households. J Econ Sustain Dev 5(24):70–79

World Bank (2024). https://data.worldbank.org/

Yousefi A, Mahdian S, Khalaj S (2016) Determinants of multidimensional poverty in Iran rural areas. J Rural Res 6(4):699–721

Zhang F, Zhang C, Yan Z, Guo S, Wang Y, Guo P (2018) An interval nonlinear multiobjective programming model with fuzzy-interval credibility constraint for crop monthly water allocation. Agric Water Manag 209:123–133

Zhang J (2016) Weighing and realizing the environmental, economic and social goals of tourism development using an analytic network process-goal programming approach. J Clean Prod 127:262–273

Development of Quantum Information Using Matrix Representation of Single and Multi-Qubit Quantum Gates

Ajanta Das, Madhubrata Bhattacharya, and Debabrata Datta

Abstract Quantum computation is an interdisciplinary field that combines computer technology with quantum mechanics. It utilises quantum superposition, entanglement, and interference to revolutionise solving complex problems. Unlike classical computing, which relies on deterministic bit manipulation, quantum computing is based on qubit that represents two basis quantum state vectors of unit length in a complex Hilbert space. In quantum mechanics, we address various operators to represent observables for dynamical variables, but in quantum computation, we address quantum gates as operators to frame quantum circuits for carrying out various computational tasks. The essential components of quantum circuits are the quantum gates and they perform unitary transformations preserving probability amplitudes and thus enabling quantum parallelism. Operators in quantum mechanics are always represented in matrix as they connect two different states and thus in an analogy quantum gates in quantum computation frame represent as matrix. The goal of this paper is to explore individual quantum gates (single and multi qubit) and their series–parallel combinations through their matrix representation, to analyse their role in computing state evolution, commutation relation, probability measurement, and expectation values. Since quantum computation is based on linear algebra, characterisation of Hilbert space using the reversibility and linearity of quantum gates is carried out. In this research work matrix representation of rotational quantum gates and universal quantum gate are also addressed. The construction of a single qubit quantum gate using rotational gates is also addressed in this paper.

Keywords Quantum gates · Matrix · Expectation values · Commutation

A. Das
Department of Physics, Heritage Institute of Technology, Kolkata, India
e-mail: ajanta.das@heritageit.edu

M. Bhattacharya
Department of Physics, The Heritage College, Kolkata, India
e-mail: madhubrata.bhattacharya@thc.edu.in

D. Datta (✉)
Department of Information Technology, Heritage Institute of Technology, Kolkata, India
e-mail: debabrata.datta@heritageit.edu

M. Ram et al. (eds.), *Advances in Mathematics for Engineering Sciences*, Mathematical Engineering, https://doi.org/10.1007/978-3-031-95693-5_7

1 Introduction

One of the most significant scientific discoveries of the twentieth century was quantum theory, which offered a cohesive framework for the creation of numerous contemporary physical theories.

Computer science, another intellectual achievement of the same age, amalgamated with it more than 50 years after it began—giving rise to the field of quantum computation. Traditional computers, whether theoretical (Turing machines) or practical (PCs), function according to classical physics. But quantum computing reveals a fundamentally different nature of reality, wherein quantum systems are capable of being in several states superposed simultaneously and exhibit interference effects during their evolution. A key objective of quantum algorithms is to outperform classical algorithms significantly in solving complex problems. The potential superiority of quantum computers over their classical counterparts has sparked widespread interest in quantum computing. This has resulted in the creation of many quantum algorithms for both discrete tasks like integer factorization and computational challenges in science and engineering, such as solving ordinary and partial differential equations, multivariate integration, path integration, eigenvalue problems, and numerical linear algebra.

Quantum gates, the fundamental building blocks of quantum circuits, can be understood qualitatively through the lens of observables and operators in quantum mechanics. In classical computation, logic gates manipulate bits deterministically, but in quantum computation, gates act as unitary operators that evolve quantum states while preserving probability amplitudes. These gates correspond to linear transformations on a quantum state vector in Hilbert space, similar to how Hermitian operators with eigenvalues that match measurable quantities are used to express observables in quantum physics.

Just as an observable like spin or energy can be represented by an operator whose action determines possible measurement outcomes, quantum gates such as the Hadamard, Pauli, and Controlled NOT (CNOT) manipulate qubits by applying reversible transformations that exploit quantum superposition and entanglement. This operator-based framework enables quantum gates to perform complex parallel computations, offering a fundamental advantage over classical logic gates.

Since operators associated with dynamical variables in quantum mechanics can be represented as matrices, quantum gates also have a natural matrix representation that governs their action on quantum states via qubits. In quantum computing, a single-qubit gate corresponds to a 2×2 unitary matrix that transforms the state of a qubit while preserving the norm of the state vector. This matrix formalism provides a powerful mathematical tool for designing and analysing quantum circuits, allowing researchers to explore the computational advantages of quantum systems through linear algebra and eigenvalue-based transformations.

Both Feynman (1982) and Manin (1980) noted separately in the early 1980s that computers constructed using quantum mechanical components would be perfect for modeling quantum mechanics (Feynman 1982a, 1986). Feynman first proposed a device that could directly simulate physics before offering some fundamental ideas for simulating time and probabilities (Feynman 1982b). Benioff (n.d., 1982a, 1982b) describes physical systems that, if the principles of quantum physics were applied, would simulate a classical turning machine; nevertheless, quantum computation is not taken into account. He built a Turing machine model using quantum mechanics. Yao (1993) shown that a quantum circuit model and a quantum Turing machine are comparable in that they can both be simulated in polynomial time. Since then, quantum circuits have emerged as the most widely used quantum computation model. Since it is extremely challenging to create quantum gates functioning on three or more qubits in existing technologies, the synthesis of quantum circuits is essential for quantum computation. It was demonstrated as early as 1995 that any quantum gate may be (roughly) broken down into a circuit made up of just CNOT gates plus a few single-qubit gates (Barenco et al. 1995). More effective synthesis algorithms for quantum circuits have recently been discovered (Shende et al. 2006). Multiple inputs can be stored on a single tape using a quantum Turing machine which execute a calculation on each input according to the quantum parallelism approach, which was first presented by Deutsch (Deutsch n.d.; Deutsch 1989) and is based on the superposition principle in quantum physics.

1.1 Research Objectives

The following are the goals of this work:

- To develop matrix representation of quantum gates along with their measurement which in turn represents probabilities and expectation values of the observables.
- To present the time evolution of matrix representation corresponding to a gate.
- To develop matrix representation of series and parallel combination of gates.

This is how the rest of the paper is structured: The foundation of quantum computations has been described in Sect. 2. In Sect. 3, matrix formulation of different quantum gates are discussed. Series and parallel combination of quantum gates are presented with an example in the Sect. 4. In Sect. 5 measurement of a quantum state is interpreted with expectation values and commutation relations of different gate are shown in Sect. 6. Finally, Sect. 7 concludes the study by mentioning the future work in this field.

2 Fundamentals of Quantum Computations

Specifically, complicated issues that are unsolvable by traditional computers may be resolved by quantum computing. Quantum computing addresses the manipulation of quantum bits by superposition, entanglement and coherence in an abstract space known as Hilbert space. Qubits leverage the principle of quantum mechanics by a probabilistic combination of both the states (qubit 0 and qubit 1) simultaneously which is known as superposition. Correlation of any quantum states formed by qubits is known as entanglement. Through superposition, it is possible to allow a single qubit to hold more information compared to that by classical bit. We represent a qubit's state as a vector in a two-dimensional complex Hilbert space having axes corresponds to one of the computational basis states qubit 0 and qubit 1 ($|0\rangle$ and $|0\rangle$) as

$$|\varphi\rangle = a|0\rangle + b|1\rangle \quad (1)$$

where the probability amplitudes of measuring $|0\rangle$ and $|1\rangle$ states are denoted by a and b respectively which may be complex numbers in general such that they satisfy the normalisation condition $a^2 + b^2 = 1$. The length of the vector signifies the amplitude of the state and its direction encodes the phase information. From the postulates of quantum mechanics, we understand that all the dynamic variables are represented as observables, all observables associate their corresponding operator which are linear, Hermitian and each such operator associate a matrix to represent them. In quantum computing, linear operators are nothing but quantum gates (QGs) and they represent on the state vectors within Hilbert space. Quantum gates, which can perform unitary transformations on up to a few qubits, are necessary for processing quantum information.

QGs perform controlled transformations, rotate and modify the superposition of qubit(s) in a precise way. Every quantum gates has input and output and they are reversible. Reversibility of a quantum gate signifies that we can reach to the input from output and this is not possible in classical gate operations. The design and sequence of these QGs are carefully articulated to guide the system's evolution through Hilbert space to achieve a computational outcome. The measurement process represents an interaction between quantum system and classical world. Before any measurement, a qubit always exists in a superposition that represents a vector in Hilbert space; however, the measurement forces a "collapse" of the superposition causing the qubit to choose randomly among the basis states ($|0\rangle$ *or* $|1\rangle$) with a probability determined by the amplitudes of the superposition. The probabilistic outcome represents the inherent uncertainty of quantum mechanics. This measurement process extracts the classical information from quantum system providing the final result of quantum computation (Nielsen and Chuang 2000). A quantum computer is a large interferometer that contains a numerous quantum circuits. The functionality of a quantum circuit lies under specific quantum algorithms and quantum gates with respect to specific quantum algorithms address the problems to be solved. In order to understand the physical interpretation or behaviour of a quantum gate, it is mandatory to explore the matrix representations. Since linear algebra serves as the foundation for quantum computation, matrix representation of a simple single qubit, two qubit and three qubit QGs can be presented by linear algebra and we can use Dirac's ket bra notation to express the linear algebra of quantum gates.

3 Quantum Gates and Their Matrix Formalism

Transformation matrices, which act on inputs to produce outputs, are the mathematical representation of quantum gates. Matrix algebra provides following properties of quantum gates from unitarity.

3.1 *Properties of Quantum Gates*

- Quantum gates are described by unitary matrices. A matrix U is said to be unitary if it satisfies $U^\dagger = U^{-1}$.
- $|det(U)| = 1$.
- The columns (rows) of U make up an orthonormal set of vectors
- For a hermitian matrix H, $U = exp(iH)$
- A quantum gate is logically reversible

In view of these properties of quantum gate, its matrix formulation is an important task because once we know the matrix formulation of quantum gate, we can use that representation to construct its operational behaviour and this representation can be mathematically structured by the following theorem.

3.2 *Theorem 1*

Let O be a quantum gate that transforms one arbitrary quantum state $|A\rangle$ to another quantum state $|B\rangle$. So mathematically, it states that $O|A\rangle = |B\rangle$. Since quantum gate O is unitary hence, the operator O can be expressed as an outer product of the states $|A\rangle and |B\rangle$. In general, following this theorem we can easily write the matrix formulation of any quantum gates say O as

$$O = \sum_i |A_i\rangle\langle B_i| \tag{2}$$

where $|A_i\rangle$ represents the ith input and $|B_i\rangle$ represents the ith output of the quantum gate under consideration. The validity of the matrix representation of a quantum gate as given in Eq. (2) is due to reversibility nature of quantum gate and maintainability of all information throughout the operation. It is also to note that the matrix formulation of single qubit gates and multi-qubit gates (Eq. (2)) is valid for those quantum gates which has a finite truth function similar to a truth table in classical logic gate. However, there exists many quantum gates like $\sqrt{X}$ gate, $\sqrt{Z}$ gate, rotation gates ($R_X(\theta)$, $R_Y(\theta)$, $R_Z(\theta)$) and many other two qubit and three qubit quantum gates whose matrix representation cannot be formulated by using Eq. (2) because they are derived from matrix representation of fundamental gates like X,

Y, Z gate, Hadamard gate, SWAP gate, controlled NOT gate, controlled SWAP, TOFFOLI gate and FREDKIN gate. Since rotation gates are constructed by a continuous rotation angle $\theta/2$ around the axes X, Y and Z on the Bloch sphere, as such no truth table for mapping input output in the qubit state exist and hence these matrix of these gates cannot be described by Eq. (2). The matrix formulation of a single qubit, two qubit and three qubit based fundamental quantum gates has been done by many researchers and can be found elsewhere in Poornima et al. (2017). In view of the absence of matrix representation of derived quantum gates ($\sqrt{X}$, $\sqrt{Z}$, $\sqrt{Y}$, controlled Hadamard, controlled reverse Hadamard, etc.) in Poornima et al. (2017), here in out research an attempt has been made to formulate their matrix representation using simple properties of matrix algebra. Rotation gates based on their task imply thatthey can produce a superposition of states rather than simple binary output. Hence, matrix representation of rotation gates around the axes takes place by Bloch sphere formalism which are presented as

$$R_X\left(\frac{\theta}{2}\right) = \begin{bmatrix} \cos\left(\frac{\theta}{2}\right) & -i\sin\left(\frac{\theta}{2}\right) \\ -i\sin\left(\frac{\theta}{2}\right) & \cos\left(\frac{\theta}{2}\right) \end{bmatrix}, R_Y\left(\frac{\theta}{2}\right) = \begin{bmatrix} \cos\left(\frac{\theta}{2}\right) & -\sin\left(\frac{\theta}{2}\right) \\ \sin\left(\frac{\theta}{2}\right) & \cos\left(\frac{\theta}{2}\right) \end{bmatrix} \textit{ and}$$
$$R_Z\left(\frac{\theta}{2}\right) = \begin{bmatrix} \exp\left(\frac{-i\theta}{2}\right) & 0 \\ 0 & \exp\left(i\frac{\theta}{2}\right) \end{bmatrix},$$

Prior to formulate the matrix representation of derived quantum gates, let us illustrate the utility of Eq. (2) to represent the matrix of Pauli gate and Hadamard gate. In quantum computing, a flip (bit-flip) and a phase flip refer to different types of transformation applied to qubit states. A flip (bit-flip) is a transformation that changes the qubit state from $|0\rangle$ to $|1\rangle$ or vice versa. The Pauli X gate performs this flip operation on qubits. A phase flip is a transformation that changes the relative phase of the $|1\rangle$ state while leaving the $|0\rangle$ state unchanged. The Pauli Z gate follows the phase flip. If a qubit in the state $|0\rangle$ is applied on Pauli Z gate, the qubit's state remain unchanged but if a qubit in the state $|1\rangle$ is applied on Pauli Z gate, the qubit's state is changed to the state $-|1\rangle$. The Pauli Y gate is a combination of the Pauli X and Pauli Z gates meaning it applies both a bit flip and a phase flip to the qubit. Therefore, if a qubit in the state $|0\rangle$ is applied as input to Pauli Y gate, the qubit's state changes to the state $i|1\rangle$ and if a qubit in the state $|1\rangle$ is applied as an input to Pauli Y gate, the qubit's state is changed to the state $-i|0\rangle$. The Hadamard gate is a single qubit operation that maps the basis state $|0\rangle$ to $\frac{|0\rangle+|1\rangle}{\sqrt{2}}$ and $|1\rangle$ to $\frac{|0\rangle-|1\rangle}{\sqrt{2}}$, thus creating an equal superposition of the two states. Using this truth table of all these fundamental quantum gates and utility of Eq. (2) we can represent the matrices of Pauli X, Y, Z gate and Hadamard gate. Table 1 presents the input output structure of the Pauli X, Y, Z gates and Hadamard gate which has been used to formulate the matrices of the Pauli X, Y and Z gates and Hadamard gate. Since matrix representation of Pauli X and Z gates are available in Poornima et al. (2017), here we only presents the matrix representation of Pauli Y gate.

Table 1 Truth table of Pauli gates and Hadamard gate

Quantum gate	Function	Input	Output	Properties
Pauli X	$Not\|x\rangle = \|1-x\rangle$ bit flip	$x = 0, 1 : \|0\rangle, \|1\rangle$	$\|1-x\rangle = \|1\rangle, \|0\rangle$	Traceless, Involuntary
Pauli Y	Phase flip and bit flip both	$\|0\rangle, \|1\rangle$	$i\|1\rangle, -i\|0\rangle$	Traceless, Involuntary
Pauli Z	Phase flip	$\|0\rangle, \|1\rangle$	$\|0\rangle, -\|1\rangle$	Traceless, Involuntary
Hadamard	Equal superposition	$\|0\rangle, \|1\rangle$	$\frac{\|0\rangle+\|1\rangle}{\sqrt{2}}, \frac{\|0\rangle-\|1\rangle}{\sqrt{2}}$	Traceless, Involuntary

3.3 Matrix Formulation of Pauli Y Gate

Usage of Eq. (2) with the help of input–output from Table 1, we can write

$$MatrixY = -i|0\rangle\langle 1|+i|1\rangle\langle 0| = -i\begin{bmatrix}1\\0\end{bmatrix}\begin{bmatrix}0 & 1\end{bmatrix} + i\begin{bmatrix}0\\1\end{bmatrix}\begin{bmatrix}1 & 0\end{bmatrix} = \begin{bmatrix}0 & -i\\ i & 0\end{bmatrix}$$

3.4 Matrix Formulation of Hadamard (H) Gate

Again using Eq. (2) and input–output of Hadamard gate from Table 1, we can write $Matrix\,H = |0\rangle\left\langle\left[\frac{|0\rangle+|1\rangle}{\sqrt{2}}\right]\right| + \left|1 >< \left[\frac{|0\rangle-|1\rangle}{\sqrt{2}}\right]\right| = \begin{bmatrix}1 & 1\\ 1 & -1\end{bmatrix}$, in general we can also write $H_{x\in 0,1} = \frac{1}{\sqrt{2}}[|0\rangle + (-1)^x|1\rangle$.

Hadamard gate is an amazing quantum gate, because of its properties like $H^2 = I$, $|\phi\rangle = H^2|\psi\rangle = I|\psi\rangle = |\psi\rangle$. Hence, in general, we have $|\phi\rangle = H^{2n}|\psi\rangle = |\psi\rangle$ and $|\phi\rangle = H^{2n+1}|\psi\rangle = H|\psi\rangle$.

3.5 Matrix Representation of Sandwiched System and Its Interpretation

Pauli Z gate does phase flip but if Z gate is sandwiched between two Hadamard gates, H it produces Pauli X gate and it is originated from the concept of a similarity transformation.

Proof Matrix representation of Pauli Z gate is obtained by

$$Z = |0\rangle\langle 0| - |1\rangle\langle 1| = \begin{bmatrix} 1 \\ 0 \end{bmatrix}\begin{bmatrix} 1 & 0 \end{bmatrix} - \begin{bmatrix} 0 \\ 1 \end{bmatrix}\begin{bmatrix} 0 & 1 \end{bmatrix} = \begin{bmatrix} 1 & 0 \\ 0 & 0 \end{bmatrix} - \begin{bmatrix} 0 & 0 \\ 0 & 1 \end{bmatrix} = \begin{bmatrix} 1 & 0 \\ 0 & -1 \end{bmatrix}. \tag{3}$$

Hadamard gate, H is represented in the matrix form as $H = \frac{1}{\sqrt{2}}\begin{bmatrix} 1 & 1 \\ 1 & -1 \end{bmatrix}$

Therefore, we have $HZH = \frac{1}{2}\begin{bmatrix} 1 & 1 \\ 1 & -1 \end{bmatrix}\begin{bmatrix} 1 & 0 \\ 0 & -1 \end{bmatrix}\begin{bmatrix} 1 & 1 \\ 1 & -1 \end{bmatrix} = \begin{bmatrix} 0 & 1 \\ 1 & 0 \end{bmatrix} = X$ (*Pauli NOT gate*).

So, matrix representation of *HZH* is nothing but the matrix representation of Pauli NOT gate.

In a similar way if we sandwich Pauli X gate between two Hadamard gates, we obtain Pauli Z gate, i.e., $HXH = Z$.

Continuing the similar strategy, matrix representation of Pauli Y gate can be obtained by a combination ofPauli X gate and phase gate, $S = \begin{bmatrix} 1 & 0 \\ 0 & i \end{bmatrix}$, which is given by

$$SXS^{\dagger} = Y \tag{4}$$

Proof

$$SXS^{\dagger} = \begin{bmatrix} 1 & 0 \\ 0 & i \end{bmatrix}\begin{bmatrix} 0 & 1 \\ 1 & 0 \end{bmatrix}\begin{bmatrix} 1 & 0 \\ 0 & i \end{bmatrix}^{\dagger} = \begin{bmatrix} 1 & 0 \\ 0 & i \end{bmatrix}\begin{bmatrix} 0 & 1 \\ 1 & 0 \end{bmatrix}\begin{bmatrix} 1 & 0 \\ 0 & -i \end{bmatrix} = \begin{bmatrix} 0 & -i \\ i & 0 \end{bmatrix} = Y$$

3.6 Matrix fFormulation of Square Root of One Qubit Gates

Square root of one qubit quantum gates is very essential to know because it allows for finer-grained control over a quantum system by providing a way to perform a partial operation, essentially "half" of the full gate operation, which can be crucial for implementing complex quantum algorithms that require precise manipulation of quantum states. In this section, we present two methodologies to derive the matrix representation of square root of Pauli X, Y, Z gates and Hadamard gate. Our first method is based on the theory of complex numbers and the second method is based on transformation rule.

3.6.1 Method I: Matrix Formulation Based on Theory of Complex Number

According to well known methods of matrix linear algebra, the generalised Euler relation for any self inverse matrix A ($A^2 = I$, or $A = A^{-1}$, I signifies the identity matrix) is given by

$$e^{\pm i\alpha A} = Icos(\alpha) \pm iAsin(\alpha)$$

For $\alpha = \pi$, we obtain $e^{\pm i\pi A} = -I$ and if $\alpha = \pi/2$, we get $e^{\pm i\frac{\pi}{2}A} = \pm iA$ and form these relations, we can easily write

$$A = i^{I-A}$$

From this last expression of matrix A, we can write that nth—root of arbitrary self-inverse gate A as

$$\sqrt[n]{A} = e^{i\frac{\pi}{2n}(I-A)}, n = 1, 2, 3, ..$$

Therefore, for n = 2, we have $\sqrt{A} = \frac{1}{\sqrt{2}i}(iI + A)$ and this relation can be used to obtain the square root of one qubit gates because they are also self-inverse (e.g. $X^2 = Y^2 = Z^2 = H^2 = S^2 = I$). Thus we have.

Matrix Representation of Square Root of Pauli Gates

$$\sqrt{X} = \frac{1}{\sqrt{2}i}(iI + X) = \frac{1}{\sqrt{2}}\begin{bmatrix} 1+i & 1-i \\ 1-i & 1+i \end{bmatrix} = \frac{1}{\sqrt{2}i}\begin{bmatrix} i & 1 \\ 1 & i \end{bmatrix}$$

$$\sqrt{Y} = \frac{1}{\sqrt{2}i}(iI + Y) = \frac{1}{\sqrt{2}i}\begin{bmatrix} i & -i \\ i & i \end{bmatrix}, \sqrt{Z} = \frac{1}{\sqrt{2}i}(iI + Z) = \begin{bmatrix} 1 & 0 \\ 0 & i \end{bmatrix} = S$$

Matrix Representation of Square Root of Hadamard (H) Gate

$$\sqrt{H} = \frac{1}{\sqrt{2}i}(iI + H) = \begin{bmatrix} \cos^2\frac{\pi}{8} + i\sin^2\frac{\pi}{8} & \frac{1-i}{2\sqrt{2}} \\ \frac{1-i}{2\sqrt{2}} & \sin^2\frac{\pi}{8} + i\cos^2\frac{\pi}{8} \end{bmatrix}$$

Matrix Representation of Square Root of Phase (S) Gate

$$\sqrt{S} = \frac{1}{\sqrt{2}}\left(e^{i\pi/4}I + e^{-i\pi/4}S\right) = \begin{bmatrix} 1 & 0 \\ 0 & e^{i\pi/4} \end{bmatrix} = T$$

3.6.2 Method II: Matrix Formulation Based on Linear Transformation

We know that if Pauli Z gate is sandwiched between two Hadamard gates we obtain Pauli X gate. Mathematically, we write

$$HZH = X$$

In a similar manner, we have already shown that $HXH = Z$. This transformation relation is so nice that we also have

$$H\sqrt{X}H = \sqrt{Z}$$

$$H\sqrt{Z}H = \sqrt{X}$$

It is possible to express the square root of Pauli Z gate as

$$\begin{aligned}\sqrt{Z} &= P\left(\pm\frac{\pi}{2}\right) = \begin{bmatrix} 1 & 0 \\ 0 & e^{\pm i\frac{\pi}{2}} \end{bmatrix} = \begin{bmatrix} 1 & 0 \\ 0 & \pm i \end{bmatrix} = |0\rangle\langle 0| \pm i|1\rangle\langle 1| \\ &= \frac{I+Z}{2} \pm i\frac{I-Z}{2} = \frac{1 \pm i}{2}I + \frac{1 \mp i}{2}Z\end{aligned}$$

Now, let us explore the relation

$$\begin{aligned}\sqrt{X} &= H\sqrt{Z}H \\ &= \frac{1 \pm i}{2}HIH + \frac{1 \mp i}{2}HZH \\ &= \frac{1 \pm i}{2}I + \frac{1 \mp i}{2}X \\ &= \frac{1}{2}\begin{bmatrix} 1 \pm i & 1 \mp i \\ 1 \mp i & 1 \pm i \end{bmatrix} + \frac{1}{2}\begin{bmatrix} 0 & 1 \mp i \\ 1 \mp i & 0 \end{bmatrix}\end{aligned}$$

Now, considering one sided sign (+ from first expression and—sign from second expression) we obtain the matrix representation of square root of X $\left(\sqrt{X}\right)$ as

$$\sqrt{X} = \frac{1}{\sqrt{2}}\begin{bmatrix} 1+i & 1-i \\ 1-i & 1+i \end{bmatrix}$$

In a similar manner, we can also formulate the matrix of square root other one qubit gates as shown in sub Sect. 3.6.1.

3.7 Matrix Formulation of Two Qubit and Three Qubit Gates

In the Hilbert space, two qubits system are defined by the tensor product of individual qubit state, such as in general, $|q_1q_2\rangle = |q_1\rangle \otimes |q_2\rangle$. Thus we have $2^2 = 4$ bases states, like $|00\rangle, |01\rangle, |10\rangle$ and $|11\rangle$. In the family of two qubit quantum gate most popular is Controlled NOT (CNOT), which has a control qubit $|q_1\rangle$ which leave unchanged and a target qubit $|q_2\rangle$, which is changed into $|q_2\rangle \oplus |q_1\rangle$. Following this protocol, the truth table (input–output) in the algebraic form can be written as $|q_1q_2\rangle = |00\rangle \rightarrow |00\rangle, |01\rangle \rightarrow |01\rangle, |10\rangle \rightarrow |11\rangle$ and $|11\rangle \rightarrow |10\rangle$. So, one can easily construct the matrix of CNOT gate using the said truth table and Eq. (2). So, we have matrix of CNOT as

$$CNOT_{matrix} = \begin{bmatrix} 1\,0\,0\,0 \\ 0\,1\,0\,0 \\ 0\,0\,0\,1 \\ 0\,0\,1\,0 \end{bmatrix}$$

The details of the calculation of CNOT gate matrix is elsewhere found in Poornima et al. (2017). Generally speaking, though, we can design a CNOT gate with qubit j as the control and qubit k as the target. This can be done by applying Hadamard gates to each qubit both before and after the CNOT, so that j is the target and k is the control. i.e.,

$$CNOT_{kj} = (H \otimes H) CNOT_{jk} (H \otimes H)$$

Using the same protocol as given in [15], one can also construct the matrix of two qubit SWAP gate. In essence, the SWAP gate switches the states of the two qubits that are used in the process and this contains three CNOT gates connected in series with each other and hence the matrix representation of SWAP gate can be represented as

$$SWAP_{matrix} = \begin{bmatrix} 1\,0\,0\,0 \\ 0\,0\,1\,0 \\ 0\,1\,0\,0 \\ 0\,0\,0\,1 \end{bmatrix}$$

3.7.1 Matrix Formulation of Controlled-Zgate

In this work, we have also carried out the matrix formulation of controlled Z (CZ) gate using Eq. (2) and truth table of CZ gate. The operation of Controlled-Z gate on input $|00\rangle$ produces the output $|00\rangle$, on input $|01\rangle$ produces the output $|01\rangle$, on input $|10\rangle$ produces the output $|10\rangle$ but when operate on input $|11\rangle$ produces the output as $-|11\rangle$. Therefore, the truth table of CZ gate can be written as shown in Table 2.

Table 2 Truth table of controlled—Z gate

Input	Output
$\|00\rangle$	$\|00\rangle$
$\|01\rangle$	$\|01\rangle$
$\|10\rangle$	$\|10\rangle$
$\|11\rangle$	$-\|11\rangle$

Using the truth table (Table 2) along with Eq. (2), the matrix elements of CZ gate is constructed which is presented as

$$\text{Matrix elements of CZ gate} = \begin{bmatrix} 1 & 0 & 0 & 0 \\ 0 & 1 & 0 & 0 \\ 0 & 0 & 1 & 0 \\ 0 & 0 & 0 & -1 \end{bmatrix}.$$

3.7.2 Matrix Formulation of Controlled Hadamard Gate

Controlled Hadamard (CH) gate is a two qubits controlled gate which functions on the basis control qubit and target qubit in the way that activation of the Hadamard gate to the target qubit occurs when the control qubit is $|1\rangle$.

To the extent of our knowledge, protocol for matrix formulation of CH gate other than its operation for quantum algorithm development is not available in the literatures of quantum computing. For this reason, we have derived the matrix formulation of CH gate using Eq. (2) and the truth table of CH gate. We are aware that a two-qubit system's computational basis states are.

$|00\rangle$, $|01\rangle$, $|10\rangle$ and $|11\rangle$. Explicit representation of these computational basis states can be written on the basis of tensor product of individual qubit states. So, we have the vectors of computational bases states as

$$|00\rangle = |0\rangle \otimes |0\rangle = \begin{bmatrix} 1 \\ 0 \end{bmatrix} \otimes \begin{bmatrix} 1 \\ 0 \end{bmatrix} = \begin{bmatrix} 1 \\ 0 \\ 0 \\ 0 \end{bmatrix}, |01\rangle = |0\rangle \otimes |1\rangle = \begin{bmatrix} 1 \\ 0 \end{bmatrix} \otimes \begin{bmatrix} 0 \\ 1 \end{bmatrix} = \begin{bmatrix} 0 \\ 1 \\ 0 \\ 0 \end{bmatrix}$$

$$|10\rangle = |1\rangle \otimes |0\rangle = \begin{bmatrix} 0 \\ 1 \end{bmatrix} \otimes \begin{bmatrix} 1 \\ 0 \end{bmatrix} = \begin{bmatrix} 0 \\ 0 \\ 1 \\ 0 \end{bmatrix} \textit{ and } |11\rangle = |1\rangle \otimes |1\rangle = \begin{bmatrix} 0 \\ 1 \end{bmatrix} \otimes \begin{bmatrix} 0 \\ 1 \end{bmatrix} = \begin{bmatrix} 0 \\ 0 \\ 0 \\ 1 \end{bmatrix}$$

We know again

$$H|0\rangle = \frac{1}{\sqrt{2}} \begin{bmatrix} 1 & 1 \\ 1 & -1 \end{bmatrix} \begin{bmatrix} 1 \\ 0 \end{bmatrix} = \frac{1}{\sqrt{2}} \begin{bmatrix} 1 \\ 1 \end{bmatrix} = \frac{1}{\sqrt{2}} [|0\rangle + |1\rangle] \textit{ and}$$

$$H|1\rangle = \frac{1}{\sqrt{2}}\begin{bmatrix} 1 & 1 \\ 1 & -1 \end{bmatrix}\begin{bmatrix} 0 \\ 1 \end{bmatrix} = \frac{1}{\sqrt{2}}\begin{bmatrix} 1 \\ -1 \end{bmatrix} = \frac{1}{\sqrt{2}}[|0\rangle - |1\rangle]$$

The truth table of Controlled Hadamard gate is presented in Table 3 and this subsection provides the details of computation of the matrix element of Controlled Hadamard gate.

Now, one can use Eq. (2) and the truth Table 3 to compute the matrix elements of CH gate, The matrix of CH gate computed results as

$$CH_{matrix} = \begin{bmatrix} 1 & 0 & 0 & 0 \\ 0 & 1 & 0 & 0 \\ 0 & 0 & \frac{1}{\sqrt{2}} & \frac{1}{\sqrt{2}} \\ 0 & 0 & \frac{1}{\sqrt{2}} & -\frac{1}{\sqrt{2}} \end{bmatrix}, CH_{matrix} = \begin{bmatrix} 1 & 0 & 0 & 0 \\ 0 & 1 & 0 & 0 \\ 0 & 0 & 1/\sqrt{2} & 1/\sqrt{2} \\ 0 & 0 & 1/\sqrt{2} & -1/\sqrt{2} \end{bmatrix}$$

It can be easily observed that the protocol used for computation of matrix elements of CH gate for two qubits can be easily handle but for higher registers qubits (say 4 qubits, 5 qubits and even 10 qubits), computation of CH gate will be a huge task. Therefore, to avoid the load of computation, we can use a generalised version of any controlled universal (U) gate (controlled U gate), where U is any arbitrary unitary operator (quantum gate) having its matrix elements as $U = \begin{bmatrix} U_{00} & U_{01} \\ U_{10} & U_{11} \end{bmatrix}$. The generalised method of expressing matrix elements of controlled U gate can be written by Eq. (5) as

$$|0\rangle\langle 0|\otimes I + |1\rangle\langle 1| \otimes U \tag{5}$$

The matrix representation of controlled U gate by Eq. (5) can be written as

$$\begin{bmatrix} 1 & 0 & 0 & 0 \\ 0 & 1 & 0 & 0 \\ 0 & 0 & U_{00} & U_{01} \\ 0 & 0 & U_{10} & U_{11} \end{bmatrix}$$

Table 3 Truth table of controlled Hadamard gate

Input	Output
$\|00\rangle$	$\|0(0 \oplus 0)\rangle = \|00\rangle$
$\|01\rangle$	$\|0(0 \oplus 1)\rangle = \|01\rangle$
$\|10\rangle$	$\|1(1 \oplus 0)\rangle = \|11\rangle = \|1\rangle H\|0\rangle = \|1\rangle(\frac{1}{\sqrt{2}}(\|0\rangle + \|1\rangle)) = \frac{1}{\sqrt{2}}[\|10\rangle + \|11\rangle]$
$\|11\rangle$	$\|1(1 \oplus 1) =\| 10\rangle = \|1\rangle H\|1\rangle = \|1\rangle(\frac{1}{\sqrt{2}}(\|0\rangle - \|1\rangle)) = \frac{1}{\sqrt{2}}[\|10\rangle - \|11\rangle]$

Therefore, using this generalised version of controlled U gate, the matrix elements of controlled Hadamard gate (CH gate) can be easily written as

$$\begin{bmatrix} 1 & 0 & 0 & 0 \\ 0 & 1 & 0 & 0 \\ 0 & 0 & H_{00} & H_{01} \\ 0 & 0 & H_{10} & H_{11} \end{bmatrix} = \begin{bmatrix} 1 & 0 & 0 & 0 \\ 0 & 1 & 0 & 0 \\ 0 & 0 & \frac{1}{\sqrt{2}} & \frac{1}{\sqrt{2}} \\ 0 & 0 & \frac{1}{\sqrt{2}} & -\frac{1}{\sqrt{2}} \end{bmatrix}$$

3.7.3 Matrix Formulation of Toffoli and Fredkin Gate

Toffoli and Fredkin gate are three qubit quantum gates. In a three qubit system, there exists $2^3 = 8$, the computational bases states which are

$$|000\rangle, |010\rangle, |011\rangle, |100\rangle, |101\rangle, |110\rangle, |001\rangle, |111\rangle$$

Vector representation of all these computational basis states are computed by the tensor product of each qubit states. Matrix elements of Toffoli gate and Fredkin gate by using the protocol as given in Eq. (2) and the truth table of each individual can be formulated and the results can be found elsewhere in [15. Thus instead of repeating the similar task, we have explored the working function of Toffoli and Fredkin gate using arbitrary three qubit states $|x\rangle, |y\rangle and |z\rangle$]. So, by postulate of quantum mechanics, we have for Toffoli gate

$$Toffoli|x, y, z\rangle = |x, y, z \oplus xy\rangle$$

Thus,

$$Toffoli|000\rangle = |00, (0 \oplus 0)\rangle = |000\rangle.$$

$$Toffoli|001\rangle = |00, (1 \oplus 0)\rangle = |001\rangle$$

$$Toffoli|100\rangle = |10, (0 \oplus 0)\rangle = |100\rangle$$

Repeating in a similar way for all other input qubit states we can obtain corresponding transformed output and finally we have the matrix representation of Toffoli gate as

$$
Toffoli = \begin{bmatrix} 1&0&0&0&0&0&0&0 \\ 0&1&0&0&0&0&0&0 \\ 0&0&1&0&0&0&0&0 \\ 0&0&0&1&0&0&0&0 \\ 0&0&0&0&1&0&0&0 \\ 0&0&0&0&0&1&0&0 \\ 0&0&0&0&0&0&0&1 \\ 0&0&0&0&0&0&1&0 \end{bmatrix}
$$

The other three qubit gate so called Fredkin gate is alternately called as controlled SWAP gate. The matrix representation of Fredkin gate can be found elsewhere in Poornima et al. (2017).

4 Series and Parallel Combination of Quantum Gates

A complicated quantum circuit primarily consists of a combination of quantum gates. The unitary operations known as quantum gates use the principles of quantum physics to control qubits. In order to use multiple quantum gates in a quantum circuit, we need to combine them either in series or in parallel or may be in both, depending on the configuration of quantum circuit for solving problems under investigation. Whatever be the configuration of the circuit, computation of matrix elements of the combinations (either series or parallel) is a challenging task and is of interest to many users. For this reason, here in this section we provide the mathematical structure of the combination of quantum gates along with the matrix elements.

4.1 Series Combination of Quantum Gates

Series combination of quantum gates is alternately called an overlap and that is why this combination of quantum gates is mathematically modelled as inner product of the matrix elements representing individual quantum gates. So, if quantum gate G_1 having matrix elements represented by matrix M_{G_1} and the quantum gate G_2 having matrix elements represented by M_{G_2} are combined in series, then the matrix elements of their combination is represented by

$$
QG_{G_2G_1} = M_{G_2}M_{G_1}
$$

If the gates represents the qubit states, say $|\psi_1\rangle$, $|\psi_2\rangle$ respectively, then the series combination can be written as $\langle\psi_2|\psi_1 >$, an inner product of the respective states.

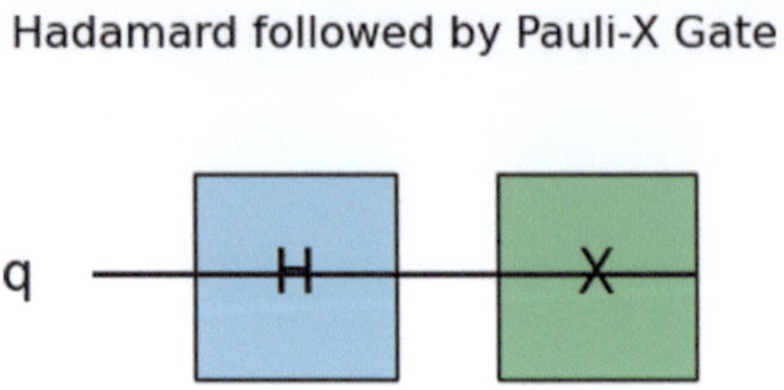

Fig. 1 Series combination of H and X gate

4.1.1 Series Combination of Hadamard (H) Gate and Pauli-X Gate

Assuming Hadamard gate(H) is in series with Pauli X gate as shown in Fig. 1.

$$H = \frac{1}{\sqrt{2}}\begin{bmatrix} 1 & 1 \\ 1 & -1 \end{bmatrix}; X = \begin{bmatrix} 0 & 1 \\ 1 & 0 \end{bmatrix}; XH = \begin{bmatrix} 0 & 1 \\ 1 & 0 \end{bmatrix}\frac{1}{\sqrt{2}}\begin{bmatrix} 1 & 1 \\ 1 & -1 \end{bmatrix} = \frac{1}{\sqrt{2}}\begin{bmatrix} 1 & -1 \\ 1 & 1 \end{bmatrix}$$

4.2 Parallel Combination of Quantum Gates

Parallel combination of quantum gates are connected by the tensor productof individual gates.

If two gates U_1 and U_2 operate on different qubits, the combined operation is:

$$U_{parallel} = U_1 \otimes U_2$$

4.2.1 Hadamard (H) and Pauli-X Gate (X)

$$H \otimes X = \frac{1}{\sqrt{2}}\begin{bmatrix} 1 & 1 \\ 1 & -1 \end{bmatrix} \otimes \begin{bmatrix} 0 & 1 \\ 1 & 0 \end{bmatrix} = \frac{1}{\sqrt{2}}\begin{bmatrix} 0 & 1 & 0 & 1 \\ 1 & 0 & 1 & 0 \\ 0 & 1 & 0 & -1 \\ 1 & 0 & -1 & 0 \end{bmatrix}$$

4.2.2 Parallel Combination of Hadamard (H) Gates

Suppose two Hadamard gates are combined in parallel as shown in Fig. 2 and act on the qubit state $|1\rangle$.

The matrix elements of the combined output can be written by the tensor product of matrix elements of both the H gates and thus the result is written as

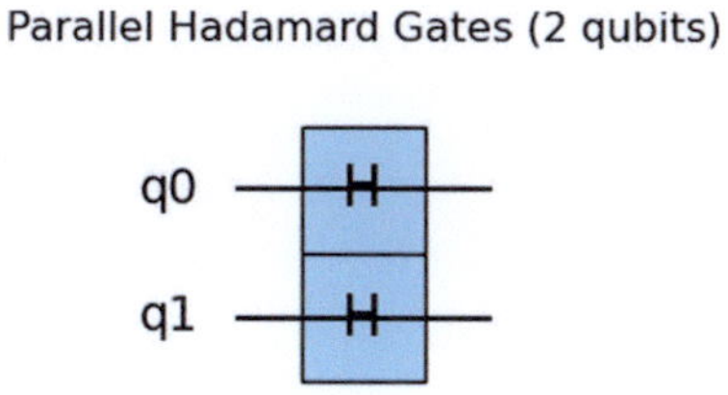

Fig. 2 Hadamard gates are in parallel and both the qubits are $|1\rangle$

$$(H|1\rangle) \otimes (H|1\rangle) = \frac{1}{2}(|0\rangle - |1\rangle) \otimes (|0\rangle - |1\rangle) = \frac{1}{2}(|00\rangle - |10\rangle - |01\rangle + |11\rangle)$$

Now, let us take the following circuit (Fig. 3) showing combination of gates in groups G1, G2 and G3. According to the diagram, $G1 = NOT \otimes Identity$, $G2 = NOT \otimes NOT$, $G3 = SWAP$. The collective qubit states are $|\psi_0$, $|\psi_1$, $|\psi_2$ and $|\psi_3$. Qubit state transformations can be written as

$$|\psi_1 = G1|\psi_0;\ |\psi_2 = G2|\psi_1 = G2G1|\psi_0\ ;\ |\psi_3 = G3|\psi_2 = G3G2G1|\psi_0$$

Hence, circuit matrix, M, for $|\psi_3 = M|\psi_0$, $M = G3G2G1$.

Calculation of matrix for each group now results as

$$G1 = NOT \otimes Identity = \begin{bmatrix} 0\,1 \\ 1\,0 \end{bmatrix} \otimes \begin{bmatrix} 1\,0 \\ 0\,1 \end{bmatrix} = \begin{bmatrix} 0\,0 & 1\,0 \\ 0\,0 & 0\,1 \\ 1\,0 & 0\,0 \\ 0\,1 & 0\,0 \end{bmatrix}$$

$$G2 = NOT \otimes NOT = \begin{bmatrix} 0\,1 \\ 1\,0 \end{bmatrix} \otimes \begin{bmatrix} 0\,1 \\ 1\,0 \end{bmatrix} = \begin{bmatrix} 0\,0 & 0\,1 \\ 0\,0 & 1\,0 \\ 0\,1 & 0\,0 \\ 1\,0 & 0\,0 \end{bmatrix}$$

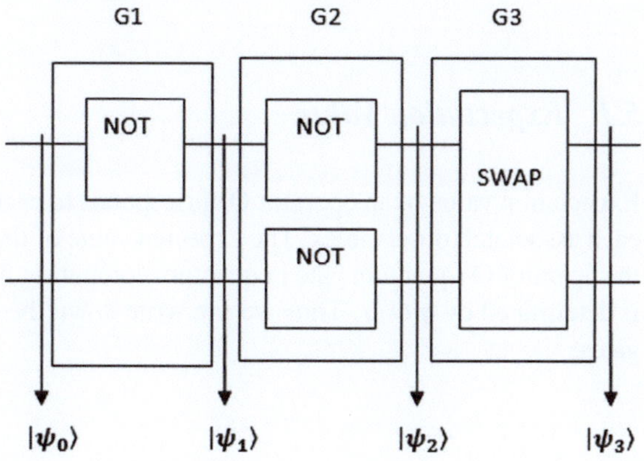

Fig. 3 Series and parallel combination of gates

$$G3 = SWAP = \begin{bmatrix} 10\ 00 \\ 00\ 10 \\ 01\ 00 \\ 00\ 01 \end{bmatrix}$$

$$M = G3G2G1 = \begin{bmatrix} 10\ 00 \\ 00\ 10 \\ 01\ 00 \\ 00\ 01 \end{bmatrix} \begin{bmatrix} 00\ 01 \\ 00\ 10 \\ 01\ 00 \\ 10\ 00 \end{bmatrix} \begin{bmatrix} 00\ 10 \\ 00\ 01 \\ 10\ 00 \\ 01\ 00 \end{bmatrix}$$

$$= \begin{bmatrix} 00\ 01 \\ 01\ 00 \\ 00\ 10 \\ 10\ 00 \end{bmatrix} \begin{bmatrix} 00\ 10 \\ 00\ 01 \\ 10\ 00 \\ 01\ 00 \end{bmatrix}$$

Hence, circuit matrix, M, for $|\psi_3 = M|\psi_0$, $M = \begin{bmatrix} 01\ 00 \\ 00\ 01 \\ 10\ 00 \\ 00\ 10 \end{bmatrix}$.

5 Measurement of a Qubit State

If a qubit is in a state $|q = \begin{bmatrix} m \\ n \end{bmatrix}$ then when the qubit is measured, the probability that $|0$ is measured is $P_\psi(0) = |0|q|^2 = \left| \begin{bmatrix} 1\ 0 \end{bmatrix} \begin{bmatrix} m \\ n \end{bmatrix} \right|^2 = |m|^2$ the probability that $|1$ is measured is $P_\psi(1) = |1|q|^2 = \left| \begin{bmatrix} 0\ 1 \end{bmatrix} \begin{bmatrix} m \\ n \end{bmatrix} \right|^2 = |n|^2$.

5.1 *Expectation Value*

Expectation value of an operator O corresponds to each dynamical variable that is to each associated observables. The expected value of the observable corresponding to the operator O (quantum gate in quantum computing field), if the qubit is in state $|q$, is determined by $q|O|q$. Thus we can write down the expectation value of quantum gates.

5.1.1 Expectation Value of Z

The expectation value of Z in state $|q$ is give by

$$E_q(Z) = q|Z\,|q = q|(|0\,0| - |1\,1|)\,|q$$
$$= q|00|q - q\big|11\big|q = |0|q\,|^2 - |1|q\,|^2 = P_{|q}\,(0) - P_{|q}\,(1)$$

5.1.2 Expectation Value of X

The expectation value of X in state $|q$ is give by

$$E_q(X) = q|X\,|q = q|HZH\,|q = H^{\dagger}q|Z\,|qH = Hq$$
$$|Z\,|qH = E_{Hq}(Z) = P_{H|q}\,(0) - P_{H|q}\,(1)$$

5.1.3 Expectation Value of Y

The expectation value of Y in state $|q$ is give by

$$E_q(Y) = q\big|SXS^{\dagger}\,|q = q\big|SHZHS^{\dagger}\,|q = H^{\dagger}S^{\dagger}q|Z\,\big|qHS^{\dagger} = HS^{\dagger}q$$
$$|Z\,\big|qHS^{\dagger} = P_{HS^{\dagger}|q}\,(0) - P_{HS^{\dagger}|q}\,(1)$$

6 Commutation Relation of Quantum Gates with Physical Interpretation

In quantum computing, commutation relation describes whether both the gates are possible to measure or not simultaneously. If two quantum gates A and B commute with each other, their order of application doesn't impact the outcome

$$[A, B] = AB - BA = 0 \tag{6}$$

However, if they do not commute, their order influences the quantum state evolution, which is crucial for quantum algorithms and error correction. In essence, commuting gates are essential to quantum computation for several reasons, particularly in circuit simplification, error correction, and measurement strategies. The knowledge of commutating gates are useful to optimize depth that also reduces noise and increases hardware efficiency. Further Commuting operators can have a common set of eigenstates, which is critical in quantum error correction (QEC) using stabilizer codes. In fact, in QEC, stabilizers are sets of commuting Pauli operators that detect and correct errors. It is worth mentioning here that when an operator U commutes with a Hamiltonian H it means that U does not change the energy eigenstates and this is useful in quantum simulation and adiabatic quantum computing, where preserving energy levels is important. In the Trotterization method (Trotter 1959) for quantum simulation, commuting terms in the Hamiltonian can be grouped together to improve efficiency. Not only commuting operators are important, search of non-commutating operators are also important. For example some quantum cryptographic protocols (e.g., BB84 quantum key distribution) rely on the non-commutativity of measurement bases to ensure security. Further more non-commuting gates introduce phase shifts or state changes that affect interference patterns.

6.1 *Commutation of Pauli Gates:*

We have $X = \begin{bmatrix} 0 & 1 \\ 1 & 0 \end{bmatrix}, Y = \begin{bmatrix} 0 & -i \\ i & 0 \end{bmatrix}, Z = \begin{bmatrix} 1 & 0 \\ 0 & -1 \end{bmatrix}$

$$[X, Y] = XY - YX = \begin{bmatrix} 0 & 1 \\ 1 & 0 \end{bmatrix}\begin{bmatrix} 0 & -i \\ i & 0 \end{bmatrix} - \begin{bmatrix} 0 & -i \\ i & 0 \end{bmatrix}\begin{bmatrix} 0 & 1 \\ 1 & 0 \end{bmatrix}$$
$$= \begin{bmatrix} i & 0 \\ 0 & -i \end{bmatrix} - \begin{bmatrix} -i & 0 \\ 0 & i \end{bmatrix} = iZ - (-iZ) = 2iZ$$

Similarly we can verify, $[Y, Z] = 2iX, [Z, X] = 2iY$.

Since Pauli matrices correspond to spin operators in quantum mechanics, their non-commutation implies that measuring spin along different axes introduces uncertainty, following Heisenberg's uncertainty principle.

6.2 *Commutation Relation of Hadamard Gate and Phase Gate*

$$H = \frac{1}{\sqrt{2}}\begin{bmatrix} 1 & 1 \\ 1 & -1 \end{bmatrix}, S = \begin{bmatrix} 1 & 0 \\ 0 & i \end{bmatrix}$$

Therefore,

$$[H, S] = HS - SH = \frac{1}{\sqrt{2}}\begin{bmatrix} 1 & 1 \\ 1 & -1 \end{bmatrix}\begin{bmatrix} 1 & 0 \\ 0 & i \end{bmatrix} - \frac{1}{\sqrt{2}}\begin{bmatrix} 1 & 0 \\ 0 & i \end{bmatrix}\begin{bmatrix} 1 & 1 \\ 1 & -1 \end{bmatrix}$$
$$= \frac{1}{\sqrt{2}}\begin{bmatrix} 1 & i \\ i & -1 \end{bmatrix} - \frac{1}{\sqrt{2}}\begin{bmatrix} 1 & 1 \\ i & -i \end{bmatrix}$$

Since, $HS \neq SH$, the Hadamard and Phase gates do not commute.

This means that applying a Hadamard gate before or after a phase shift changes the quantum state differently. This property is essential for quantum interference and algorithms like the quantum Fourier transform.

6.3 CNOT and Rotation Gate Commutation

Commutation relation of CNOT and rotation gates are carried out by using standard commutation relation and Sect. 6.3 presents rules

$CNOT = \begin{bmatrix} 1\,0 & 0\,0 \\ 0\,1 & 0\,0 \\ 0\,0 & 0\,1 \\ 0\,0 & 1\,0 \end{bmatrix}$, ThePauli-X rotation gate is given by: $R_x(\theta) = \begin{bmatrix} \cos\frac{\theta}{2} & -i\sin\frac{\theta}{2} \\ -i\sin\frac{\theta}{2} & \cos\frac{\theta}{2} \end{bmatrix}$.

Since $R_x(\theta)$ is applied to the target qubit, the two-qubit operation is:

$$R_x(\theta) \otimes \boldsymbol{I} = \begin{bmatrix} cos\frac{\theta}{2} \;\; -isin\frac{\theta}{2} & 0\,0 \\ -isin\frac{\theta}{2} \;\; cos\frac{\theta}{2} & 0\,0 \\ 0\,0 & cos\frac{\theta}{2} \;\; -isin\frac{\theta}{2} \\ 0\,0 & -isin\frac{\theta}{2} \;\; cos\frac{\theta}{2} \end{bmatrix}$$

$$CNOT(R_x(\theta) \otimes I) = \begin{bmatrix} 1\,0 & 0\,0 \\ 0\,1 & 0\,0 \\ 0\,0 & 0\,1 \\ 0\,0 & 1\,0 \end{bmatrix} \begin{bmatrix} \cos\frac{\theta}{2} \;\; -i\sin\frac{\theta}{2} & 0\,0 \\ -i\sin\frac{\theta}{2} \;\; \cos\frac{\theta}{2} & 0\,0 \\ 0\,0 & \cos\frac{\theta}{2} \;\; -i\sin\frac{\theta}{2} \\ 0\,0 & -i\sin\frac{\theta}{2} \;\; \cos\frac{\theta}{2} \end{bmatrix}$$

$$= \begin{bmatrix} \cos\frac{\theta}{2} \;\; -i\sin\frac{\theta}{2} & 0\,0 \\ -i\sin\frac{\theta}{2} \;\; \cos\frac{\theta}{2} & 0\,0 \\ 0\,0 & -i\sin\frac{\theta}{2} \;\; \cos\frac{\theta}{2} \\ 0\,0 & \cos\frac{\theta}{2} \;\; -i\sin\frac{\theta}{2} \end{bmatrix}$$

$$(R_x(\theta) \otimes I)CNOT = \begin{bmatrix} \cos\frac{\theta}{2} \;\; -i\sin\frac{\theta}{2} & 0\,0 \\ -i\sin\frac{\theta}{2} \;\; \cos\frac{\theta}{2} & 0\,0 \\ 0\,0 & \cos\frac{\theta}{2} \;\; -i\sin\frac{\theta}{2} \\ 0\,0 & -i\sin\frac{\theta}{2} \;\; \cos\frac{\theta}{2} \end{bmatrix} \begin{bmatrix} 1\,0 & 0\,0 \\ 0\,1 & 0\,0 \\ 0\,0 & 0\,1 \\ 0\,0 & 1\,0 \end{bmatrix}$$

$$= \begin{bmatrix} \cos\frac{\theta}{2} \;\; -i\sin\frac{\theta}{2} & 0\,0 \\ -i\sin\frac{\theta}{2} \;\; \cos\frac{\theta}{2} & 0\,0 \\ 0\,0 & \cos\frac{\theta}{2} \;\; -i\sin\frac{\theta}{2} \\ 0\,0 & -i\sin\frac{\theta}{2} \;\; \cos\frac{\theta}{2} \end{bmatrix}$$

This confirms that $CNOT(R_x(\theta) \otimes \boldsymbol{I}) \neq (R_x(\theta) \otimes \boldsymbol{I})CNOT$.

Non-commutation means that the sequence of applying a rotation and an entangling gate (like CNOT) affects the final state. If we apply the rotation before entangling the qubits, the entangled state will reflect the rotated basis. If we entangle first and then rotate, each qubit's rotation will behave differently depending on the entanglement.

6.4 CNOT and $(I \otimes X)$ Gate Commutation

$$CNOT = \begin{bmatrix} 1 & 0 & 0 & 0 \\ 0 & 1 & 0 & 0 \\ 0 & 0 & 0 & 1 \\ 0 & 0 & 1 & 0 \end{bmatrix} \text{ and } (I \otimes X) = \begin{bmatrix} 1 & 0 \\ 0 & 1 \end{bmatrix} \otimes \begin{bmatrix} 0 & 1 \\ 1 & 0 \end{bmatrix} = \begin{bmatrix} 0 & 1 & 0 & 0 \\ 1 & 0 & 0 & 0 \\ 0 & 0 & 0 & 1 \\ 0 & 0 & 1 & 0 \end{bmatrix}$$

$$CNOT(I \otimes X) = \begin{bmatrix} 1 & 0 & 0 & 0 \\ 0 & 1 & 0 & 0 \\ 0 & 0 & 0 & 1 \\ 0 & 0 & 1 & 0 \end{bmatrix} \begin{bmatrix} 0 & 1 & 0 & 0 \\ 1 & 0 & 0 & 0 \\ 0 & 0 & 0 & 1 \\ 0 & 0 & 1 & 0 \end{bmatrix} = \begin{bmatrix} 0 & 1 & 0 & 0 \\ 1 & 0 & 0 & 0 \\ 0 & 0 & 1 & 0 \\ 0 & 0 & 0 & 1 \end{bmatrix}$$

$$(I \otimes X)CNOT = \begin{bmatrix} 0 & 1 & 0 & 0 \\ 1 & 0 & 0 & 0 \\ 0 & 0 & 0 & 1 \\ 0 & 0 & 1 & 0 \end{bmatrix} \begin{bmatrix} 1 & 0 & 0 & 0 \\ 0 & 1 & 0 & 0 \\ 0 & 0 & 0 & 1 \\ 0 & 0 & 1 & 0 \end{bmatrix} = \begin{bmatrix} 0 & 1 & 0 & 0 \\ 1 & 0 & 0 & 0 \\ 0 & 0 & 1 & 0 \\ 0 & 0 & 0 & 1 \end{bmatrix}$$

Therefore, we can write the commutation relation of CNOT gate and Pauli X gate as $CNOT(I \otimes X) = (I \otimes X)CNOT$.

7 Conclusion

In this study, we have looked upon the matrix algebra required to understand the operational behaviour of quantum gates. We have presented the methodology of constructing the matrix elements of quantum gates. We have also carried out a survey of historical backgrounds of quantum computing along with its necessity in the field of computing. We have derived the matrix elements of controlled not gate, controlled Z gate and controlled Hadamard gate. Matrix elements of three qubit quantum gates such as Toffoli (Controlled-Controlled NOT) gate and Fredkin (Controlled Controlled Swap) gate are already available in Poornima et al. (2017) and hence we have excluded these two from our study. Utility of linear transformation required for combining two gates is also presented. Continuation of our study will be towards the time evolution of quantum gates and formulation of corresponding Hamiltonian gate. Hamiltonian gate later on can be used to solve Schrödinger equation with a specified potential which can be used to explain quantum Fourier transformation and quantum phase estimation.

References

Barenco A, Bennet C, Cleve R, DiVincenzo DP, Margolus N, Shor P, Sleator T, Smolin JA, Weinfurter H (1995) Elementary gates for quantum computation. Phys Rev A 52:34573467

Benioff P (1982) Quantum mechanical Hamiltonian model of turning machines. J Stat Phys 29:515–546

Benioff P (1982b) Quantum mechanical Hamiltonian model of Turning machines that dissipate no energy. Phys Rev Lett 48:1581–1585

Benioff P, The computer as a physical system: a microscopic quantum mechanical Hamiltonian model of computers as represented by Turning machines. J Stat Phys 22(5):563–591

Deutsch D (1989) Quantum computational networks. Proc Royal Soc London A 425:7390

Deutsch D, Quantum theory, the Church-Turing principle and the universal quantum computer. Proceedings of the royal society of London A400:97117

Feynman RP (1982a) Simulating physics with computers. Int J Theor Phys 21(6/7):467–488

Feynman RP (1982) Simulating physics with computers. Int J Theor Phys 21:467–488

Feynman RP (1985) Quantum mechanical computers. Foundation of Physics, Vol 16, pp 507–531(1986). (Originally appeared in optics news, February 1985)

Nielsen MA, Chuang IL (2000) Quantum computation and quantum information. Cambridge University Press, Cambridge

Poornima A, Naghabhushana NM, Ujjinimatad R (2017) Matrix representation of quantum gates. Int J Comput Appl 159(8):1–6

Shende VV, Bullock AS, Markov IL (2006) Synthesis of quantum-logic circuits. IEEE Trans Comput Aided Design Integr Circuits Syst 25:10001010

Trotter HF (1959) On the product of semi-groups of operators. Proc Am Math Soc 10:545

Yao AC (1993) Quantum circuit complexity. In: Proceedings of the 34th Annual IEEE symposium on foundations of computer science, pp 352361

Computational Approach to Solve Ricci Solitons on Kenmotsu Manifolds

Shivani Sundriyal

Abstract This chapter aims to study the Ricci-soliton on the Kenmotsu manifold. Initially, we studied the Ricci Flow, Kenmotsu manifolds, M-projective curvature tensor, pseudo-projective curvature tensor, and semi-symmetric properties of Ricci tensor. We find some results satisfying some mathematical conditions like $\hat{\mathcal{S}}(\xi, X_1).\hat{\mathcal{R}} = 0$ on Ricci tensor, $\hat{\mathcal{R}}(\xi, X_1).\hat{P} = 0$ on Pseudo-Projective, and $\hat{\mathcal{R}}(\xi, X_1).\hat{M} = 0$ on M-projective curvature tensors.

1 Introduction

In 1958, Boothby and Wang (1958), studied the topological properties of an odd-dimensional differentiable manifold. Sasaki (1960), did tremendous work in the field of contact geometry. He used the tensor calculus to characterize the postulates of an odd-dimensional differentiable manifold equipped with contact structures known as the contact manifolds. The several classes of contact manifolds have been characterized by many researchers who studied their properties with different connections. In this series, Kenmotsu (1972), introduced the concept of the Kenmotsu manifold by considering a class of contact metric manifolds satisfying certain tensorial relations. Later, many geometers studied the various properties of the Kenmotsu manifold and its sub-manifolds along with various connections Sato (1976), Matsumoto et al. (1998), Chaubey and Ojha (2010), Chaubey et al. (2012), Sundriyal and Haseeb (2024).

A. Hayden introduced a metric connection on a Riemannian manifold, in Hyden (1932). A linear connection ∇ is said to be a metric on a manifold $\mathbb{M}^n$ of dimension n if $\nabla g = 0$; otherwise, it is non-metric. In 1970, a semi-symmetric metric connection on the Riemannian manifold was introduced by Yano (1970). Later many other geometers like Sengupta et al. (2000), Agashe (1992) and Chaubey and Ojha (2008), Chaubey and Ojha (2012) studied various and important properties of semi-symmetric metric and non-metric connections on several differentiable manifolds.

S. Sundriyal (✉)
Graphic Era Hill University, Dehradun, India
e-mail: shivani.sundriyal5@gmail.com

M. Ram et al. (eds.), *Advances in Mathematics for Engineering Sciences*, Mathematical Engineering, https://doi.org/10.1007/978-3-031-95693-5_8

They also defined some new types of connections on the Riemannian manifold. For the detailed understanding of connections I refer Chaubey and Yildiz (2019), Chaubey and Ojha (2010), Chaubey et al. (2012).

The notion of a semi-symmetric non-metric connection on a Riemannian manifold was introduced by S.K Chaubey and Yildiz in Chaubey and Yildiz (2019). Most recently, S. Sundriyal and J. Upreti in 2021, applied this connection on the H.S.U structure manifold Sundriyal and Upreti (2021) and later on in 2022, with the Kenmotsu manifold.

Hamilton (1988) used Ricci flow and solitons to get the canonical metric on a smooth manifold. The introduction of Ricci flow made studying manifolds with positive curvature easier. Perelman used Ricci flow to establish the Poincare conjecture Perelman (2002), Perelman (2003). The Ricci soliton refers to the limit of Ricci flow solutions. An almost Ricci soliton is a simplified version of the Einstein metric. A Riemannian metric g on a Riemannian manifold $\mathbb{M}^n$ of size n is considered a nearly Ricci soliton if a vector field X on $\mathbb{M}^n$ satisfies the equation Hadimani and Prakasha (2016).

$$\mathcal{L}_X g + 2\mathcal{S} + 2\alpha g = 0 \tag{1}$$

It will be shrinking, steady, or expanding according to $\alpha < 0, \alpha = 0, \alpha > 0$, respectively.

The M-projective curvature tensor was first given by R.S. Mishra and G.P. Pokhariyal in 1971 Pokhariyal and Mishra (1971). This curvature tensor is further studied by many geometers Mandal and Das (2020), Prakasha and Mirji (2017), Ojha (1975), Ojha (1986), Chaubey and Ojha (2010), Chaubey et al. (2012) on various manifolds. For an n-dimensional manifold, the M-projective curvature tensor can be obtain from,

$$M(X_1, X_2)X_3 = \mathcal{R}(X_1, X_2)X_3 - \frac{1}{2(n-1)}\{\mathcal{S}(X_2, X_3)X_1 - \mathcal{S}(X_1, X_3)X_2\} - \frac{1}{2(n-1)}\{g(X_2, X_3)\mathcal{Q}X_1 - g(X_1, X_3)\mathcal{Q}X_2\} \tag{2}$$

for all vector fields $\forall X_1, X_2, X_3 \in \chi(\mathbb{M})$, where $\mathcal{Q}$ is the Ricci operator.

The pseudo-projective curvature tensor on a Riemannian manifold of dimensions greater than two was introduced by B. Prasad, in 2002 Prasad (2002) and it is formulated as,

$$P(X_1, X_2)X_3 = a\mathcal{R}(X_1, X_2)X_3 + b\{\mathcal{S}(X_2, X_3)X_1 - \mathcal{S}(X_1, X_3)X_2\} - \frac{r}{n}\left(\frac{a}{(n-1)} + b\right)\{g(X_2, X_3)X_1 - g(X_1, X_3)X_2\} \tag{3}$$

for all vector fields $\forall X_1, X_2, X_3 \in \chi(\mathbb{M})$, where a, b, and c are non-vanishing constants.

Definition 1 (*Kenmotsu Manifolds*) A manifold with the almost contact Riemannian structure (ϕ, ξ, η, g) is an almost Kenmotsu manifold if the following conditions are satisfied,

$$\phi^2 X_1 = -X_1 + \eta(X_1)\xi, \quad \eta(\xi) = 1, \quad \phi\xi = 0, \quad \eta \circ \phi = 0 \tag{4}$$

$$g(X_1, X_2) = g(\phi X_1, \phi X_2) + \eta(X_1)\eta(X_2) \tag{5}$$

$$g(X_1, \phi X_2) = -g(\phi X_1, X_2), \, and \, g(X_1, \xi) = \eta(X_1) \tag{6}$$

for all $X_1, X_2 \in \chi(\mathbb{M})$, where ϕ is $(1, 1)-$ tensor field, ξ is a vector field and η is a 1-form. An almost Kenmotsu manifold is called a Kenmotsu manifold if it satisfies,

$$(\nabla_{X_1}\phi)X_2 = -g(X_1, \phi X_2)\xi - \eta(X_2)\phi X_1 \tag{7}$$

$$(\nabla_{X_1}\eta)X_2 = g(X_1, X_2) - \eta(X_1)\eta(X_2) \tag{8}$$

$$\nabla_{X_1}\xi = X_1 - \eta(X_1)\xi \tag{9}$$

where ∇ is the Levi-Civita connection on a Riemannian metric g.

An odd-dimensional say $(n = 2m + 1)$, Kenmotsu manifold is normal but not Sasakian. Also, every Kenmotsu manifold is non-compact and $div\xi = n - 1$. Moreover, the Riemannian curvature tensor $\mathcal{R}$, Ricci tensor $\mathcal{S}$, and Ricci operator $\mathcal{Q}$ for Kenmotsu manifolds are given by,

$$\mathcal{R}(X_1, X_2)\xi = \eta(X_1)X_2 - \eta(X_2)X_1, \tag{10}$$

$$\mathcal{R}(\xi, X_1)X_2 = \eta(X_2)X_1 - g(X_1, X_2)\xi, \tag{11}$$

$$\mathcal{R}(\xi, X_1)\xi = X_1 - \eta(X_1)\xi, \tag{12}$$

$$\mathcal{S}(X_1, \xi) = -(n-1)\eta(X_1), \tag{13}$$

$$\mathcal{Q}\xi = -(n-1)\xi, \tag{14}$$

Definition 2 (*Semi-symmetric non-metric connection*) In 2019, S.K Chaubey familarized semi-symmetric non-metric connection in Chaubey and Yildiz (2019). This linear connection $\hat{\nabla}$ on a Riemannian manifold $(\mathbb{M}^n, g)$ can be formulated as,

$$\hat{\nabla}_{X_1}X_2 = \nabla_{X_1}X_2 + \frac{1}{2}\{\eta(X_2)X_1 - \eta(X_1)X_2\} \tag{15}$$

where ∇ is the Levi-Civita connection on a Riemannian metric g. The torsion tensor $\hat{T}$ on $\mathbb{M}^n$ concerning $\hat{\nabla}$ satisfies the equations,

$$\hat{T}(X_1, X_2) = \eta(X_2)X_1 - \eta(X_1)X_2 \tag{16}$$

where η is 1-form and ξ is associated vector field, also

$$\eta(X_1) = g(X_1, \xi) \tag{17}$$

The metric g holds the relation

$$(\hat{\nabla}_{X_1}g)(X_2, X_3) = \frac{1}{2}\{2\eta(X_1)g(X_2, X_3) - \eta(X_2)g(X_1, X_3) - \eta(X_3)g(X_1, X_2)\} \quad (18)$$

For any arbitrary vector fields X_1, X_2, and the vector field ξ, it is known that

$$(\hat{\mathcal{L}}_\xi g)(X_1, X_2) = (\mathcal{L}_\xi g)(X_1, X_2) + 2\{\eta(\xi)g(X_1, X_2) - \eta(X_1)\eta(X_2)\} \quad (19)$$

where $\hat{\mathcal{L}}_\xi$ and $\mathcal{L}_\xi$ are the Lie derivatives along with the vector field ξ corresponding to $\hat{\nabla}$ and ∇, respectively.

The curvature tensor $\hat{\mathcal{R}}$ corresponding to the semi-symmetric non-metric connection on an n-dim Kenmotsu manifold is given by Sundriyal and Upreti (2021),

$$\hat{\mathcal{R}}(X_1, X_2)X_3 = \mathcal{R}(X_1, X_2)X_3 + \frac{3}{4}\{\eta(X_2)X_1 - \eta(X_1)X_2\}\eta(X_3) + \frac{1}{2}\{g(X_1, X_3)X_2 - g(X_2, X_3)X_1\} \quad (20)$$

The Ricci curvature corresponding to $\hat{\nabla}$ on an n-dimensional Kenmotsu manifold is

$$\hat{\mathcal{S}}(X_2, X_3) = \mathcal{S}(X_2, X_3) - \frac{(n-1)}{2}g(X_2, X_3) + \frac{3(n-1)}{2}\eta(X_2)\eta(X_3) \quad (21)$$

The Ricci curvature $\hat{\mathcal{S}}(X_2, X_3)$ is symmetric for this connection on the Kenmotsu manifold.

The Ricci operator $\hat{\mathcal{Q}}$ for the Kenmotsu manifold concerning $\hat{\nabla}$ is given by,

$$\hat{\mathcal{Q}}X_1 = \mathcal{Q}X_1 - \frac{(n-1)}{2}X_1 + \frac{3(n-1)}{4}\eta(X_1)\xi \quad (22)$$

Also, the constant curvature $\hat{r}$ is given by,

$$\hat{r} = r - \frac{(2n-3)(n-1)}{4} \quad (23)$$

From equations (21) and (22), we can deduce the following,

$$\hat{\mathcal{S}}(X_2, \xi) = -\frac{3(n-1)}{4}\eta(X_2) \quad (24)$$

$$\hat{\mathcal{Q}}\xi = -\frac{3(n-1)}{4}\xi \quad (25)$$

Definition 3 (*Solitons*) A Riemannian metric g on an n-dimensional Riemannian manifold $\mathbb{M}^n$ is called an almost Ricci soliton if there exists a complete vector field X_2 on $\mathbb{M}^n$, that satisfies the following equation,

$$\mathcal{L}_E g + 2\mathcal{S} + 2\alpha g = 0$$

where $\mathcal{L}$ is the Lie derivative, $\mathcal{S}$ is the Ricci tensor and α is a smooth function. If α is a constant then the metric g satisfying the above equation is called a Ricci soliton. A Ricci soliton is expanding, steady, and shrinking according to $\alpha > 0, \alpha = 0$ and $\alpha < 0$ respectively.

A Riemannian manifold $\mathbb{M}^n$ with Riemannian metric g is called an almost η-Ricci soliton if there exists a smooth vector field X_2 such that,

$$\mathcal{L}_E g + 2\mathcal{S} + 2\alpha g + 2\beta\eta \otimes \eta = 0$$

where α and β are smooth functions. If α and β are constants, then the metric g is known as η-Ricci soliton. A Ricci soliton has constant curvature for a compact manifold of dimension 2 or 3.

2 Ricci Semi-symmetry with Respect to ∇ and $\hat{\nabla}$

In this section, we characterize the Ricci semi-symmetry in Kenmotsu manifolds concerning $\hat{\nabla}$. Now,

$$(\hat{\mathcal{R}}(X_1, X_2) \cdot \hat{\mathcal{S}})(U, V) = -\hat{\mathcal{S}}(\hat{\mathcal{R}}(X_1, X_2)U, V) - \hat{\mathcal{S}}(U, \hat{\mathcal{R}}(X_1, X_2)V) \tag{26}$$

Now again, using (20) in (26), we have

$$\begin{aligned}(\hat{\mathcal{R}}(X_1, X_2) \cdot \hat{\mathcal{S}})(U, V) = {} & (\mathcal{R}(X_1, X_2) \cdot \mathcal{S})(U, V) - \frac{3}{4}\{\mathcal{S}(X_1, V)\eta(X_2)\eta(U) + \mathcal{S}(U, X_1)\eta(X_2)\eta(V) - \mathcal{S}(U, X_2)\eta(X_1)\eta(V) \\ & - \mathcal{S}(X_2, V)\eta(X_1)\eta(U)\} - \frac{1}{2}\{g(X_1, U)\mathcal{S}(X_2, V) - g(X_2, U)\mathcal{S}(X_1, V) + g(X_1, V)\mathcal{S}(U, X_2) \\ & - g(X_2, V)\mathcal{S}(U, X_1)\} - \frac{3}{8}(n-1)\{g(X_1, U)\eta(X_2)\eta(V) + g(X_1, V)\eta(X_2)\eta(U) \\ & - g(X_2, U)\eta(X_1)\eta(V) - g(X_2, V)\eta(X_1)\eta(U)\}\end{aligned} \tag{27}$$

Now suppose $(\hat{\mathcal{R}}(X_1, X_2) \cdot \hat{\mathcal{S}})(U, V) = (\mathcal{R}(X_1, X_2) \cdot \mathcal{S})(U, V)$, then Right hand side of equation (27) will be equal to zero.

Now put $X_1 = U = e_i$; $1 \le i \le (2n + 1)$, we have

$$\frac{3(n-1)}{2}\{g(\mathcal{R}(e_i, X_2)e_i, \xi)\eta(V) + g(\xi, (e_i, X_2)V)\eta(e_i)\}$$
$$-\frac{3}{4}\{\mathcal{S}(e_i, V)\eta(X_2)\eta(e_i) + \mathcal{S}(e_i, e_i)\eta(U)\eta(X_2) - \mathcal{S}(e_i, X_2)\eta(e_i)\eta(V) - \mathcal{S}(X_2, V)\eta(e_i)\eta(e_i)\}$$
$$-\frac{1}{2}\{g(e_i, e_i)\mathcal{S}(X_2, V) - g(X_2, e_i)\mathcal{S}(e_i, V) + g(e_i, V)\mathcal{S}(e_i, X_2) - g(X_2, V)\mathcal{S}(e_i, e_i)\}$$
$$-\frac{3(n-1)}{8}\{n\eta(X_2)\eta(V) - g(X_2, U)\} = 0$$

this will simplify as,

$$\frac{3(n-1)}{2}\{\mathcal{S}(X_2, \xi)\eta(V) + \mathcal{S}(\xi, X_2)\eta(U)\} - \frac{3}{4}\{r\eta(X_2)\eta(V) - \mathcal{S}(X_2, V)\}$$
$$-\frac{1}{2}\{n\mathcal{S}(X_2, V) - rg(X_2, V)\} - \frac{3(n-1)}{8}\{n\eta(X_2)\eta(V) - g(X_2, U)\} = 0$$

Hence we get,

$$\frac{3-2n}{4}\mathcal{S}(X_2, V) = -\frac{4r + 3(n-1)}{8}g(X_2, V) + \frac{6r + 3n(n-1)}{8}\eta(X_2)\eta(V) \tag{28}$$

Hence we can state the following:

Theorem 1 *The Ricci symmetries concerning* $\hat{\nabla}$ *and* ∇ *are equivalent if and only if the manifold is* $\eta-$*Einstein with respect to the* ∇.

3 Ricci-Soliton on Kenmotsu Manifold Concerning $\hat{\nabla}$

Let ξ is a killing vector field on n-dimensional Kenmotsu manifold equipped with $\hat{\nabla}$, then we have,

$$(\hat{\mathcal{L}}_\xi g)(X_1, X_2) = 2\{g(X_1, X_2) - \eta(X_1)\eta(X_2)\} \tag{29}$$

Now the Ricci-soliton for Kenmotsu manifold concerning $\hat{\nabla}$ is formulated as,

$$(\hat{\mathcal{L}}_\xi g)(X_1, X_2) + 2\hat{\mathcal{S}}(X_1, X_2) + 2\lambda g(X_1, X_2) = 0 \tag{30}$$

As ξ is killing, so

$$2\{g(X_1, X_2) - \eta(X_1)\eta(X_2)\} + 2\hat{\mathcal{S}}(X_1, X_2) + 2\lambda g(X_1, X_2) = 0 \tag{31}$$

Hence, we have

$$\hat{\mathcal{S}}(X_1, X_2) = -(\lambda + 1)g(X_1, X_2) + \eta(X_1)\eta(X_2) \tag{32}$$

Hence, the following theorem is stated.

Theorem 2 *An n-dimensional Kenmotsu manifold is an η-Einstein manifold concerning a semi-symmetric non-metric connection if ξ is killing.*

Now, put $X_2 = \xi$ in equation (32), we have

$$\hat{\mathcal{S}}(X_1, \xi) = -\lambda\eta(X_1) \tag{33}$$

Now, put the value of $\hat{\mathcal{S}}$ from equation (21) in (32), we get

$$\mathcal{S}(X_1, X_2) = \left(\frac{n-3-2\lambda}{2}\right) g(X_1, X_2) + \left(\frac{7-3n}{4}\right) \eta(X_1)\eta(X_2) \tag{34}$$

Now put $X_2 = \xi$ in equation (34),

$$\mathcal{S}(X_1, \xi) = \left(\frac{1-n-4\lambda}{4}\right) \eta(X_1) \tag{35}$$

From equation (13), we have

$$\lambda = \frac{3}{4}(n-1) \tag{36}$$

We can state the following theorem based on the above discussion.

Theorem 3 *The Ricci soliton for Kenmotsu manifold concerning $\hat{\nabla}$, is expanding for all $n > 1$.*

4 Ricci Soliton on Kenmotsu Manifold Concerning $\hat{\nabla}$ satisfying $\hat{\mathcal{S}}(\xi, X_1).\hat{\mathcal{R}} = 0$

Let $(\mathbb{M}^n, g)$ be a Kenmotsu manifold concerning the connection $\hat{\nabla}$, and satisfying $\hat{\mathcal{S}}(\xi, X_1).\hat{\mathcal{R}} = 0$.

From the definition, we have

$$\begin{aligned}(\hat{\mathcal{S}}(\xi, X_1).\hat{\mathcal{R}})(U, V)W = {} & \hat{\mathcal{S}}(\xi, \hat{\mathcal{R}}(U, V)W)X_1 - \hat{\mathcal{S}}(X_1, \hat{\mathcal{R}}(U, V)W)\xi + \hat{\mathcal{S}}(\xi, U)\hat{\mathcal{R}}(X_1, V)W - \\ & \hat{\mathcal{S}}(X_1, U)\hat{\mathcal{R}}(\xi, V)W + \hat{\mathcal{S}}(\xi, V)\hat{\mathcal{R}}(U, X_1)W - \hat{\mathcal{S}}(X_1, V)\hat{\mathcal{R}}(U, \xi)W \\ & + \hat{\mathcal{S}}(\xi, W)\hat{\mathcal{R}}(U, V)X_1 - \hat{\mathcal{S}}(X_1, W)\hat{\mathcal{R}}(U, V)\xi \\ & = 0\end{aligned} \tag{37}$$

Using equations (20), (32), (33), and (37), and then taking the inner product with ξ, we have

$$-(\lambda+1)\begin{bmatrix} 3g(U,W)\eta(V)\eta(X_1) - 3g(V,W)\eta(X_1)\eta(U) - g(X_1,\hat{\mathcal{R}}(U,V)W) + \frac{3}{2}g(X_1,U)g(V,W) \\ -\frac{3}{2}g(X_1,V)g(U,W) + \frac{3}{2}g(X_1,U)\eta(V)\eta(W) - \frac{3}{2}g(X_1,V)\eta(U)\eta(W) \end{bmatrix} = 0 \tag{38}$$

Now put $X_1 = U = e_i$ in equation (38) and summing over $i = 1, 2, ..., n$. we have,

$$-(\lambda+1)\left[-\hat{\mathcal{S}}(V,W) + \frac{3}{2}(n+1)\eta(V)\eta(W) + \frac{3}{2}(n-3)g(V,W)\right] = 0 \tag{39}$$

Now put $V = W = \xi$ in equation (39),

$$-(\lambda+1)(\lambda+3n-3) = 0 \tag{40}$$

from (40), $\lambda = -1$ or $\lambda = -3(n-1)$. Hence, we state the following theorem.

Theorem 4 *A Ricci soliton on a Kenmotsu manifold of dim $n > 1$ satisfying $\hat{\mathcal{S}}(\xi, X_1).\hat{\mathcal{R}} = 0$, is always shrinking.*

5 Ricci Soliton on Kenmotsu Manifold Concerning $\hat{\nabla}$ satisfying $\hat{\mathcal{R}}(\xi, X_1).\hat{P} = 0$

The pseudo-projective curvature tensor for the Kenmotsu manifold $(\mathbb{M}^n, g)$ concerning $\hat{\nabla}$ is given by (3),

$$\hat{P}(X_1,X_2)X_3 = a\hat{\mathcal{R}}(X_1,X_2)X_3 + b\{\hat{\mathcal{S}}(X_2,X_3)X_1 - \hat{\mathcal{S}}(X_1,X_3)X_2\} - \frac{\hat{r}}{n}\left(\frac{a}{(n-1)} + b\right)\{g(X_2,X_3)X_1 - g(X_1,X_3)X_2\} \tag{41}$$

for all vector fields $\forall X_1, X_2, X_3 \in \chi(\mathbb{M})$, where a, b, and c are non-vanishing constants.

Now, taking $X_3 = \xi$ in equation (41) and using equations (20) and (33) we have

$$\hat{P}(X_1,X_2)\xi = \left[\frac{3}{4}a + (\lambda+1)b + \frac{\hat{r}}{n}\left(\frac{a}{(n-1)} + b\right)\right]\{\eta(X_1)X_2 - \eta(X_2)X_1\} \tag{42}$$

Again,

$$\eta\left(\hat{P}(X_1,X_2)X_3\right) = \left[\frac{3}{2}a + (\lambda+1)b + \frac{\hat{r}}{n}\left(\frac{a}{(n-1)} + b\right)\right]\{g(X_1,X_3)\eta(X_2) - g(X_2,X_3)\eta(X_1)\} \tag{43}$$

Let the manifold $(\mathbb{M}^n, g)$ concerning $\hat{\nabla}$, and satisfies $\hat{\mathcal{R}}(\xi, X_1).\hat{P} = 0$.

From the definition, we have

$$\begin{aligned}(\hat{\mathcal{R}}(\xi,X_1).\hat{P})(U,V)W &= \hat{\mathcal{R}}(\xi,X_1).\hat{P}(U,V)W - \hat{P}(\hat{\mathcal{R}}(\xi,X_1)U,V)W - \hat{P}(U,\hat{\mathcal{R}}(\xi,X_1)V)W - \hat{P}(U,V)\hat{\mathcal{R}}(\xi,X_1)W \\ &= 0\end{aligned} \tag{44}$$

Using equations (20), (41), (42), and (44), and then taking the inner product with ξ, we have

$$-\frac{3}{2}ag(X_1, \hat{\mathcal{R}}(U, V)W) + \frac{9}{4}a(g(X_1, V)g(U, W) - g(X_1, U)g(V, W)) = 0 \tag{45}$$

Now put $X_1 = U = e_i$ in equation (45) and summing over $i = 1, 2, ...n$. we have,

$$-\frac{3}{2}a\hat{\mathcal{S}}(V, W) + \frac{9}{4}(1-n)g(V, W) = 0 \tag{46}$$

Now put $V = W = \xi$ in equation (46)

$$\lambda = -\frac{3}{2}(n-1) \tag{47}$$

Hence, we can state,

Theorem 5 *A Ricci soliton on a Kenmotsu manifold of dim $n > 1$ satisfying $\hat{\mathcal{R}}(\xi, X_1).\hat{P} = 0$, is always shrinking.*

6 Ricci Soliton on Kenmotsu Manifold Concerning $\hat{\nabla}$ satisfying $\hat{\mathcal{R}}(\xi, X_1).\hat{M} = 0$

The M-projective curvature tensor $\hat{M}$ for the Kenmotsu manifold $(\mathbb{M}^n, g)$ concerning $\hat{\nabla}$ is given by (2),

$$\hat{M}(X_1, X_2)X_3 = \hat{\mathcal{R}}(X_1, X_2)X_3 - \frac{1}{2(n-1)}\{\hat{\mathcal{S}}(X_2, X_3)X_1 - \hat{\mathcal{S}}(X_1, X_3)X_2\} - \frac{1}{2(n-1)}\{g(X_2, X_3)\hat{Q}X_1 - g(X_1, X_3)\hat{Q}X_2\} \tag{48}$$

for all vector fields $\forall X_1, X_2, X_3 \in \chi(\mathbb{M})$.

Now, taking $X_3 = \xi$ in equation (48) and using equations (20) and (33) we have

$$\hat{M}(X_1, X_2)\xi = \left[\frac{3n-3-4\lambda}{4(n-1)}\right]\{\eta(X_1)X_2 - \eta(X_2)X_1\} \tag{49}$$

Again,

$$\eta\left(\hat{M}(X_1, X_2)X_3\right) = \left[\frac{3n-4-2\lambda}{2(n-1)}\right]\{g(X_1, X_3)\eta(X_2) - g(X_2, X_3)\eta(X_1)\} \tag{50}$$

Let the manifold $(\mathbb{M}^n, g)$ concerning $\hat{\nabla}$, and satisfies $\hat{\mathcal{R}}(\xi, X_1).\hat{M} = 0$.

From the definition, we have

$$(\hat{\mathcal{R}}(\xi, X_1).\hat{M})(U, V)W = \hat{\mathcal{R}}(\xi, X_1).\hat{M}(U, V)W - \hat{M}(\hat{\mathcal{R}}(\xi, X_1)U, V)W - \hat{M}(U, \hat{\mathcal{R}}(\xi, X_1)V)W - \hat{M}(U, V)\hat{\mathcal{R}}(\xi, X_1)W$$
$$= 0 \tag{51}$$

Using equations (20), (49), (50), and (51), and then taking the inner product with ξ, we have

$$
-\frac{3}{2}g(X_1,\hat{\mathcal{R}}(U,V)W)-\frac{3(3n-2)}{4(n-1)}a(g(X_1,U)g(V,W)-g(X_1,V)g(U,W))+\frac{3}{4(n-1)}(g(V,W)\eta(U)\eta(X_1)
$$
$$
-g(U,W)\eta(V)\eta(X_1))+\frac{3(3n-2-2\lambda)}{8(n-1)}(g(U,X_1)\eta(V)\eta(W)-g(V,X_1)\eta(U)\eta(W))=0 \tag{52}
$$

Now put $X_1=U=e_i$ in equation (52) and summing over $i=1,2,...n$. we have,

$$
-\frac{3}{2}\hat{\mathcal{S}}(V,W)-\frac{3(3n-2)}{4}g(V,W)+\frac{3}{4(n-1)}(g(V,W)-\eta(V)\eta(W))+\frac{3(3n-2-2\lambda)}{8}\eta(V)\eta(W)=0 \tag{53}
$$

Now put $V=W=\xi$ in equation (53)

$$
\lambda=\frac{3(n-2)}{2} \tag{54}
$$

Hence, we have the following theorem.

Theorem 6 *A Ricci soliton on a Kenmotsu manifold of dim $n>1$ satisfying $\hat{\mathcal{R}}(\xi,X_1).\hat{M}=0$, is always expanding.*

7 Example

Example 1 Let us consider a 5-dimensional manifold $\mathbb{M}=\{(l,m,n,o,p)\in\mathbb{R}^5\}$ where (l,m,n,o,p) are the standard coordinates of $\mathbb{R}^5$. We choose the linearly independent vector fields, at each point of $\mathbb{M}$.

$$
\mu_1=\frac{\partial}{\partial l},\ \mu_2=e^{-l}\frac{\partial}{\partial m},\ \mu_3=e^{-l}\frac{\partial}{\partial n},\ \mu_4=e^{-l}\frac{\partial}{\partial o},\ \mu_5=e^{-l}\frac{\partial}{\partial p}
$$

also, $l\neq 0$.

Let g be the Riemannian metric defined by,

$$
g(\mu_i,\mu_j)=\begin{cases}1 \text{ if } i=j\\ 0 \text{ if } i\neq j\end{cases};\ i,j=1,2,3,4,5.
$$

Let η be the 1-form defined by,

$$
\eta(A)=g(A,\mu_1)
$$

for any $A\in\chi(\mathbb{M})$. Using the linearity of ϕ and g, we have

$$
\phi^2A=-A+\eta(A)\mu_1 \text{ and } \eta(\mu_1)=1
$$

Also,

$$
g(\phi A,\phi B)=g(A,B)-\eta(A)\eta(B)
$$

For any $A, B \in \chi(\mathbb{M})$. Thus $\mu_1 = \xi$, the structure (ϕ, ξ, η, g) defines an almost Kenmotsu manifold. For Levi-Civita connection ∇, we have

$$[\mu_1, \mu_2] = -\mu_2, [\mu_1, \mu_3] = -\mu_3, [\mu_1, \mu_4] = -\mu_4, [\mu_1, \mu_5] = -\mu_5,$$

$$[\mu_2, \mu_3] = [\mu_2, \mu_4] = [\mu_2, \mu_5] = [\mu_3, \mu_4] = [\mu_3, \mu_5] = [\mu_4, \mu_5] = 0$$

The Riemannian connection ∇ of the metric g is given by using Koszul's formula and also consider $\mu_1 = \xi$, we get

$$\begin{array}{lllll}
\nabla_{\mu_1}\mu_1 = 0 & \nabla_{\mu_1}\mu_2 = 0 & \nabla_{\mu_1}\mu_3 = 0 & \nabla_{\mu_1}\mu_4 = 0 & \nabla_{\mu_1}\mu_5 = 0 \\
\nabla_{\mu_2}\mu_1 = \mu_2 & \nabla_{\mu_2}\mu_2 = -\mu_1 & \nabla_{\mu_2}\mu_3 = 0 & \nabla_{\mu_2}\mu_4 = 0 & \nabla_{\mu_2}\mu_5 = 0, \\
\nabla_{\mu_3}\mu_1 = \mu_3 & \nabla_{\mu_3}\mu_2 = 0 & \nabla_{\mu_3}\mu_3 = -\mu_1 & \nabla_{\mu_3}\mu_4 = 0 & \nabla_{\mu_3}\mu_5 = 0 \\
\nabla_{\mu_4}\mu_1 = \mu_4 & \nabla_{\mu_4}\mu_2 = 0 & \nabla_{\mu_4}\mu_3 = 0 & \nabla_{\mu_4}\mu_4 = -\mu_1 & \nabla_{\mu_4}\mu_5 = 0, \\
\nabla_{\mu_5}\mu_1 = \mu_5 & \nabla_{\mu_5}\mu_2 = 0 & \nabla_{\mu_5}\mu_3 = 0 & \nabla_{\mu_5}\mu_4 = 0 & \nabla_{\mu_5}\mu_5 = -\mu_1.
\end{array} \tag{55}$$

From the above values, it is clear that (ϕ, ξ, η, g) is a 5-dimensional Kenmotsu manifold.

Using the results from equation (55), we can obtain the for $\hat{\nabla}$,

$$\begin{array}{lllll}
\hat{\nabla}_{\mu_1}\mu_1 = 0 & \hat{\nabla}_{\mu_1}\mu_2 = -\frac{\mu_2}{2} & \hat{\nabla}_{\mu_1}\mu_3 = -\frac{\mu_3}{2} & \hat{\nabla}_{\mu_1}\mu_4 = -\frac{\mu_4}{2} & \hat{\nabla}_{\mu_1}\mu_5 = -\frac{\mu_5}{2} \\
\hat{\nabla}_{\mu_2}\mu_1 = \frac{3}{2}\mu_2 & \hat{\nabla}_{\mu_2}\mu_2 = -\mu_1 & \hat{\nabla}_{\mu_2}\mu_3 = 0 & \hat{\nabla}_{\mu_2}\mu_4 = 0 & \hat{\nabla}_{\mu_2}\mu_5 = 0, \\
\hat{\nabla}_{\mu_3}\mu_1 = \frac{3}{2}\mu_3 & \hat{\nabla}_{\mu_3}\mu_2 = 0 & \hat{\nabla}_{\mu_3}\mu_3 = -\mu_1 & \hat{\nabla}_{\mu_3}\mu_4 = 0 & \hat{\nabla}_{\mu_3}\mu_5 = 0 \\
\hat{\nabla}_{\mu_4}\mu_1 = \frac{3}{2}\mu_4 & \hat{\nabla}_{\mu_4}\mu_2 = 0 & \hat{\nabla}_{\mu_4}\mu_3 = 0 & \hat{\nabla}_{\mu_4}\mu_4 = -\mu_1 & \hat{\nabla}_{\mu_4}\mu_5 = 0, \\
\hat{\nabla}_{\mu_5}\mu_1 = \frac{3}{2}\mu_5 & \hat{\nabla}_{\mu_5}\mu_2 = 0 & \hat{\nabla}_{\mu_5}\mu_3 = 0 & \hat{\nabla}_{\mu_5}\mu_4 = 0 & \hat{\nabla}_{\mu_5}\mu_5 = -\mu_1.
\end{array}$$

Using the results from equation (55), we can obtain the components of the Riemannian curvature tensors concerning ∇ as follows:

$$\begin{array}{llll}
\mathcal{R}(\mu_1, \mu_2)\mu_1 = \mu_2, & \mathcal{R}(\mu_1, \mu_2)\mu_2 = -\mu_1, & \mathcal{R}(\mu_1, \mu_3)\mu_1 = \mu_3, & \mathcal{R}(\mu_1, \mu_3)\mu_3 = -\mu_1, \\
\mathcal{R}(\mu_1, \mu_4)\mu_1 = \mu_4, & \mathcal{R}(\mu_1, \mu_4)\mu_4 = -\mu_1, & \mathcal{R}(\mu_1, \mu_5)\mu_1 = \mu_5, & \mathcal{R}(\mu_1, \mu_5)\mu_5 = -\mu_1, \\
\mathcal{R}(\mu_2, \mu_3)\mu_2 = \mu_3, & \mathcal{R}(\mu_2, \mu_3)\mu_3 = -\mu_2, & \mathcal{R}(\mu_2, \mu_4)\mu_2 = \mu_4, & \mathcal{R}(\mu_2, \mu_4)\mu_4 = -\mu_2, \\
\mathcal{R}(\mu_2, \mu_5)\mu_2 = \mu_5, & \mathcal{R}(\mu_2, \mu_5)\mu_5 = -\mu_2, & \mathcal{R}(\mu_3, \mu_4)\mu_3 = \mu_4, & \mathcal{R}(\mu_3, \mu_4)\mu_4 = -\mu_3, \\
\mathcal{R}(\mu_3, \mu_5)\mu_3 = \mu_5, & \mathcal{R}(\mu_3, \mu_5)\mu_5 = -\mu_3, & \mathcal{R}(\mu_4, \mu_5)\mu_4 = \mu_5, & \mathcal{R}(\mu_4, \mu_5)\mu_5 = -\mu_4.
\end{array}$$

Hence, the non-vanishing Riemannian curvature tensors concerning $\hat{\nabla}$ as follows:

$$\begin{array}{llll}
\hat{\mathcal{R}}(\mu_1, \mu_2)\mu_1 = \frac{3}{4}\mu_2, & \hat{\mathcal{R}}(\mu_1, \mu_2)\mu_2 = -\frac{3}{2}\mu_1, & \hat{\mathcal{R}}(\mu_1, \mu_3)\mu_1 = \frac{3}{4}\mu_3, & \hat{\mathcal{R}}(\mu_1, \mu_3)\mu_3 = -\frac{3}{2}\mu_1, \\
\hat{\mathcal{R}}(\mu_1, \mu_4)\mu_1 = \frac{3}{4}\mu_4, & \hat{\mathcal{R}}(\mu_1, \mu_4)\mu_4 = -\frac{3}{2}\mu_1, & \hat{\mathcal{R}}(\mu_1, \mu_5)\mu_1 = \frac{3}{4}\mu_5, & \hat{\mathcal{R}}(\mu_1, \mu_5)\mu_5 = -\frac{3}{2}\mu_1, \\
\hat{\mathcal{R}}(\mu_2, \mu_3)\mu_2 = \frac{3}{2}\mu_3, & \hat{\mathcal{R}}(\mu_2, \mu_3)\mu_3 = -\frac{3}{2}\mu_2, & \hat{\mathcal{R}}(\mu_2, \mu_4)\mu_2 = \frac{3}{2}\mu_4, & \hat{\mathcal{R}}(\mu_2, \mu_4)\mu_4 = -\frac{3}{2}\mu_2, \\
\hat{\mathcal{R}}(\mu_2, \mu_5)\mu_2 = \frac{3}{2}\mu_5, & \hat{\mathcal{R}}(\mu_2, \mu_5)\mu_5 = -\frac{3}{2}\mu_2, & \hat{\mathcal{R}}(\mu_3, \mu_4)\mu_3 = \frac{3}{2}\mu_4, & \hat{\mathcal{R}}(\mu_3, \mu_4)\mu_4 = -\frac{3}{2}\mu_3, \\
\hat{\mathcal{R}}(\mu_3, \mu_5)\mu_3 = \frac{3}{2}\mu_5, & \hat{\mathcal{R}}(\mu_3, \mu_5)\mu_5 = -\frac{3}{2}\mu_3, & \hat{\mathcal{R}}(\mu_4, \mu_5)\mu_4 = \frac{3}{2}\mu_5, & \hat{\mathcal{R}}(\mu_4, \mu_5)\mu_5 = -\frac{3}{2}\mu_4.
\end{array} \tag{56}$$

So, the Ricci tensor with respect to ∇ will be,

$$\mathcal{S}(\mu_1, \mu_1) = \mathcal{S}(\mu_2, \mu_2) = \mathcal{S}(\mu_3, \mu_3) = \mathcal{S}(\mu_4, \mu_4) = \mathcal{S}(\mu_5, \mu_5) = -4.$$

So, the scalar curvature r with respect to ∇, of the manifold will be,

$$r = -20.$$

Hence, the Ricci curvature for $\hat{\nabla}$ as follows:

$$\hat{\mathcal{S}}(\mu_1, \mu_1) = -3, \hat{\mathcal{S}}(\mu_2, \mu_2) = \hat{\mathcal{S}}(\mu_3, \mu_3) = \hat{\mathcal{S}}(\mu_4, \mu_4) = \hat{\mathcal{S}}(\mu_5, \mu_5) = -6. \quad (57)$$

And hence, the scalar curvature of $\hat{\nabla}$ is

$$\hat{r} = -27.$$

Now, using the values from (56) and (57), in $\hat{\mathcal{S}}(\xi, X_1).\hat{\mathcal{R}} = 0$, $\hat{\mathcal{R}}(\xi, X_1).\hat{P} = 0$, and $\hat{\mathcal{R}}(\xi, X_1).\hat{M} = 0$, we get the Theorems 4, 5 and 6.

References

Agashe NS (1992) A semi-symmetric non-metric connection on a Riemannian manifold. Indian J Pure Appl Math 23:399–409

Boothby WM, Wang HC (1958) On contact manifolds. Ann Math 68(2):721–734

Chaubey SK, Ojha RH (2008) On a semi-symmetric non-metric and quarter symmetric metric connexions, Tensor N.S. 70(2):202–203

Chaubey SK, Ojha RH (2010) On the m-projective curvature tensor of a Kenmotsu manifold. Differ Geom Dyn Syst 12:52–60

Chaubey SK, Ojha RH (2012) On a semi-symmetric non-metric connection. Filomat 26(2):269–275

Chaubey SK, Yildiz A (2019) Riemannian manifolds admitting a new type of semisymmetric nonmetric connection. Turkish J Mathem 43(4):1887–1904

Chaubey SK, Prakash S, Nivas S (2012) Some properties of M-projective curvature tensor in Kenmotsu manifolds. Bull Math Anal Appl 4(3):48–56

Hadimani BS, Prakasha DG (2016) Ricci Solitons on Para-Sasakian manifolds. IOSR J Mathem 12:39–43

Hamilton RS (1988) The Ricci flow on surfaces. American Mathematical Society, Contemporary Mathematics

Hyden A (1932) Sub-spaces of a space with torsion. Pro London Mathem Soc 2(1):27–50

Kenmotsu K (1972) A class of almost contact Riemannian manifolds. Tohoku Math J 2(2):93–103

Mandal A, Das A (2020) On M-projective curvature tensor of Sasakian manifolds admitting Zamkovoy connection. Adv Math Sci J 9(10):8929–8940

Matsumoto K, Mihai I, Shahid MH (1998) Certain submanifolds of a Kenmotsu manifold, Monogr Geom Topology Int. Press, Cambridge, MA, 25:183–193

Ojha RH (1975) A note on the M-projective curvature tensor. Indian J Pure Appl Math 8(12):1531–1534

Ojha RH (1986) M-projectively flat Sasakian manifolds. Indian J Pure Appl Math 17(4):481–484

Perelman G (2002) The entropy formula for the Ricci flow and its geometric applications, 1–39

Perelman G (2003) Ricci flow with surgery on three-manifolds, 1–22

Pokhariyal GP, Mishra RS (1971) Curvature tensors and their relativistic significance II, Yokohama Math J

Prakasha DG, Mirji K (2017) On the M-projective curvature tensor of a (κ, μ)- contact metric manifold. Ser Math Inform 32(1):117–128

Prasad B (2002) A pseudo projective curvature tensor on a Riemannian manifold. Bull Calcutta Math Soc 94(3):163–166

Sasaki S (1960) On differentiable manifolds with certain structures which are closely related to almost contact structure, I. Tohoku Math J 2(12):459–476

Sato I (1976) On a structure similar to the almost contact structure. Tensor(N.S) 30:219-224

Sengupta J, De UC, Binh T (2000) On a type of semi-symmetric non-metric connection on a Riemannian manifold. Indian J Pure Appl Mathem 31(12):1659–1670

Sundriyal S, Haseeb A (2024) Generalized Tanaka-Webster connection on β-Kenmotsu manifolds. Int J Maps Mathem 7(2):258–272

Sundriyal S, Upreti J (2021) On a type of semi-symmetric non-metric connection in HSU-unified structure manifold. Int Electron J Geometry 14(2):383–390

Yano K (1970) On semi-symmetric metric connections. Revue Roumaine de Mathematiques Pures et Appliquees 15:1579–1586

A Study on the Performance Analysis of Handoff Requirement in a Wireless Vehicular Communication Network for Urban Areas: Case of Artificial Neural Network (ANN)

Chandan Singh Ujarari, Shshank Chaube, Rakesh Chandra Bhadula, and Yaghvendra Kumawat

Abstract Wireless communication has become a necessity in our lives in recent times. Queuing theory is essential in designing communication protocols. One common challenge in identifying issues with wireless devices involves transferring signals from one base station to another suitable base station, especially in the context of vehicular wireless communication. To emphasize the importance of a mathematical approach in practical applications, we propose an innovative service handoff scheme that ensures seamless transitions between different service providers for Safe-as-a-Service (Safe-aaS) provisioning in road transportation. It is also crucial to recognize that a user's request for safety services may encompass distances that overlap with the service areas of multiple Service Support Providers (SSPs). In this context, we have analyzed a real-life scenario in the city of Dehradun, Uttarakhand, India. Our study specifically focused on the necessary handoff times at six different signal tower locations across various areas. Our findings provide insights into the efficient utilization of wireless communication systems in urban settings concerning handoff requirements. To enhance the quality of service (QoS), it is essential to integrate advanced technology into the systems. We conducted a thorough comparison of our study using artificial neural networks (ANNs). To validate our system model, we employed the Levenberg–Marquardt Backpropagation training algorithm within the ANN framework.

Keywords Wireless vehicular communication · QOS · Handoff · Safe-aaS · ANN

C. S. Ujarari (✉) · R. C. Bhadula
Department of Mathematics, Graphic Era Hill University, Dehradun, Uttarakhand, India
e-mail: ujarari0000@gmail.com

S. Chaube
Symbiosis Institute of Technology, Hyderabad Campus, Symbiosis International (Deemed University), Pune, India

Y. Kumawat
Department of Mathematics, Jaipur Engineering College and Research Centre, Jaipur, India

M. Ram et al. (eds.), *Advances in Mathematics for Engineering Sciences*, Mathematical Engineering, https://doi.org/10.1007/978-3-031-95693-5_9

1 Introduction

Wireless communication involves transmitting data or information between two or more points without using an electrical conductor. This is done using electromagnetic waves or signals, typically in the radio frequency (RF) spectrum, with air serving as the medium.

Basics Terms in Wireless Communication

(a) **Propagation of Radio Waves**– Propagation of radio waves is the tendency of radio waves to travel through Earth's atmosphere.
(b) **Modes of Propagation of Radio Waves:** Direct Propagation, Ground Wave Propagation, Sky Wave Propagation
(c) **Multiple Propagation:** Multiple propagation consists of two key effects that are critical to understanding signal behaviour. First, radio waves consistently reach the receiver from multiple directions, which impacts the quality of the signal. Second, "fading" occurs, representing the fluctuation in signal strength due to various influencing factors. In wireless systems, fading is primarily caused by multipath propagation and adverse weather conditions, especially rain. This understanding is essential for optimizing wireless communications.

Applications of Wireless Communication

In mobile communication we can categorize its applications are cell phones, smartphones and other mobile devices from which we make voice calls, send text messages and access the internet through cellular networks. The Wireless Networking involved with Wi-fi technology enables wireless networking and allows devices such as laptops, tablets and smart home devices to connect to internet and communicate with each other within a Local Area Network (LAN). We can see also its application in Bluetooth technology which enable short range wireless communication between devices such as connecting wireless headphones to a smart phone. The satellite communication is very important application which facilitate wireless communication over long distances enabling services such as satellite TV, GPS, navigation and satellite internet access while the Wireless Sensor Networks played also an important role in our life in which Sensors and actuators deployed in various environment and collect data, monitor conditions and control processes without the need for wired connections.

2 Review of Literature

Good wireless communication can affect the working scenario of any organization or country. In this context, upgrading is required day by day. A prioritization scheme for access control in cellular networks was first introduced in Bartolini (2001). To achieve an appropriate number of frequencies in each cell there is a competition between low-priority and high-priority users. The scheme given in Bartolini (2001) focuses on giving priority to handoff calls over initial calls to prevent the forced termination of call-in progress. The low-priority and high-priority users correspond to initial access requests and handoff requests respectively. In Huang et al. (2001), a new handoff scheme was proposed to improve the call completion probability of initial and handoff calls in mobile cellular systems. The authors observed in this study that the prioritized handoff calls may not optimize the better call completion probability. An overview of the existing handover schemes for current and future system networks was then presented in Bauer and Rees (2002). In this study, the authors classified the different schemes under two main classes: a vertical class and a horizontal class. The vertical class is based on the concept that the scheme adopts whereas the horizontal class works on how mobile hosts identify the new potential base stations.

In this continuation, Zeng and Agrawal (2002) presented two handoff schemes: one is with a priority procedure, and while the other one is without a priority procedure. These schemes are designed for integrated wireless mobile networks in which they categorized service calls into four classes: originating voice calls, originating data calls, data handoff request calls, and voice handoff request calls. Two separate queues were discussed in this study for the two presented handoff schemes. Leu and Mark (2002) and Soh and Kim (2003) introduced an innovative class of fast handoff algorithms and accurate positioning method. These studies involve a local averaging technique to effectively eliminate only the fast-fading component from the received signal strength and they discussed how the innovative features can be leveraged to enhance mobility predictions. A new admission control scheme called Dynamic Thermal Balance Routing (DTBR) was introduced by Li et al. (2004). The authors presented a detailed comparative study based on mathematical and simulation models, quantitatively demonstrating that DTBR can provide Quality of Service (QoS) guarantees for each traffic type, leading to improved efficiency. A new dynamic guard channel (OGC) scheme was introduced by Wei et al. (2004). This scheme combines a differentiated Quality of Service (QoS) model with a priority handoff mechanism. It provides fairness among different QoS services and adjusts to various traffic loads by assuming a stationary distribution for mean arrival times and adopting a passive feedback mechanism. Barcelo (2004) introduced a state-dependent rejection scheme (SRS) structure designed to analyze existing handoff schemes and facilitate the creation of new ones. In this article, Markov analysis of Sounding Reference Signals (SRS) is considered for drawing initial conclusions about the evaluation of handoff strategies. This model demonstrates the ability of SRS to adapt to various mobility and load scenarios. Chang and Leu (2004) proposed a method for improving the quality of service (QoS) in multimedia wireless systems based on the prioritization of handover requests and introduced an algorithm of Signal strength for multimedia communication (SSMC). The authors analyzed well their algorithm and then its performance was compared with other methods by simulation.

Although various handoff schemes existed at that time, some classifications and fundamental criteria for these schemes had yet to be established. In this way,

Diederich and Zitterbart (2005) proposed the basic functions of handoff prioritization scheme and described one proposal for a classification for handoff schemes. To reduce and avoid interruption after handoff, they gave importance to the sessions performing a handoff into a cell compared to newly arriving sessions in that cell. Iraqi and Baoutaba (2005) explained how the call dropping probability (CDP) is very important in contexts of QoS parameter. The CDP represents the probability of a hand off failure due to insufficient available resources in the target cell. Mehbodniya and Chitizadeh (2005) presented a new vertical handoff algorithm in multitier (overlay) networks. The most important issue of this study was to optimize vertical handoff process as it leads to reduction of network signaling and mobile device power. On the other hand, it improves network quality of service (QoS) and grade of service (GoS) it uses pattern reorganization to estimate the user's position. Wang et al. (2005) introduced a vertical handoff approach to analyse the handoff delay. The vertical handoff procedure was separated into two steps: pre-handoff and handoff. They also divided the vertical handoff system into decision and action modules. Lin and Lin (2005) presented a handoff ordering method constructed based on packet success rate (PSR) for multimedia communication in wireless networks. The priority of a handoff request is considered in the context of the current PSR, the PSR degradation rate, and the minimum PSR requirement of its service class. The method focused on improving the QoS and in reducing the new call blocking probability during the handoff process.

However, the strengths and weaknesses of the several handoff schemes have already been presented by the researchers in their study. In this context Ghaderi and Boutaba (2006) conducted a survey on admission control schemes for cellular networks, offering a detailed description of numerous admission control methods. In this article, the performances and the complexities of the admission control schemes are compared. Pati (2007) focussed his study on the prioritization of handoff calls to reserve the optimal number of guard channels at each base station and gave a distributive adaptive guard channel reservation (DAGCR) scheme in this context. The scheme was then evaluated for a wide range of offered loads and tested well in managing the blocking performance of the networks. To develop an advanced 5G network, Wang et al. (2014) proposed an innovative cellular network model that distinguishes between outdoor and indoor environments. In this article, they point out various challenges of the 4G networks like network congestion in highly populated regions. This perspective paves the way for addressing these challenges and improving connectivity in urban settings. Furthermore, Ujarari and Kumar (2024) conducted a priority-based analytical study of a handoff scheme for a wireless network operating on a specific channel. In this article, the authors examine how the quality of service (QoS) can be improved by addressing the issues of handoff failure probability in wireless networks. The stochastic geometry is then applied to determine the average user rate affected by handoff delays, Arshad et al. (2016). In this study, they developed a flexible handoff program based on single handoff skipping to deal with the handoff delays. Ni et al. (2021) proposed a perceptive mobile networks (PMNs) uplink sensing scheme to determine sensing uncertainty and to enhance sensing accuracy. Through this scheme, all measurements across the temporal, spatial, and frequency domains were collectively examined and processed. A hierarchical cooperative game model for networks was developed by Gao et al. (2021) to explore how device-to-device (D2D) pairs cooperate with one another. Roy and Misra (2021) presented an advanced Safe-Passe handoff scheme for provisioning Safety-as-a-Service (Safe-aaS) to the end-users in a 5G environment.

3 Model Description

In this model, we explore the handover system within cellular networks, particularly in a 5G-enabled Intelligent Transportation System (ITS) environment that incorporates a Safety-as-a-Service (Safe-aaS) infrastructure. In this Safe-aaS architecture, two types of sensor nodes are involved one is static, and the other one is mobile. To enhance vehicular communication the vehicles are to be employed with mobile sensor nodes while the static sensor nodes are to be deployed at identified locations. To functionalize this architecture there are four essential layers: the device, edge, decision, and decision virtualization. The data collection process is done by the node in the Device Layer and then data is transmitted to the edge layer or the cloud. The Edge Layer processes this data, which is then forwarded to the decision layer. Here, a synthesis of data from multiple sensors leads to the decision-making step (Fig. 1).

Finally, the system effectively matches the end user's safety service requests with the decisions generated in the decision virtualization layer. When an end-user seeks a safety service, the Service Switching Point (SSP) evaluates the distance associated with the requested service and calculates the potential number of handoff requests. This structured approach not only enhances safety but also optimizes communication and response times in critical situations. The requirement of handoff to the end user

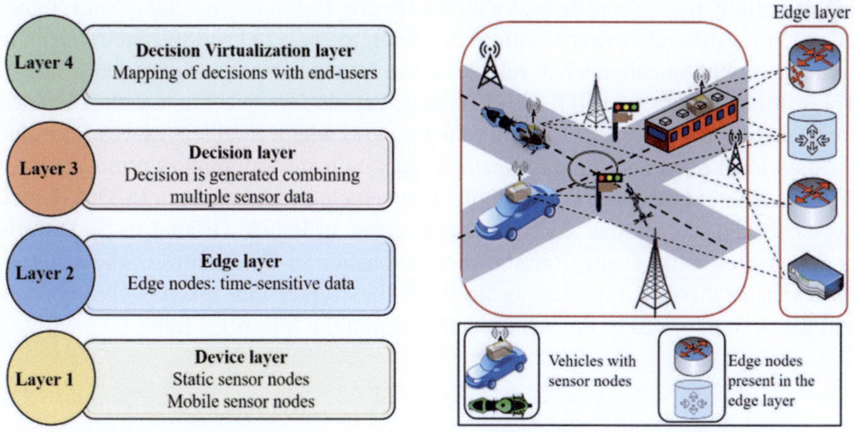

(a) Layers of Safe-aaS

(b) Dynamic selection of edge node

Fig. 1 DENSE: The System Archiecture

can be shown with a two-tuple (λ_r, λ_n), where λ_r and λ_n denotes whether handoff is required or not respectively. This requirement of handoff This requirement of handoff can be handled with the following Mathematical expression.

$$\lambda = \begin{cases} \lambda = \begin{cases} 1 & \textit{handoff possible} \\ 0 & \textit{handof not possible} \end{cases} \\ \lambda_n = \mathrm{x}, \quad \textit{handoff not required} \end{cases} \tag{1}$$

Therefore, the overall time required (T_j^{total}) by the jth end-user to reach his/her destination is given as:

$$T_j^{total} = \Delta t \left(\frac{\alpha v_j^t - (1-\alpha) v_j^{(t-1)}}{\Delta v_j^t} \right) \tag{2}$$

where $v_j^{(t-1)}$ denotes the jth end-user velocity at the instant $(t - 1)$. Further the change in velocity in time interval Δt is represented by Δv_j^t while α shows the handoff delaying parameter whose value can vary between 0 and 1. In case handoff is required to serve the jth end-users request, the time left to reach the destination of the jth end-user (i.e. required handoff time) is given by:

$$T_j^{left} = (T_j^{total} - T_{ji}) \tag{3}$$

4 Model Problem

In this section, the authors discuss a model for city Dehradun in Uttarakhand (India) in which six different tower locations are taken to analyze handoff requirements in vehicular communication when vehicles move from one location to another location. These locations are as ISBT, Clock Tower, Parade Ground, Sudhowala(Chakrata road), Rispana, Garhi Catt. Consider the scenario where a vehicle moves from one location to another, and the subsequent impact on handoff timing as it moves to yet another different location. For example: one vehicle is moving from ISBT to Clock Tower then it has another four choices to move to Parade Ground or Sudhowala or Rispana or Garhi Cantt from the current location of Clock Tower. The different measures for these six locations as distances, the average time taken by the vehicles, and the average speed of the vehicles are given in the following Table 1.

Table 1 Average Time taken by vehicles and their average speed

From	To	Distance	Averagetime (hour) taken by vehicles to cover distance	Average speed (Km/Hr)
Dehradun ISBT	Clock Tower	D_1=7.1 km	T_1= 17 min = 0.283 h	$V_1^{t-1} = 25.088$
Dehradun ISBT	Parade Ground	D_2= 8.6 km (NH 7)	T_2= 20 min = 0.333 h	$V_2^{t-1} = 25.825$
Dehradun ISBT	Sudhowala (Chakrata Road)	D_3=12 km (NH 7)	T_3= 28 min = 0.467 h	$V_3^{t-1} = 25.695$
Dehradun ISBT	Rispana	D_4=6.6 km (NH 7)	T_4=16.5 min = 0.275 h	$V_4^{t-1} = 24.000$
Dehradun ISBT	Garhi Cantt	D_5=10.7 km	T_5 = 25 min = 0.417 h	$V_5^{t-1} = 25.659$
Clock tower	Parade Ground	D_6= 4.7 km	T_6 = 13.5 min = 0.225 h	$V_1^{t1} = 20.889$
Clock tower	Rispana	D_7= 5.1	T_7= 13.2 min = 0.22 h	$V_1^{t2} = 23.182$
Clock tower	Sudhowala (Chakrata Road)	D_8= 11.3 km	T_8= 25.6 min = 0.427 h	$V_1^{t3} = 26.463$
Clock tower	Garhi Cantt	D_9=5.3 km	T_9= 14.2 min = 0.237 h	$V_1^{t4} = 22.746$
Parade ground	Rispana	D_{10}=4.8 km	T_{10}= 14 min = 0.233 h	$V_2^{t1} = 20.601$
Parade ground	Sudhowala (Chakrata Road)	D_{11}= 12.3	T_{11}= 28 min = 0.467 h	$V_2^{t2} = 26.338$
Parade ground	Garhi Cantt	D_{12}=4.6 km	T_{12}=13.5 min = 0.225 h	$V_2^{t3} = 20.444$
Sudhowala (Chakrata Road)	Rispana	D_{13}=16 km	T_{13}= 37 min = 0.617 h	$V_3^{t1} = 25.932$
Sudhowala (Chakrata Road)	Garhi Cantt	D_{15}=13.3 km	T_{15}= 31 min = 0.517 h	$V_3^{t2} = 25.725$
Rispana	Garhi Cantt	D_{14}=8.6 km	T_{14}= 22 min = 0.367 h	$V_4^{t1} = 23.43.3$

5 Result Analysis

In this section, we discuss the calculations of the total time required (T_j^{total}) by the jth end-user to reach its destination and the required handoff time to switch the base station to get a good quality of service in the context of wireless communication while moving between any three consecutive locations. The following Table 2 shows all the essential calculations.

Table 2 Calculations of total time required

Movement of a Vehicle	Time slots		Time interval Δt in hours	Speed of vehicle km/hr	$\Delta v_j^t = \lvert v_j^t - v_j^{(t-1)} \rvert$ km/hr	$T_j^{total} = \Delta t\left(\frac{\alpha v_j^t-(1-\alpha)v_j^{(t-1)}}{\Delta v_j^t}\right)$ in hour	$T_j^{left} = (T_j^{total} - T_{ji})$ in hour
	From	To					
From ISBT to Clock Tower	11:00 am	11:17 am	–	$V^{t-1} = V_1^{t-1} =$ 25.088	–	–	–
Clock Tower to Parade Ground	11:17 am	11:37 am	20 min = 0.333 h	$V^t = V_1^{t1} =$ 20.889	4.199	0.1665	0.1235
Clock Tower to Rispana	11:17 am	$11:30':12''$ am	13 min 12 s = 0.22 h	$V^t = V_1^{t2} =$ 23.182	1.906	0.1100	0.067
Clock Tower to Sudhowala	11:17 am	$11:30':30''$ am	13 min 30 s = 0.225 h	$V^t = V_1^{t3} =$ 26.463	1.375	0.1125	0.0695
ISBT to Parade Ground	11:00 am	11:20 am	–	$V^{t-1} = V_2^{t-1} =$ 25.825	–	–	–
Parade Ground to Rispana	11:20 am	11:34 am	14 min = 0.233 h	$V^t = V_2^{t1} =$ 20.601	5.224	0.1165	0.0755
Parade Ground to Sudhowala	11:20 am	11:48 am	28 min = 0.467 h	$V^t = V_2^{t2} =$ 26.338	0.513	0.2335	0.1925
Parade Ground to Garhi Cantt	11:20 am	$11:33':30''$ am	13 min 30 s = 0.225 h	$V^t = V_2^{t3} =$ 20.444	5.381	0.1125	0.0715

(continued)

Table 2 (continued)

Movement of a Vehicle	Time slots		Time interval Δt in hours	Speed of vehicle km/hr	$\Delta v_j^t = \lvert v_j^t - v_j^{(t-1)} \rvert$ km/hr	$T_j^{total} = \Delta t\left(\frac{\alpha v_j^t - (1-\alpha) v_j^{(t-1)}}{\Delta v_j^t}\right)$ in hour	$T_j^{left} = (T_j^{total} - T_{ji})$ in hour
	From	To					
ISBT to Sudhowala	11:00 am	11:28 am	–	$V^{t-1} = V_3^{t-1} =$ 25.695	–	–	–
Sudhowala to Rispana	11:28 am	12:05 pm	37 min = 0.616 h	$V^t = V_3^{t1} =$ 25.932	0.237	0.3080	0.2630
Sudhowala to Garhi Cantt	11:28 am	11:59 am	31 min = 0.516 h	$V^t = V_3^{t2} =$ 25.725	0.03	0.2580	0.213
ISBT to Rispana	11:00 am	11 : 16′ : 30″ am	–	$V^{t-1} = V_4^{t-1} =$ 24.000	–	–	–
Rispana to Garhi Cantt	11:16′:30″ am	11 : 38′ : 30″ am	28 min = 0.467 h	$V^t = V_4^{t1} =$ 23.433	0.567	0.2335	0.1945

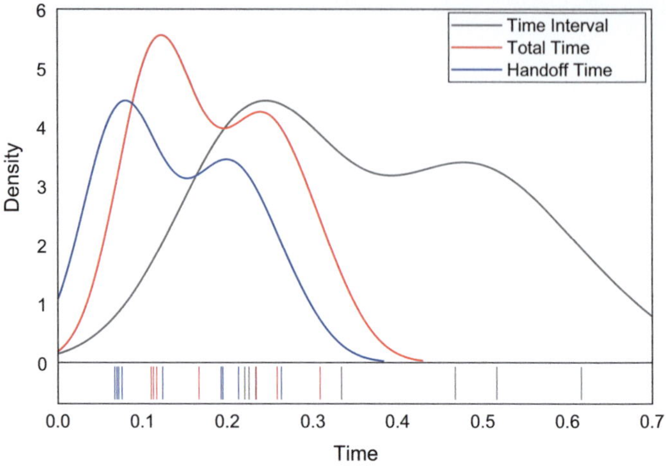

Fig. 2 Density for time interval, total time and handoff time versus time

The Fig. 2 represents how the time interval, total time required, and handoff time varies with the time continuously.

The Predictions of a vehicle's movement from one location to another can be analyzed using Artificial Neural Networks (ANN). Let's consider a scenario where a vehicle needs to travel from point A to point B, and then from point B to one of 20 possible destinations (point C) within a city. In this context, we can generate handoff predictions, which are illustrated in the Table 3.

We trained our data through ANN by taking time intervals and instant velocity as the random variables. The following different measurement graphs given in Figs. 3, 4, and 5, show the best fit of our trained data.

The Fig. 6 the variations in time interval, total time required and handoff time versus time through ANN analysis.

Table 3 Calculation of required handoff time

Time interval Δt in hours	Speed of vehicle km/hr (V^t)	Speed of vehicle km/hr (V^{t-1})	$\Delta v_j^t = \lvert v_j^t - v_j^{(t-1)} \rvert$ km/hr	T_j^{total}	Required Handoff time T_j^{left}
0.426952863	30.06206522	25.088	4.974065223	0.213476432	0.170476432
0.324170225	28.16814497	25.088	3.080144974	0.162085112	0.119085112
0.347703531	32.43540512	25.088	7.34740512	0.173851766	0.130851766
0.40842297	33.70564063	25.088	8.617640628	0.204211485	0.161211485
0.491820166	28.53356926	25.088	3.445569256	0.245910083	0.202910083
0.298326488	32.57945747	25.088	7.491457468	0.149163244	0.106163244
0.451340955	27.23986603	25.088	2.151866033	0.225670477	0.182670477
0.421721668	33.53050504	25.088	8.442505036	0.210860834	0.167860834
0.486252337	27.24741544	25.088	2.159415442	0.243126168	0.200126168
0.209576789	31.29388447	25.088	6.205884469	0.104788394	0.061788394
0.307060696	29.2954112	25.088	4.207411204	0.153530348	0.110530348
0.39879615	33.2608359	25.088	8.172835902	0.199398075	0.156398075
0.284450468	30.53402707	25.088	5.446027072	0.142225234	0.099225234
0.26911492	30.40634905	25.088	5.318349049	0.13455746	0.09155746
0.413338565	33.89343851	25.088	8.80543851	0.206669283	0.163669283
0.387371875	29.17827632	25.088	4.090276316	0.193685938	0.150685938
0.377182596	30.04499201	25.088	4.956992009	0.188591298	0.145591298
0.39813139	34.67177258	25.088	9.583772582	0.199065695	0.156065695
0.214266402	26.38068018	25.088	1.292680181	0.107133201	0.064133201
0.304635443	34.75662501	25.088	9.668625007	0.152317721	0.109317721

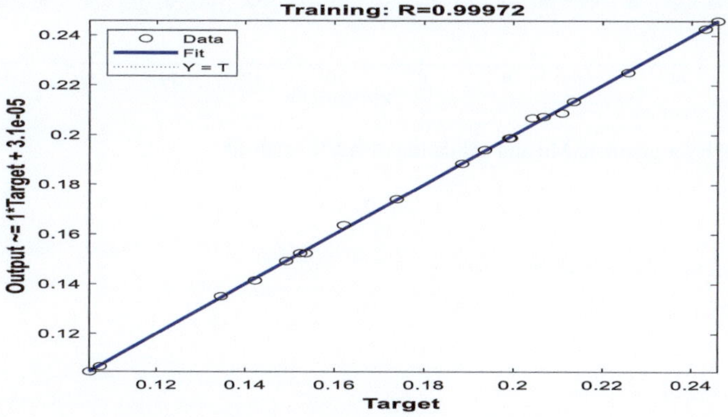

Fig. 3 Graph of regression coefficient of ANN analysis

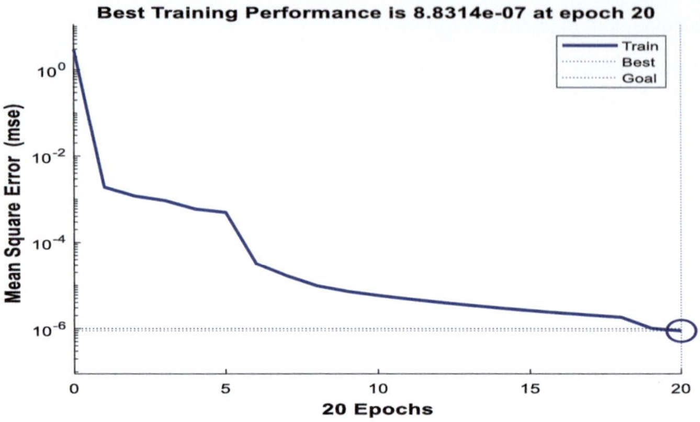

Fig. 4 Graph for training performance of ANN analysis

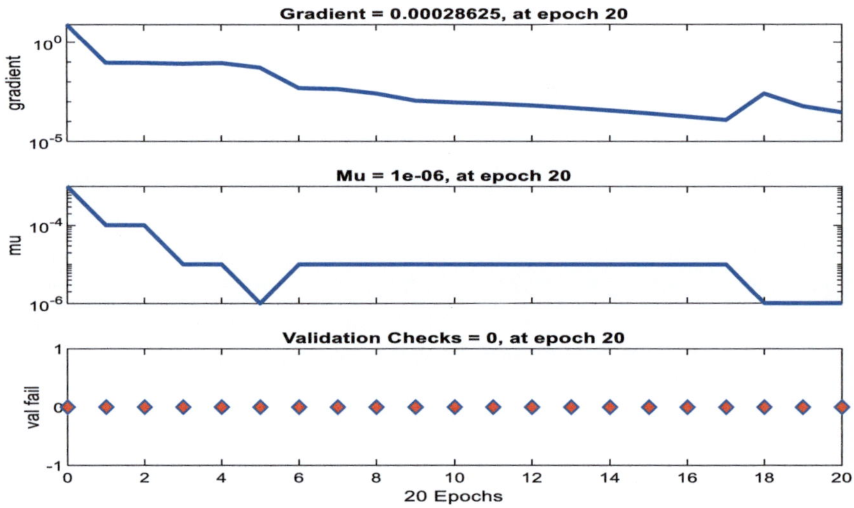

Fig. 5 Graph for gradient, Mu and validation of ANN Analysis

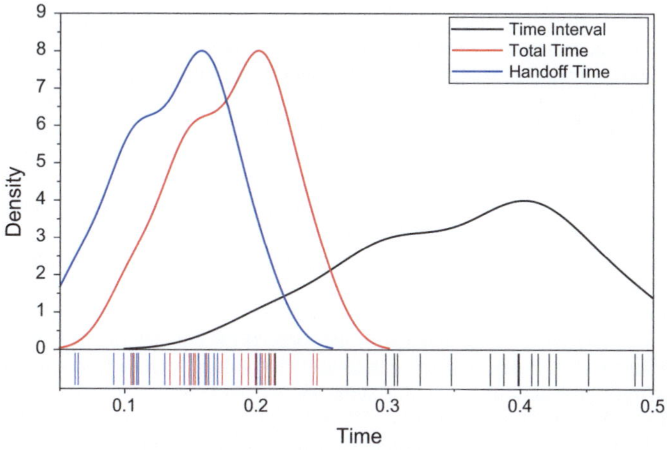

Fig. 6 Density for time interval, total time and handoff time versus time of ANN analysis

6 Conclusion

The Safe-aaS infrastructure is designed with the end user in mind, aiming to provide tailored safety-related decisions that significantly enhance their experience. In the context of a 5G-enabled Intelligent Transport System (ITS), the integration of multiple Service Switching Points (SSPs) can facilitate the delivery of comprehensive services. Although the current range of these SSPs is limited, this presents an opportunity for innovation in addressing real-world challenges within cellular wireless communication systems. By focusing on a single end-user as the starting point, we can analyze various locations at different distances and routes. Utilizing tools like Google Maps allows us to effectively estimate the time and distance between these points, thereby calculating the total travel time across the routes. Additionally, we can pinpoint critical locations where network connectivity may be compromised which is essential for proactive problem solving. The results derived from both analytical methods and ANN (Artificial Neural Network) analysis demonstrate a strong alignment, validating our system model's performance. This cohesive approach not only identifies potential issues but also sets the foundation for developing enhanced service strategies in the future, ultimately leading to a more robust and user-centric experience.

References

Arshad R, ElSawy H, Sorour S, Al-Naffouri TY, Alouini MS (2016) Handover management in dense cellular networks: a stochastic geometry approach. In: 2016 IEEE international conference on communications (icc), pp 1–7. IEEE

Barceló F (2004) Performance analysis of handoff resource allocation strategies through the state-dependent rejection scheme. IEEE Trans Wireless Commun 3(3):900–909

Bartolini N (2001) Handoff and optimal channel assignment in wireless networks. Mobile Networks Appl 6:511–524

Bauer CI, Rees SJ (2002) Classification of handover schemes within a cellular environment. In: The 13th IEEE international symposium on personal, indoor and mobile radio communications (Vol 5, pp 2199–2203). IEEE

Chang RS, Leu SJ (2004) Handoff ordering using signal strength for multimedia communications in wireless networks. IEEE Trans Wireless Commun 3(5):1526–1532

Diederich J, Zitterbart M (2005) Handoff prioritization schemes using early blocking. IEEE Commun Surveys Tutorials 7(2):26–45

Gao X, Niyato D, Yang K, An J (2021) Cooperative scheme for backscatter-aided passive relay communications in wireless-powered D2D networks. IEEE Internet Things J 9(1):152–164

Ghaderi M, Boutaba R (2006) Call admission control in mobile cellular networks: a comprehensive survey. Wirel Commun Mob Comput 6(1):69–93

Huang CJ, Lai WK, Wang CC, Jin YJ, Chen HW (2001) A ratioed channel assignment scheme for initial and handoff calls in mobile cellular systems. Comput Commun 24(3–4):308–318

Iraqi Y, Baoutaba R (2005) Handoff and call dropping probabilities in wireless cellular networks. In: 2005 International conference on wireless networks, communications and mobile computing (Vol 1, pp 209–213). IEEE

Leu AE, Mark BL (2002) Modeling and analysis of fast handoff algorithms for microcellular networks. In: Proceedings. 10th IEEE international symposium on modeling, analysis and simulation of computer and telecommunications systems, pp 321–328. IEEE

Li B, Li L, Li B, Sivalingam KM, Cao XR (2004) Call admission control for voice/data integrated cellular networks: performance analysis and comparative study. IEEE J Sel Areas Commun 22(4):706–718

Lin TN, Lin PC (2005) Handoff ordering using link quality estimator for multimedia communications in wireless networks. In: GLOBECOM'05. IEEE Global telecommunications conference, 2005. (Vol 2, pp 6-pp). IEEE

Mehbodniya A, Chitizadeh J (2005) An intelligent vertical handoff algorithm for next generation wireless networks. In: Second IFIP international conference on wireless and optical communications networks, 2005. WOCN 2005, pp 244–249. IEEE

Ni Z, Zhang JA, Huang X, Yang K, Yuan J (2021) Uplink sensing in perceptive mobile networks with asynchronous transceivers. IEEE Trans Signal Process 69:1287–1300

Pati HK (2007) A distributed adaptive guard channel reservation scheme for cellular networks. Int J Commun Syst 20(9):1037–1058

Roy C, Misra S (2021) Safe-Passé: Dynamic handoff scheme for provisioning safety-as-a-service in 5G-enabled intelligent transportation system. IEEE Trans Intell Transp Syst 22(8):5415–5425

Soh WS, Kim HS (2003) QoS provisioning in cellular networks based on mobility prediction techniques. IEEE Commun Mag 41(1):86–92

Ujarari CS, Bhadula RC, Chaube S, Mishra RK (2024) Performance analysis of a queuing dynamics model in a supermarket to optimize the efficiency: case of Monte Carlo simulation. In: 2024 International conference on electrical electronics and computing technologies (ICEECT) (Vol 1, pp 1–6). IEEE

Wang CX, Haider F, Gao X, You XH, Yang Y, Yuan D, Hepsaydir E (2014) Cellular architecture and key technologies for 5G wireless communication networks. IEEE Commun Mag 52(2):122–130

Wang Y, Ho PH, Shen S, Li S, Naik S (2005) Modularized two-step vertical handoff scheme in integrated WWAN and WLAN. In: HPSR. 2005 workshop on high performance switching and routing, pp 520–524. IEEE

Wei Y, Lin C, Ren F, Raad R, Dutkiewicz E (2004) Dynamic handoff scheme in differentiated QoS wireless multimedia networks. Comput Commun 27(10):1001–1011

Zeng QA, Agrawal DP (2002) Modeling and efficient handling of handoffs in integrated wireless mobile networks. IEEE Trans Veh Technol 51(6):1469–1478

Mathematical Insights on Influenza Virus: Unveiling Dynamics and Informing Control Strategies

Pooja Khurana and Deepak Kumar

Abstract Influenza virus continues to pose a significant global health threat, causing seasonal epidemics and occasional pandemics. Mathematical modeling is vital to understand the dynamics of influenza virus transmission and shaping effective control strategies. This chapter provides an overview of mathematical insights into influenza virus dynamics, including compartmental models, evolutionary models, and agent-based models. These mathematical approaches allow researchers to simulate the spread of influenza within populations, assess the impact of interventions, and predict the emergence of novel strains. Challenges in model validation, data integration, and interdisciplinary collaboration are discussed, along with future directions for advancing the field. By integrating mathematical modeling with epidemiological data and molecular biology techniques, researchers can gain a deeper understanding of influenza virus transmission dynamics and develop more effective strategies for prevention, control, and pandemic preparedness. This chapter underscores the importance of mathematical insights in unraveling the complexities of influenza virus dynamics and guiding evidence-based public health interventions.

Keywords Influenza disease · Mathematical model · Applied mathematics

1 Introduction

Influenza, a contagious respiratory disease caused by influenza viruses, continues to pose significant public health challenges worldwide. The dynamic nature of influenza viruses, characterized by frequent antigenic changes and the emergence of novel strains, necessitates a multifaceted approach for understanding virus dynamics, predicting outbreaks, and developing effective control strategies (Shao et al. 2017). In

P. Khurana · D. Kumar (✉)
Department of Applied Sciences, Manav Rachna International Institute of Research and Studies, Faridabad 121004, India
e-mail: deepakman12@gmail.com

P. Khurana
e-mail: pooja.set@mriu.edu.in

M. Ram et al. (eds.), *Advances in Mathematics for Engineering Sciences*, Mathematical Engineering, https://doi.org/10.1007/978-3-031-95693-5_10

recent times, mathematical modeling has become an effective method for clarifying the intricate dynamics of influenza virus transmission, development, and intervention (Bedford et al. 2024; Zitzmann et al. 2018; Ewald et al. 2020).

This chapter seeks to deliver an extensive summary of the diverse array of mathematical models and computational techniques that have been employed to study influenza virus dynamics. By integrating mathematical modeling with epidemiological data, genomic sequencing, and clinical observations, researchers have gained valuable insights into the mechanisms driving influenza virus spread, the impact of vaccination and antiviral interventions, and the emergence of drug-resistant strains (Chen-Charpentier 2024).

Firstly, the etiology and pathophysiology of influenza virus has been discussed followed by exploring compartmental models, which offer a simplified yet informative framework for representing the transmission dynamics of influenza viruses within populations. These models divide the population into compartments based on infection status (susceptible, exposed, infected, recovered) and incorporate parameters governing transmission, immunity, and intervention measures.

Next, the agent-based models simulating the interactions between individual hosts and viruses within a spatially structured environment have been addressed. These models capture heterogeneity in contact patterns, transmission routes, and immune responses, allowing researchers to explore the effects of population demographics, social mixing, and spatial dynamics on influenza virus spread.

In addition to exploring transmission dynamics, this chapter examined evolutionary models that elucidate the processes driving the genetic diversity and antigenic evolution of influenza viruses. These models integrate genetic and antigenic data with epidemiological information to infer viral phylogenies, predict the emergence of novel strains, and assess the effectiveness of vaccination strategies.

Finally, the challenges and future directions in the field of mathematical modeling of influenza virus has been discussed, including the need for real-time data integration, model validation, and collaboration between mathematical modelers, epidemiologists, virologists, and public health officials.

Overall, this chapter highlights the critical role of mathematical modeling in advancing our understanding of influenza virus dynamics and informing evidence-based strategies for prevention, control, and preparedness in the face of seasonal epidemics and pandemic threats.

1.1 Introduction to Influenza Virus

Influenza virus, often referred to simply as "the flu," is an extremely transmissible respiratory virus that that triggers regular seasonal epidemics and can occasionally lead to global pandemics, presenting a significant global health challenge. Influenza viruses, which are part of the Orthomyxoviridae family, are divided into types A, B, and C, with type A being the primary cause of seasonal flu outbreaks in humans (Forrest and Webster 2010). The virus is recognized for its swift evolution via antigenic drift and shift, resulting in the development of new strains capable of evading existing immunity. Influenza virus spreads easily through respiratory droplets, direct contact, and contaminated surfaces, leading to a wide range of symptoms from mild respiratory illness to severe complications and death, particularly among high-risk populations (Moghadami 2017). A thorough understanding of the influenza virus's biology, transmission dynamics, and evolutionary patterns is critical for designing effective prevention and control strategies. This knowledge is essential for creating vaccination programs, developing antiviral treatments, and implementing public health measures to decrease the incidence and impact of influenza-related illnesses and fatalities.

1.2 Etiology and Risk Factors

The etiology of influenza virus is attributed to several strains, primarily belonging to the influenza A, B, and C viruses. Influenza A viruses are further classified based on surface glycoproteins hemagglutinin (HA) and neuraminidase (NA), contributing to the virus's antigenic diversity (Bouvier and Palese 2008). The virus is transmitted through respiratory droplets, inhalation of aerosols, or direct contact with contaminated surfaces. While influenza virus infection can affect individuals of all ages, certain populations are at higher risk, including young children, elderly adults, pregnant women, individuals with underlying medical conditions, and immunocompromised individuals. Environmental factors like overcrowded living conditions and inadequate ventilation can elevate the risk of influenza transmission (Huang et al. 2019). A thorough review of these elements offers valuable insights into the multifaceted nature of influenza virus infection and informs evidence-based approaches for prevention and management.

1.3 Pathophysiology

The pathophysiology of influenza virus infection involves a complex interplay between the virus, the host immune response, and various physiological systems. Upon exposure to the virus, typically through inhalation of respiratory droplets containing infectious viral particles, influenza virus targets epithelial cells lining the respiratory tract. The virus enters host cells by attaching to sialic acid receptors on the cell's exterior, subsequently leading to internalization and replication inside the host cell. Once inside the host cell, influenza virus hijacks cellular machinery to produce viral proteins and genetic material, leading to the generation of new viral particles. This mechanism harms the infected cells and activates an inflammatory reaction, leading to the secretion of cytokines and chemokines that draw immune cells to the infection location. The host immune response plays a critical role in combating influenza virus infection. Innate immune cells such as macrophages, dendritic cells, and natural killer cells recognize viral antigens and mount an initial defense by producing antiviral cytokines and initiating the adaptive immune response. Adaptive immunity, mediated by B cells and T cells, plays a crucial role in clearing the virus and providing long-term protection against reinfection. B cells produce antibodies that neutralize the virus and facilitate its clearance, while T cells recognize and eliminate virus-infected cells. However, the immune response to influenza virus can also contribute to disease pathology. Excessive inflammation and cytokine release, known as a cytokine storm, can cause tissue damage and contribute to the development of severe symptoms such as fever, cough, sore throat, muscle aches, and respiratory distress. In certain situations, especially among high-risk individuals, the immune response can become dysfunctional, resulting in issues like pneumonia, acute respiratory distress syndrome (ARDS), and multiple organ failure. In addition to the direct effects of viral replication and immune response, influenza virus infection can also increase susceptibility to secondary bacterial infections, further exacerbating disease severity. The disruption of the respiratory epithelium and impaired mucociliary clearance provide an ideal environment for bacterial colonization and secondary infections (Kalil and Thomas 2019; Taubenberger and Morens 2008; Fukuyama and Kawaoka 2011).

2 Mathematical Models

Mathematical models play a critical role in understanding the complex dynamics of influenza virus transmission, evolution, and immune responses. By simulating how the virus spreads within populations, these models help unravel the factors driving seasonal outbreaks and pandemics. They provide important perspectives on the efficiency of public health measures, including vaccination initiatives, antiviral therapies, and social distancing protocols. Additionally, models can predict the impact of viral mutations and the potential emergence of drug-resistant strains, enabling better preparedness and targeted responses. Table 1 captures the key features, applications, and limitations of each model type as they pertain to influenza virus research, with a focus on their utility in predicting, controlling, and understanding influenza transmission.

Table 1 Overview of mathematical models for influenza virus research: features, applications, and limitations

Model type	Key features	Applications	Limitations
SIR model	Simplest model (Susceptible-Infectious-Recovered), without an exposed compartment	Basic disease dynamics understanding, theoretical studies	Lacks realism for diseases with incubation periods, such as influenza
SEIR model	Adds exposed (E) compartment to SIR model to account for incubation period	Influenza transmission dynamics, particularly incubation and latent period modeling	Assumes fixed progression rates, may oversimplify complex disease progression
SEIRS model	Extends SEIR by allowing loss of immunity (recovered individuals return to susceptible pool)	Recurrent epidemic and seasonal influenza patterns simulation	Requires assumptions about immunity loss rates, limited by difficulty in modeling real-world immune response variability
SIRS model	Permanent immunity after recovery (recovered to susceptible transition is excluded)	Endemic influenza modeling	Simplifies the immune response, ignores immune waning or vaccination impacts
Agent-based models (ABMs)	Models' individuals (agents) with heterogeneous behaviors and immune responses	Household transmission, antiviral strategies, behavioral interventions, zoonotic transmission	Computationally intensive, requires detailed agent-level data, stochastic nature increases variability in predictions
Individual-based models (household transmission)	Agents simulate interactions within families, workplaces, etc.	Models influenza spread within households and communities	Requires detailed demographic and contact pattern data, outcomes highly sensitive to assumptions

(continued)

Table 1 (continued)

Model type	Key features	Applications	Limitations
Agent-based model of antiviral treatment	Simulates drug distribution and use within a population	Studying antiviral treatment efficacy, drug-resistant strain emergence	Complex pharmacokinetic modeling required, sensitivity to treatment adherence and distribution assumptions
Agent-based model of behavioral interventions	Explores social distancing, mask-wearing, hygiene practices	Evaluating public health strategies during influenza outbreaks	Difficult to model individual behavior changes accurately, reliance on behavioral data
Evolutionary models	Analyze genetic changes and evolution of viral strains over time	Predicts antigenic evolution and vaccine strain selection	Requires extensive genetic data, limited by assumptions of genetic drift and mutation rates
Phylogenetic models	Reconstruct evolutionary relationships between viral strains	Traces viral lineage and mutation history, guides vaccine strain updates	Phylogenetic trees depend on available sequence data, uncertainties in inferring ancestral relationships
Phylogeographic models	Combine phylogenetic and spatial data to model viral transmission between regions	Tracks geographic spread of influenza virus strains	Sensitive to geographic sampling biases, migration rate assumptions may not reflect true viral movement patterns
Phylodynamic models	Integrates phylogenetic data with epidemiological trends	Estimates transmission rates, effective population size, and epidemic dynamics	Complex modeling, requires integration of high-resolution genetic and epidemiological data, high computational demands

2.1 Compartmental Models

They are mathematical frameworks that divide a population into distinct compartments based on their infection status and disease progression. These models provide a structured approach to understanding the dynamics of influenza transmission within a population and the impact of interventions such as vaccination, antiviral treatment, and social distancing measures.

SIR Model

The simplest compartmental model for infectious diseases is the SIR model, which does not include an exposed compartment. This model assumes that individuals move directly from the susceptible to the infectious compartment upon infection. While less realistic for diseases with an incubation period like influenza, the SIR model can still provide valuable insights into disease dynamics (Osthus et al. 2017). The equations for the SIR model are:

$$\frac{dS}{dt} = -\beta S.I$$

$$\frac{dI}{dt} = \beta S.I - \gamma I$$

$$\frac{dR}{dt} = \gamma I$$

where, $S(t), I(t)$ and $R(t)$ represent the number or proportion of individuals in the susceptible, infectious, and recovered compartments at time t, respectively. β represents the rate of transmission, indicating the likelihood of transmission per contact between a susceptible and an infectious individual. γ represents the rate of recovery, depicting the rate at which infected people heal and move to the recovered group.

SEIR Model

This model is used to simulate the transmission dynamics of infectious diseases, including influenza virus (Etbaigha et al. 2018). In this model, individuals in the population are divided into four compartments based on their infection status: Susceptible (S), Exposed (E), Infectious (I), and Recovered (R). The dynamics of the SEIR model can be described by the following set of ordinary differential equations (ODEs):

$$\frac{dS}{dt} = -\beta S.I$$

$$\frac{dE}{dt} = \beta S.I - \sigma E$$

$$\frac{dI}{dt} = \sigma E - \gamma I$$

$$\frac{dR}{dt} = \gamma I$$

where, $S(t)$ denotes the population of susceptible individuals at time t.

$E(t)$ indicates the number of individuals who have been exposed to the infection at time t but are not yet infectious. $I(t)$ denotes the count of individuals who are currently infectious at time t. $R(t)$ reflects the number of individuals who have recovered from the infection at time t. β is the transmission rate, representing the rate at which susceptible individuals become exposed through contact with infectious individuals. σ indicates the speed at which individuals who are exposed become infectious, γ denotes the rate of recovery, indicating how quickly infectious individuals recover and gain immunity.

SEIRS Model

The SEIRS model extends the SEIR model by including a section for those who have recovered from the infection but gradually lose immunity and return the susceptible group. This model allows for the simulation of recurrent epidemics and seasonal patterns of influenza transmission (Bjornstad et al. 2020). The equations for the SEIRS model are similar to the SEIR model, with an additional term representing the loss of immunity:

$$\frac{dS}{dt} = -\beta S.I + \omega R$$

$$\frac{dE}{dt} = \beta S.I - \sigma E$$

$$\frac{dI}{dt} = \sigma E - \gamma I$$

$$\frac{dR}{dt} = \gamma I - \omega R$$

where, ω represents the rate of loss of immunity.

SIRS Model

The SIRS model further simplifies the SEIRS model by assuming that individuals who recover from the infection gain permanent immunity and do not return to the susceptible pool. This model is often used to study endemic diseases where individuals acquire natural immunity after infection. The equations for the SIRS model are:

$$\frac{dS}{dt} = -\beta S.I + \gamma R$$

$$\frac{dI}{dt} = \beta S.I - \gamma I$$

$$\frac{dR}{dt} = \gamma I - \gamma R$$

2.2 *Agent Based Models*

These models interactions between individual hosts and viruses within a spatially organized environment. These models represent each individual as an agent with specific attributes and behaviors, allowing for heterogeneity in contact patterns, transmission routes, and immune responses (Krauland et al. 2023; Arduin et al. 2017; Roche et al. 2011). Equations in agent-based models are typically stochastic and capture interactions between agents.

Individual-Based Model of Immune Response Heterogeneity

In this model, individuals are represented as agents with heterogeneous immune responses to influenza virus infection. The model incorporates variations in immune status, including differences in pre-existing immunity, vaccination history, and immune function (Sambaturu et al. 2018). Equations can describe the dynamics of immune responses, such as antibody production, T cell activation, and cytokine release, in response to infection or vaccination. These equations may include parameters representing immune cell kinetics, antigenic stimulation, and immune memory formation (Zarnitsyna et al. 2016).

Individual-Based Model of Household Transmission

In this model, each individual in a population is represented as an agent with attributes such as age, location, and infection status. Agents interact within households, workplaces, schools, and other community settings, with transmission occurring through close contact. The model incorporates parameters such as household size, contact rates, and susceptibility to infection to simulate the spread of influenza within households and communities (Tsang et al. 2023, 2016; Rolfes et al. 2023).

Agent-Based Model of Antiviral Treatment Strategies

This model replicates the use of antiviral medications for both the treatment and prevention of influenza virus infections in a community. Agents may receive antiviral treatment based on predefined criteria such as symptoms, risk factors, or contact with infectious individuals. Equations can describe the pharmacokinetics and pharmacodynamics of antiviral drugs, including drug absorption, distribution, metabolism, and excretion. The model tracks the distribution of antiviral drugs, adherence to treatment regimens, and the emergence of drug-resistant strains over time (Du et al. 2020).

Agent-Based Model of Behavioural Interventions

This model examines how behavioural interventions like social distancing, mask usage, and hygiene measures affect the spread of the influenza virus in communities. Agents adjust their behaviours in response to perceived infection risk, public health messages, and social norms. Equations can describe individual decision-making processes, including risk perception, information processing, and social influence.

The model evaluates the effectiveness of different intervention strategies in reducing transmission and mitigating the impact of influenza outbreaks (Valle et al. 2013; Ge et al. 2011). Several studies have proposed enhancements to the traditional SIR model by introducing additional compartments to better represent the complexities of disease dynamics. Roche et al. (2011) added additional classes to better represent realistic agent states and incorporating a stochastic approach using a normal distribution for transitions. They categorized into nine classes: Susceptible (S), In Contact (C), Exposed (E), Infectious (I), Quarantined (Q), Not Quarantined (NQ), Dead (D), Recovered (R), and Immunized (M). Initially, all agents are Susceptible, and they may move into the In Contact class if exposed to an infectious agent. Those infected transition to the Exposed class during the incubation period, then to the Infectious class once they become contagious. Depending on their interaction with healthcare systems, infected agents may be quarantined or not. The model also accounts for eventual outcomes, including death, recovery, or immunization.

Agent-Based Model of Zoonotic Transmission

This model aims to simulate the transmission of influenza viruses from animal hosts to humans, including avian influenza viruses or swine influenza viruses. Agents represent animals, humans, and environmental factors, with interactions occurring through animal-human contact, environmental contamination, and interspecies transmission events. Equations can describe the dynamics of viral replication, mutation, and adaptation within animal hosts, as well as the likelihood of zoonotic transmission to humans and subsequent human-to-human spread (Hackl and Dubernet 2019; Laskowski et al. 2011).

Tabasi et al., modelled the agent-based model wherein he incorporated the biological ageing (senescence) of the sand flies. They defined the "age-specific mortality rates" for both the lavaes and adults of the sand flies, using agent-based model (ABM), with the probability of death increasing as sand flies age. Following Styer et al. a logistic mortality model is used:

$$DMR_{age} = \frac{\rho \times e^{age \times \sigma}}{1 + \frac{\rho \times \varepsilon}{\sigma}\left(e^{age \times \sigma} - 1\right)}$$

The model calculates the daily mortality rate (DMR) of sand flies, ρ represents factoring in the baseline DMR, σ is an aging component that increases exponentially with age, and ε signifies the degree of mortality deceleration (Tabasi et al. 2020; Styer et al. 2007).

2.3 Evolutionary Models

Evolutionary models elucidate the processes driving the genetic diversity and antigenic evolution of influenza viruses. These models integrate genetic and antigenic data with epidemiological information to infer viral phylogenies, predict the emergence of novel strains, and assess the effectiveness of vaccination strategies (Morris et al. 2018). For example, the Wright-Fisher model simulates the genetic drift of viral populations over time and can be described by equations governing allele frequencies and genetic diversity (Wright 1945).

Phylogenetic Model

Phylogenetic models reconstruct the evolutionary relationships between influenza virus strains based on genetic sequence data (Lam et al. 2013). These models use methods such as maximum likelihood or Bayesian inference to estimate phylogenetic trees that represent the ancestral relationships between viral lineages (Liu et al. 2022, Hayati et al. 2020). Equations for phylogenetic models typically involve likelihood calculations to evaluate the fit of different evolutionary models to the observed sequence data. For example, the likelihood function for a simple nucleotide substitution model may be represented as:

$$L(\theta) = \prod_i P(y_i|\theta)$$

where, $L(\theta)$ denotes the likelihood of the data conditioned on the model parameters (θ)

y_i represents the observed sequence data for each viral lineage i

$P(y_i|\theta)$ indicates the probability of observing the sequence data yiy_iyi given the specified model parameters (θ).

Phylogeographic Model

Phylogeographic models combine phylogenetic information with spatial data to infer the spatiotemporal dynamics of influenza virus transmission. These models use methods such as continuous-time Markov chains or structured coalescent processes to estimate the rates of viral migration between different geographic regions (Dellicour et al. 2019). Equations for phylogeographic models may involve transition rate matrices to model the movement of viral lineages between spatial compartments, with parameters representing migration rates and spatial distances between locations.

Multiple rasters were generated by transforming the original cell values for each environmental factor, using the formula: $v_t = 1+k\left(\frac{v_0}{v_{max}}()\right)$ where v_t the transformed cell value, and v_0 is the original value, and $v_{\max}$ is the maximum cell value in the raster. k is the scaling parameter that was used to adjust and test varying levels of conductance or resistance, with a minimum value of "1." Three different scaling values, $k = 10, 100$ and 1000 were tested for each of the three environmental factors.

Phylodynamic Model

Phylodynamic models integrate phylogenetic information with epidemiological data to infer the evolutionary and epidemiological dynamics of influenza virus transmission. These models typically combine phylogenetic trees with demographic and epidemiological parameters to estimate transmission rates, population sizes, and other epidemiological parameters (Ratmann et al. 2012). Equations for phylodynamic models may involve stochastic branching processes to simulate the co-evolution of viral lineages and host populations over time. For example, the birth–death skyline model represents changes in effective population size over time as a piecewise constant function, with equations describing birth and death rates in each time interval. They utilized a simulation-based method known as "Approximate Bayesian Computation" to conduct phylodynamic inference and evaluate the "goodness-of-fit" for intricate models. The aim was to assess the posterior distribution of epidemiological and evolutionary parameters θ by estimating the likelihood $l\left(\frac{x}{\theta}\right)$ of observed incidence in populations and phylogenetic data x.

The use of prior distributions $\pi(\theta)$ enabled the integration of existing knowledge or constraints on the range of possible model parameters

$$\pi\left(\frac{\theta}{x}\right) \propto l\left(\frac{x}{\theta}\right)\pi(\theta)$$

Antigenic Cartography Models

Antigenic cartography models map antigenic relationships between influenza virus strains based on serological data from neutralization assays. These models use multidimensional scaling techniques to visualize the antigenic landscape of influenza viruses in low-dimensional space, allowing researchers to infer antigenic distances between viral strains (Anderson et al. 2018). Equations for antigenic cartography models may involve distance-based metrics to quantify the antigenic similarity between strains, such as Euclidean distance (*ED*) or Mahalanobis distance.

$$ED_{i,y} = \frac{Number\ of\ a\min o\ acid\ changes\ in\ antigenic\ site}{Total\ number\ of\ a\min o\ acid\ in\ antigenic\ site} \times 20$$

$$AD_{i,y} = \frac{ED_{sa} + ED_{sb1} + ED_{ca1} + ED_{ca2} + ED_{sb2}}{5}$$

wherein, i and y denote the HA sequences from viral strain of Influenza ED_{xx} corresponds to the antigenic sites: $Sa,\ Sb_1,\ Ca_1,\ Ca_2\ and\ Sb_2$.

3 Graphical Results and Finding

Graphical Representation involves analysing numerical data through visual means. It represents the connection among data, ideas, information, and concepts through a figure. It is simple to comprehend and is a learning strategy

$$\frac{dS}{dt} = -\beta S.I$$

$$\frac{dI}{dt} = \beta S.I - \gamma I$$

$$\frac{dR}{dt} = \gamma I$$

SIR Model

The Figs. 1 and 2 shows a decline in the susceptible class over time, while the removal class increases due to timely treatment. Swine flu (Fig. 1) and covid-19 (Fig. 2), a respiratory illness in pigs caused by type A influenza viruses, often leads to outbreaks. The viruses that typically spread among pigs are called swine influenza viruses. Initially, the infected class rises, but it eventually approaches zero following treatment (Table 2).

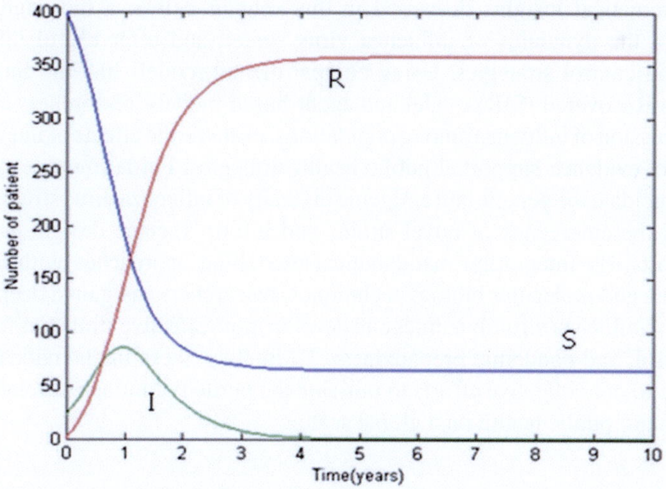

Fig. 1 In this graphical representation, S (susceptible individuals) is shown in blue, I (infected individuals) is represented in green, and R (recovered individuals) is depicted in red, each illustrating their respective changes over time (Kumar et al. 2016)

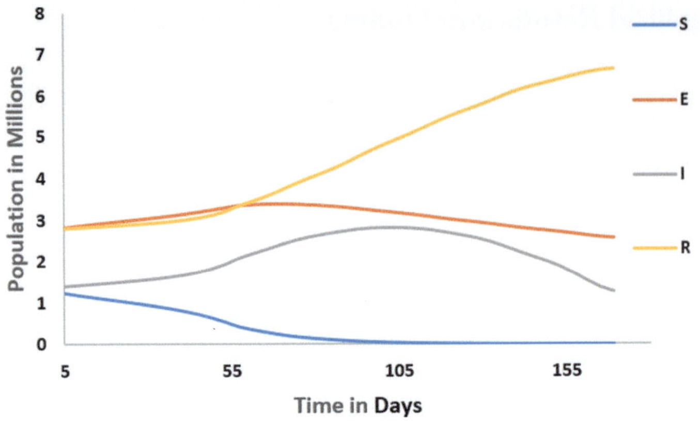

Fig. 2 In this graphical representation, S (susceptible individuals), E (exposed individuals), I (infected individuals) is represented and R (recovered individuals), each illustrating their respective changes over time. (Sharma et al. 2022)

Table 2 Numerical values of the parameters (Kumar et al. 2016)

Parameters	β	γ	*S(0)*	*I(0)*	*R(0)*
Values	0.012531 per year	2.46269 per year	399 per year	23 per year	0.00125 per year

4 Conclusion

The mathematical insights discussed in this chapter deliver a thorough comprehension of the dynamics of influenza virus spread and offer useful information for guiding control strategies. Using compartmental models like the Susceptible-Infectious-Recovered (SIR) model and agent-based models, researchers can model the transmission of influenza among populations, evaluate the effects of interventions, and inform evidence-supported public health strategies. Furthermore, evolutionary models elucidate the genetic and antigenic diversity of influenza virus strains, helping to predict the emergence of novel strains and inform vaccine design and surveillance efforts. By integrating mathematical modelling approaches with epidemiological data and molecular biology techniques, researchers can gain a deeper understanding of influenza virus dynamics and develop more effective strategies for prevention, control, and pandemic preparedness. Ultimately, these mathematical insights contribute to our collective efforts to mitigate the burden of influenza-related illness and safeguard public health on a global scale.

5 Challenges and Future Directions

The field of mathematical modelling of influenza virus transmission faces several challenges and holds promising future directions. One key challenge is the need for real-time data integration to inform model development and validation. Integrating diverse data sources, including epidemiological surveillance data, viral genetic sequences, and host immune responses, is essential to enhance the precision and dependability of mathematical models. Additionally, model validation against real-world data remains a critical challenge, as it requires comprehensive and high-quality data on influenza virus dynamics and transmission parameters. Collaboration between mathematical modelers, epidemiologists, virologists, and public health officials is essential to address these challenges effectively. By combining expertise from multiple disciplines, researchers can develop models that better capture the complex dynamics of influenza virus transmission and inform evidence-based public health interventions. Future directions in the field should prioritize interdisciplinary collaboration, data sharing mechanisms, and the development of predictive models that can guide timely and effective responses to influenza outbreaks and pandemics. By addressing these challenges and fostering collaboration, the field of mathematical modeling of influenza virus transmission can contribute to improved influenza control and pandemic preparedness efforts.

Competing Interests The authors declare no conflicts of interest that pertain to the content of the chapter.

References

Anderson CS, McCall PR, Stern HA, Yang H, Topham DJ (2018) Antigenic cartography of H1N1 influenza viruses using sequence-based antigenic distance calculation. BMC Bioinform 19(1):1–1. https://doi.org/10.1186/s12859-018-2042-4

Arduin H, Domenech de Celles M, Guillemot D, Watier L, Opatowski L (2017) An agent-based model simulation of influenza interactions at the host level: insight into the influenza-related burden of pneumococcal infections. BMC Infect Dis 17:1–2. https://doi.org/10.1186/s12879-017-2464-z

Bedford T, Marc AS, Philippe L, Gytis D, Victoria G, Alan JH, John WM, Russell CA, Smith DJ, Andrew R (2014) Integrating influenza antigenic dynamics with molecular evolution. elife 3:e01914. https://doi.org/10.7554/elife.01914

Bjornstad ON, Shea K, Krzywinski M, Altman N (2020) The SEIRS model for infectious disease dynamics. Nat Methods 17(6):557–559. https://doi.org/10.1038/s41592-020-0856-2

Bouvier NM, Palese P (2008) The biology of influenza viruses. Vaccine 26. https://doi.org/10.1016/j.vaccine.2008.07.039

Chen-Charpentier B (2024) SIRS epidemic models with delays, partial and temporary immunity and vaccination. AppliedMath 4(2):666–689. https://doi.org/10.3390/appliedmath4020036

Dellicour S, Lequime S, Vrancken B, Gill MS, Bastide P, Gangavarapu K, Matteson NL, Tan Y, du Plessis L, Fisher AA, Nelson MI (2019) Phylogeographic and phylodynamic approaches to epidemiological hypothesis testing. bioRxiv 788059. https://doi.org/10.1101/788059

Du Z, Nugent C, Galvani AP, Krug RM, Meyers LA (2020) Modeling mitigation of influenza epidemics by baloxavir. Nat Commun 11(1):2750. https://doi.org/10.1038/s41467-020-16585-y

Etbaigha F, Willms A, Poljak Z (2018) An SEIR model of influenza A virus infection and reinfection within a farrow-to-finish swine farm. PLOS One 13(9). https://doi.org/10.1371/journal.pone.0202493

Ewald J, Sieber P, Garde R, Lang SN, Schuster S, Ibrahim B (2020) Trends in mathematical modeling of host–pathogen interactions. Cell Mol Life Sci 77:467–480. https://doi.org/10.1007/s00018-019-03382-0

Forrest HL, Webster RG (2010) Perspectives on influenza evolution and the role of research. Anim Health Res Rev 11(1):3–18. https://doi.org/10.1017/s1466252310000071

Fukuyama S, Kawaoka Y (2011) The pathogenesis of influenza virus infections: the contributions of virus and host factors. Curr Opin Immunol 23(4):481–486. https://doi.org/10.1016/j.coi.2011.07.016

Ge Y, Liu L, Chen B, Qiu X, Huang K (2011) Agent-based modeling for influenza H1N1 in an artificial classroom. Syst Eng Procedia 2:94–104. https://doi.org/10.1016/j.sepro.2011.10.012

Hackl J, Dubernet T (2019) Epidemic spreading in urban areas using agent-based transportation models. Future Internet 11(4):92. https://doi.org/10.3390/fi11040092

Hayati M, Biller P, Colijn C (2020) Predicting the short-term success of human influenza virus variants with machine learning. Proc R Soc B 287(1924):20200319. https://doi.org/10.1098/rspb.2020.0319

Huang QS, Bandaranayake D, Wood T, Newbern EC, Seeds R, Ralston J, Waite B, Bissielo A, Prasad N, Todd A, Jelley L (2019) Risk factors and attack rates of seasonal influenza infection: results of the southern hemisphere influenza and vaccine effectiveness research and surveillance (SHIVERS) seroepidemiologic cohort study. J Infect Dis 219(3):347–357. https://doi.org/10.1093/infdis/jiy443

Kalil AC, Thomas PG (2019) Influenza virus-related critical illness: pathophysiology and epidemiology. Crit Care 23:1–7. https://doi.org/10.1186/s13054-019-2539-x

Krauland MG, Zimmerman RK, Williams KV, Raviotta JM, Harrison LH, Williams JV, Roberts MS (2023) Agent-based model of the impact of higher influenza vaccine efficacy on seasonal influenza burden. Vaccine: X 13:100249. https://doi.org/10.1016/j.jvacx.2022.100249

Kumar V, Kumar D, Pooja (2016) SIR model of swine flu in shimla. In: Advanced computing and communication technologies: proceedings of the 9th ICACCT, 2015. Springer Singapore, pp 297–303. https://doi.org/10.1007/978-981-10-1023-1_30

Lam TT, Chong YL, Shi M, Hon CC, Li J, Martin DP, Tang JW, Mok CK, Shih SR, Yip CW, Jiang J (2013) Systematic phylogenetic analysis of influenza A virus reveals many novel mosaic genome segments. Infect Genet Evol 18:367–378. https://doi.org/10.1016/j.meegid.2013.03.015

Laskowski M, Demianyk BC, Witt J, Mukhi SN, Friesen MR, McLeod RD (2011) Agent-based modeling of the spread of influenza-like illness in an emergency department: a simulation study. IEEE Trans Inf Technol Biomed 15(6):877–889. https://doi.org/10.1109/TITB.2011.2163414

Liu H, Wu C, Pang Z, Zhao R, Liao M, Sun H (2022) Phylogenetic and phylogeographic analysis of the highly pathogenic H5N6 avian influenza virus in China. Viruses 14(8):1752. https://doi.org/10.3390/v14081752

Moghadami M (2017) A narrative review of influenza: a seasonal and pandemic disease. Iran J Med Sci 42(1):2. https://pmc.ncbi.nlm.nih.gov/articles/PMC5337761/

Morris DH, Gostic KM, Pompei S, Bedford T, Luksza M, Neher RA, Grenfell BT, Lässig M, McCauley JW (2018) Predictive modeling of influenza shows the promise of applied evolutionary biology. Trends Microbiol 26(2):102–118. https://doi.org/10.1016/j.tim.2017.09.004

Osthus D, Hickmann KS, Caragea PC, Higdon D, Del Valle SY (2017) Forecasting seasonal influenza with a state-space SIR model. Ann Appl Stat 11(1):202. https://doi.org/10.1214/16-AOAS1000

Ratmann O, Donker G, Meijer A, Fraser C, Koelle K (2012) Phylodynamic inference and model assessment with approximate Bayesian computation: influenza as a case study. PLoS Comput Biol 8(12). https://doi.org/10.1371/journal.pcbi.1002835

Roche B, Drake JM, Rohani P (2011) An agent-based model to study the epidemiological and evolutionary dynamics of influenza viruses. BMC Bioinform 12:1. https://doi.org/10.1186/1471-2105-12-87

Rolfes MA, Talbot HK, McLean HQ, Stockwell MS, Ellingson KD, Lutrick K, Bowman NM, Bendall EE, Bullock A, Chappell JD, Deyoe JE (2023) Household transmission of influenza a viruses in 2021–2022. JAMA 329(6):482–489. https://doi.org/10.1001/jama.2023.0064

Sambaturu N, Mukherjee S, Lopez-García M, Molina-Paris C, Menon GI, Chandra N (2018) Role of genetic heterogeneity in determining the epidemiological severity of H1N1 influenza. PLoS Comput Biol 14(3). https://doi.org/10.1371/journal.pcbi.1006069

Shao W, Li X, Goraya MU, Wang S, Chen JL (2017) Evolution of influenza a virus by mutation and re-assortment. Int J Mol Sci 18(8):1650. https://doi.org/10.3390/ijms18081650

Sharma B, Khurana P, Kumar D (2022) Covid-19 analysis: a critique of a paper by Q. Lin. In: AIP conference proceedings, vol 2576, no 1. AIP Publishing. https://doi.org/10.1063/5.0105700

Styer LM, Carey JR, Wang JL, Scott TW (2007) Mosquitoes do senesce: departure from the paradigm of constant mortality. Am J Trop Med Hyg 76:111–7. https://pubmed.ncbi.nlm.nih.gov/17255238/

Tabasi M, Alesheikh AA, Sofizadeh A et al (2020) A spatio-temporal agent-based approach for modeling the spread of zoonotic cutaneous leishmaniasis in northeast Iran. Parasit Vectors 13:572. https://doi.org/10.1186/s13071-020-04447-x

Taubenberger JK, Morens DM (2008) The pathology of influenza virus infections. Annu Rev Pathol Mech Dis 3:499–522. https://doi.org/10.1146/annurev.pathmechdis.3.121806.154316

Tsang TK, Lau LL, Cauchemez S, Cowling BJ (2016) Household transmission of influenza virus. Trends Microbiol 24(2):123–133. https://doi.org/10.1016/j.tim.2015.10.012

Tsang TK, Wang C, Fang VJ, Perera RA, So HC, Ip DK, Leung GM, Peiris JM, Cauchemez S, Cowling BJ (2023) Reconstructing household transmission dynamics to estimate the infectiousness of asymptomatic influenza virus infections. Proc Natl Acad Sci 120(33). https://doi.org/10.1073/pnas.2304750120

Valle SY, Mniszewski SM, Hyman JM (2013) Modeling the impact of behavior changes on the spread of pandemic influenza. In: Modeling the interplay between human behavior and the spread of infectious diseases. pp 59–77. https://doi.org/10.1007/978-1-4614-5474-8_4

Wright S (1945) The differential equation of the distribution of gene frequencies. Proc Natl Acad Sci 31(12):382–389. https://doi.org/10.1073/pnas.31.12.382

Zarnitsyna VI, Handel A, McMaster SR, Hayward SL, Kohlmeier JE, Antia R (2016) Mathematical model reveals the role of memory CD8 T cell populations in recall responses to influenza. Front Immunol 7:165. https://doi.org/10.3389/fimmu.2016.00165

Zitzmann C, Kaderali L (2018) Mathematical analysis of viral replication dynamics and antiviral treatment strategies: from basic models to age-based multi-scale modeling. Front Microbiol 9:1546. https://doi.org/10.3389/fmicb.2018.01546

Using Mathematical Approaches for Predictive Modelling of Herbal-Drug Interactions

Pooja Khurana, Deepak Kumar, and Sanjeev Kumar

Abstract Diabetes Mellitus (DM) and Alzheimer's Disease (AD) are two main burden chronic diseases with a remarkably high prevalence rate worldwide. They share pathophysiological characteristics such as insulin resistance, oxidative stress and chronic inflammation. Medicinal plants have drawn increasing attention over the years as possible therapeutic agents for the management of both diseases owing to their bioactive compounds, which target various mechanisms and have fewer side effects compared to the current pharmaceuticals. This chapter outlines the findings of two earlier studies aimed at determining the potential application of mathematical models in evaluating the neuroprotective and metabolic impacts of herbal therapies. The beneficial compounds and diverse therapeutic actions of plants like Curcuma longa, Tinospora cordifolia and Ginkgo biloba are analyzed. This chapter aims to bridge traditional plant medicine with contemporary medical therapies via literature data integration. The conclusions note that predictive modeling helps in clarifying the mechanisms at work, rationalizing dosages, and favoring better treatment results. The findings show the dual therapeutic advantages of herbal medicines to DM and AD by being able to slow down disease progression and bring about metabolic restoration while paving the road for personalized treatment strategies.

Keywords Alzheimer disease · Diabetes mellitus · Herbal approach · Predictive modeling · Mathematical models

P. Khurana · D. Kumar (✉)
Department of Applied Sciences, Manav Rachna International Institute of Research and Studies, Faridabad 121004, India
e-mail: deepakman12@gmail.com

P. Khurana
e-mail: pooja.set@mriu.edu.in

S. Kumar
Maharaja Suhail Dev University, Azamgarh 276001, Uttar Pradesh, India
e-mail: sanjeevibs@yahoo.co.in

M. Ram et al. (eds.), *Advances in Mathematics for Engineering Sciences*, Mathematical Engineering, https://doi.org/10.1007/978-3-031-95693-5_11

1 Introduction

Non communicable diseases (NCDs) are on the rise as a major threat to global health, driving a surge in morbidity and mortality around the world (WHO 2020). As per the International Diabetes Federation and Alzheimer's Association, Diabetes mellitus (DM) & Alzheimer's Disease (AD) need to be well understood as they are most common and chronic NCDs (International Diabetes Federation 2021; Alzheimer's Association 2022). DM is characterized by hyperglycemia (higher levels of blood glucose) resulting from absolute or relative insulin deficiency, insulin resistance or insensitivity. This can develop into life threatening pathological consequences such as myocardial damage, renal failure and neuropathy (Goyal et al. 2023). On the other hand, AD is the primary cause of dementia, a neurodegenerative disorder with a clinical course of steadily worsening cognitive dysfunction and brain amyloid-beta plaques, tau tangles (McGirr et al. 2020). Despite the advances in drug development, treating these diseases presents a major challenge. This is due to the intricate and only partly elucidated molecular rationale of them, that no effective therapy exists.

With the increasing incidence of chronic diseases several years back, the interest of herbal medicine has started to utilized as a complimentary treatment of chronic diseases. The herbal medicine is primarily derived from natural plant sources, some of these bioactive plant constituents include alkaloids, flavonoids glycosides, terpenoids and phenolic acids act as herbal remedies. These compounds are generally multi-functional and often display antioxidant, anti-inflammatory and even neuroprotective actions that may endow them as a natural alternative treatment option of DM and AD owing to their holistic, synergistic impact (Bommakanti et al. 2023). Gymnemic acid from Gymnema sylvestre helps in blood sugar stabilization and the cognitive enhancing properties of Bacopasides (Bacopa monnieri) are known for their neuroprotective capacity (Muddapur et al. 2024; Nishanth et al. 2023). However, the application of herbal medicine is limited due to its low scientific evidence regarding efficacy and safety, and the techniques for standardizing its therapeutic mechanisms are still many years from being developed.

Predictive modeling has evolved as an alternative solution for these specific challenges by performing artificial intelligence based predictive assessment of herbal bioactives toward their pharmacological utility (Zeng and Jia 2024) These models provide a means of quantifying and better understand disease dynamics, the effect of herbal treatment (Arora et al. 2021). With these models, herbal compounds' mechanism of action may be elucidated, maximum dose level can be determined and beneficial therapeutic potentials can be discovered to understand the disease management in a more effective manner.

Combining predictive modeling with herbal medicine is in keeping with the vision of precision medicine as it is to enable personalized treatment for each individual patient depending on his/her specific biological/clinical profile. This is particularly crucial for diseases such as DM or AD where different underlying disease mechanisms will need personalized therapeutic intervention. Thus, for cases such as insulin resistance and amyloid-beta accumulation (Type 3 Diabetes) illustrates the importance of tackling several targets at once. Herbal medicines that are polypharmacological in nature are ideally suited to this type of action. This chapter aims to explore the role of predictive modeling in assessing herbal interventions for DM and AD.

2 Mathematical Models

Mathematical modeling has become one of the fundamental methods for forecasting molecular interactions and recognizing possible drug candidates based on herbs' bioactivities in recent decades. It therefore contributes to realizing the therapeutic potential of herbal medicines by illuminating the relationship between bioactive compounds from herbs and their biological targets. This helps in reveal their mode of actions, efficacies and side-effect profiles (Ismail et al. 2018). Such models are very important in solving the problem of complex chronic diseases, such as diabetes mellitus and Alzheimer's disease, in which effective treatment is generally aimed at modulating the pathways or targets of multiple agents. By doing so, they deliver meaningful quantitative understanding of pharmacodynamics, systems-wide effects and varying responses to herbal therapies proximal to predictive targets.

Pharmacokinetic (PK) and pharmacodynamic (PD) models are the main basis used to represent the different facets of absorption, distribution, metabolism, and elimination of herbal substances within the body. These models have been applied to optimize dose schedules and predict therapeutic effectiveness; thus, they represent a very important step towards the fine-tuning of herbal medicine in prediction and treatment. PK and PD models provide a structured approach in studying the time-course of drug effects and their interaction with biological systems; consequently, this is the basis for tailoring treatments for different populations of patients (Gabrielsson and Weiner 2001). The systems biology perspective posits that graph theory-based systems biology modeling enables one to perform quantitative analyses of interactions at multiple targets. Applying adjacency matrices and network analysis allows such models to uncover the link between herbal compounds and biological pathways and identify critical nodes for the modulation of diseases (Lee 2015; Cai et al. 2018).For diseases e.g., DM and AD, the network-based models could provide the basis on which certain combinations of bioactive compounds synergistically interface multiple pathways with the intention of augmenting knowledge from computational network pharmacology (Adnan et al. 2022; Li et al. 2017; Arora and Khurana 2023; Wang et al. 2022).

Population-based models also simulate the differential responses of individuals from various demographic backgrounds to herbal treatments, considering genetic factors, metabolic aspects, and environmental influences. Such models are the very basis for personalized medicine such as population pharmacokinetics, which seeks to evaluate their differential responses at various ages to tailor therapeutic approaches more effectively (Kaur et al. 2022).

2.1 Model on Alzheimer

The neurodegenerative disorder Alzheimer's influences thinking, memory, behaviour, and diverse activities. Apart from affecting memory, it also poses a severe socioeconomic challenge. Therefore, various drugs have been synthesized to tackle Alzheimer's, such as cholinesterase inhibitors, antioxidants, statins, antagonists, and a few combinations. Despite this; however, the general trajectory of Alzheimer's disease in progress continues to linger, imposing significant impact on society. Herbal treatments provide an alternative to conventional medications that have not very much served well to treat the disease.

Mathematical models have been employed to examine the progression and handling of Alzheimer's disease across various factors, including amyloid-beta aggregation, tau-phosphorylation, micro-glia, astrocytes, and macro-phages (Mukherjee et al. 2008). Puri and Li used mathematical modeling to describe the pathogenesis of Alzheimer's disease (Hao and Friedman 2016), while another model has been proposed that represents neuronal death in terms of Alzheimer's (Puri and Li 2010). Such models provide a way of assembling a coherent framework by which optimal doses of bioactive compounds from medicinal plants could be determined in order to inhibit amyloid-beta accumulation from the brain and to enhance its clearance from the brain, and these may thus valorize as new and improved treatment modalities in the fight against Alzheimer's disease.

The extracellular amyloid beta peptide fulfills the subsequent equation:

$$\frac{d{A^0}_\beta}{dt} = \lambda_N \frac{N}{N_0} + \lambda_A \frac{A}{A_0} + \lambda_D \frac{D}{D_0} - \left(d_{A^0\beta M}(M) + d_{A^0\beta DE}(DE)\right) + \frac{{A^0}_\beta}{{A^0}_\beta + k{A^0}_\beta} \tag{1}$$

Five plant extracts with four dosage variations (250 mg/day, 500 mg/day, 750 mg/day, and 1000 mg/day), were considered for studying the best therapeutic effect of herbal drugs in amyloid beta regulation in the brain for the period of five months.

The dosage equation is as follows:

$$\frac{d{A^0}_\beta}{dt} = \lambda_N \frac{N}{N_0} + \lambda_A \frac{A}{A_0} + \lambda_D \frac{D}{D_0} - \left(d_{A^0\beta M}(M) + d_{A^0\beta DE}(DE)\right) + \frac{H{A^0}_\beta}{{A^0}_\beta + k{A^0}_\beta} \tag{2}$$

Initial value ${A^0}_\beta = 92$.

Parametric description

$\lambda_N \frac{N}{N_0}$: The cleavage of amyloid precursor protein (APP) by $\beta-$secretase and $\gamma-$secretase leads to the release of amyloid-beta ($A\beta$) peptides.

$\lambda_A \frac{A}{A_0}$: Activated astrocytes contribute to the production and release of amyloid-beta (Aβ) peptides.

A_0: The standard concentration of astrocyte cells within brain tissue.

$\lambda_D \frac{D}{D_0}$: Dendritic cells facilitate the secretion of amyloid-beta ($A\beta$) peptides.

D_0: The standard concentration of dendritic cells within brain tissue.

$d_{A^0 \beta M}(M)$: Microglial cells mediate the clearance of amyloid-beta ($A\beta$) peptides.

$d_{A^0 \beta DE}(DE)$: Degrading enzymes mediate the clearance of amyloid-beta ($A\beta$) peptides.

H: Quantity of dose.

k: Michael's-Menten coefficient.

2.1.1 General Solution of First Order Differential Equations-Euler's Method

Numerical methods can be employed to determine the standard situation for addressing first-order differential equations. The differential equation $\frac{dA^0{}_\beta}{dt} = f(A^0{}_\beta, t)$ is replaced with the following approximation $\frac{A^0{}_\beta(t_0+\Delta t)-A^0{}_\beta(t_0)}{\Delta t} = f(t_0, A^0{}_\beta(t_0))$.

For sufficiently small Δt. This can be arranged to

$$A^0{}_\beta(t_0 + \Delta t) = A^0{}_\beta(t_0) + \Delta t f(t_0, A^0{}_\beta(t_0))$$

So that given a step size Δt and initial condition $A^0{}_\beta(t_0)$.

2.1.2 Graphical Representation on Alzheimer Modelling

Medicinal plants, rich in bioactive compounds, hold significant potential for managing various health conditions. The capacity of bioactive substances to lower beta-amyloid levels has been thoroughly investigated and reported. These include curcumin, steroids, glycosides, anthocyanins, terpenoids, asiatic acid, and alkaloids. To identify effective treatments, it is crucial to investigate their pharmacokinetic profiles and physicochemical attributes, including water solubility, gastrointestinal absorption, lipophilicity, bioavailability scores, and other related properties.

Lipinski's rule provides a standard framework for evaluating the drug-likeness of molecules by assessing their suitability as oral pharmaceutical agents. According to this rule, a molecule with molecular weight (MW) exceeding 500 g/mol, more than 5 hydrogen-bond donors, more than 10 hydrogen-bond acceptors, or a log P value greater than 5 is unlikely to exhibit optimal oral bioavailability. The natural compounds selected for Alzheimer's treatment demonstrated favorable pharmacokinetics and physicochemical characteristics, with no significant violations of Lipinski's rule, except for the active components bacoside and asiaticoside found in Bacopa monniera and Centella asiatica, respectively.

Furthermore, the efficacy of various plants was evaluated at dosages of 250, 500, 750, and 1000 mg/day over a five-month period. Among these, Curcuma longa exhibited optimal results at a dose of 1000 mg/day, Bacopa monniera at 750 mg/day, Convolvulus pluricaulis at 1000 mg/day, Withania somnifera at 500 mg/day, and Centella asiatica at 750 mg/day. These findings are visually represented in Figs. 1, 2, 3, 4, and 5. Based on the results, Withania somnifera emerges as the most effective among the five plants, demonstrating significant amyloid beta clearance at the lowest tested dose of 500 mg/day.

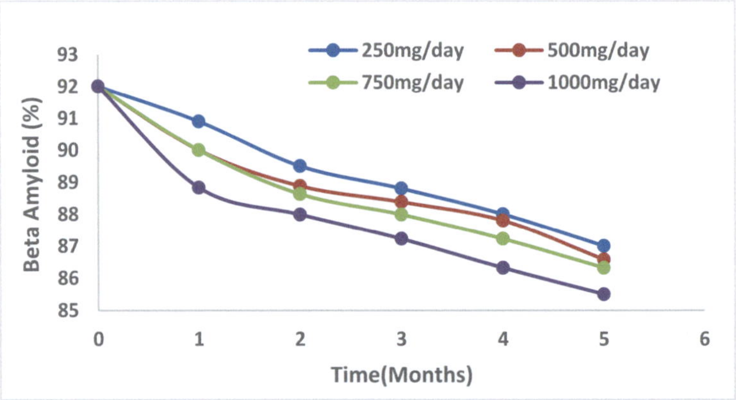

Fig. 1 The influence of curcumin on amyloid beta levels over time across various dosages

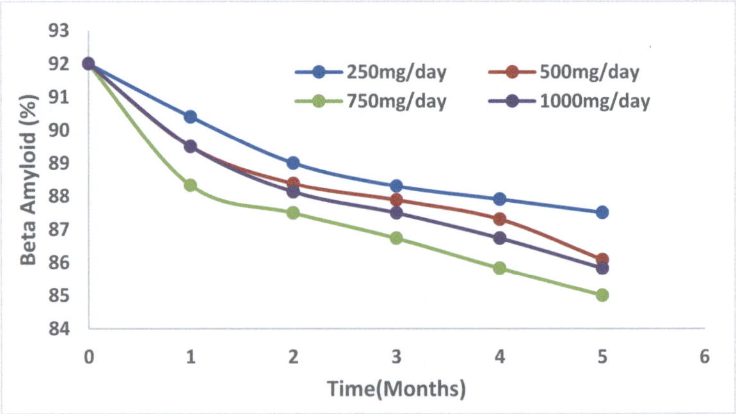

Fig. 2 The influence of bacoside on amyloid beta levels over time across various dosages

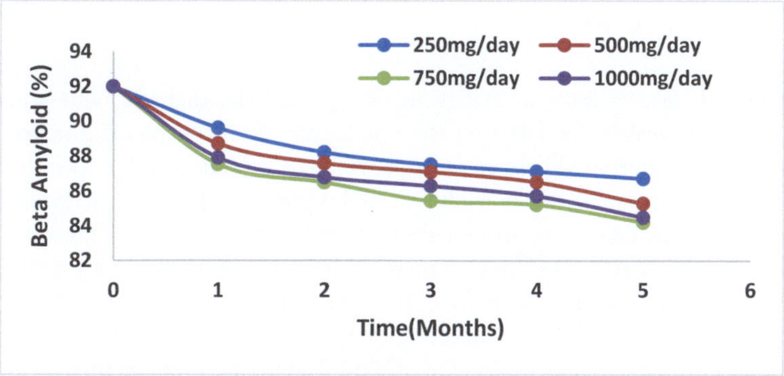

Fig. 3 The influence of asiaticside on amyloid beta levels over time across various dosages

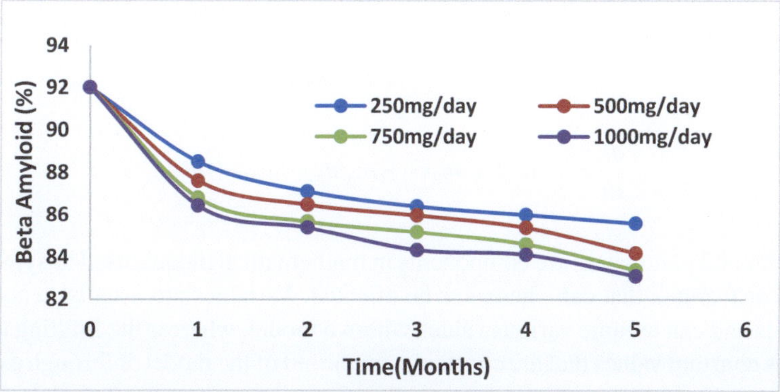

Fig. 4 The influence of convolvine on amyloid beta levels over time across various dosages

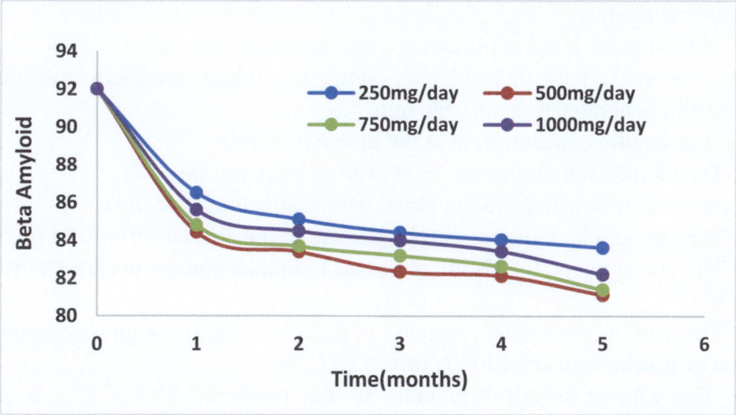

Fig. 5 The influence of withaferin a on amyloid beta levels over time across various dosages

2.2 Model on Diabetes

Mathematical models have been developed to elucidate diabetic behaviour, with many of these models focusing on the connection between blood glucose levels and sugar concentration. These models prove to be effective under specific conditions. In one such model, we consider variables such as blood sugar levels (BSL), insulin generation $G(t)$, insulin generalization variable $X(t)$ and plasma insulin concentration $I(t)$. After meeting certain criteria, a comprehensive mathematical model is constructed to better understand and predict the dynamics of diabetes (Tyagi et al. 2024; Huard et al. 2022; Chudtong et al. 2021, Xie 2023, Lopez et al. 2020, Choudhury et al. 2018; Sugiyanto et al. 2020). The model is characterized as

$$\frac{dG}{dt} = -(m_1 + a_1)G + m_2 I + m_1 G_b \tag{3}$$

$$\frac{dX}{dt} = -m_2 X + m_3 I - m_3 I_b + m_6 I_b \tag{4}$$

$$\frac{dI}{dt} = -m_3 I + m_4 G + m_4 m_5 - m_6 I + m_6 I_b \tag{5}$$

Variables and parameters are components in mathematical models used to symbolize values or features that can change or be affected. Variables are usually denoted by symbols and can assume various values within a model, whereas the limiting factors possess constant values that are constant with the aid of the model or through external input. Variables and parameters are crucial elements of mathematical models utilized across different domains.

Below is a description of all of the values of variables and parameters utilized in mathematical models:

$G(t)$: blood sugar level in plasma at a specific time $t(mg/dl)$

$X(t)$: A variable in the distant bodily compartment that serves as a representation of the insulin generalization variable min^{-1}.

$I(t)$: The insulin concentration at the time $t(\mu U/ml)$

G_b: The blood's initial glucose level prior to injection (mg/dl)

I_b: The insulin concentration in blood before injection $(\mu U/ml)$

m_1: The rate at which tissues absorb glucose in the absence of insulin (min^{-1}).

m_2: The rate at which the ability of tissues to uptake glucose decreases over time (min^{-1}).

m_3: The insulin-independent increase in tissue's capacity for glucose uptake as a function of insulin concentration I_b $(min^{-2}(\mu U/ml))$.

m_4: The rate at which beta cells in the pancreas release insulin due to a glucose infusion and elevated glucose levels beyond a specific threshold. h $[(\mu\text{U/ml})\ \text{min}^{-2}(\text{mg/dl})^{-1}]$

m_5: The lowest blood sugar level needed to trigger the release of insulin from the beta cells in the pancreas.

m_6: The trend of insulin's gradual elimination from circulatory systems.

a_1: the rate at which ingesting active compounds from medicinal plants alters blood glucose levels.

The amount of three factors above stated $N(t) = G(t) + X(t) + I(t)$.

2.2.1 Mathematical Analysis

Theorem 1 *The system of Eqs.* (1), (2) and (3) *obtained the equilibrium points* E_{q0}

$$G_0 = \frac{m_2 m_4 g_b + m_2 m_4 m_5 + m_2 m_6 i_b}{(m_1 + a_1)(m_3 + m_6) - m_2 m_4} + g_b$$

$$X_0 = \frac{(m_1 + a_1)m_3(m_4 g_b + m_4 m_5 + m_6 i_b) + (m_6 i_b - m_3 i_b)((m_1 + a_1)(m_3 + m_6 - m_2 m_4)}{m_2((m_1 + a_1)(m_3 + m_6) - m_2 m_4)}$$

$$I_0 = \frac{(m_1 + a_1)m_4 g_b + (m_1 + a_1)m_4 m_5 + (m_1 + a_1)m_6 i_b}{(m_1 + a_1)(m_3 + m_6) - m_2 m_4}$$

Theorem 2 *The equilibrium point exists if* $G_0, X_0,$ *and* I_0 *is real and non-negative.*

Proof Existence of the equilibrium point above exist if G_0, X_0, and I_0 is real and and non-negative.

1. G_0 is real and non-negative, which $(m_1 + a_1)(m_3 + m_6) - m_2 m_4 > 0$.
2. X_0 is real and non-negative, which $(m_1 + a_1)(m_3 + m_6) - m_2 m_4 > 0$.
3. I_0 is real and non-negative, which $(m_1 + a_1)(m_3 + m_6) - m_2 m_4 > 0$.

The discovered equilibrium point will be used to discuss the stability of the diabetes mellitus model.

Theorem 3 *The equilibrium point,* E_{q0}*, system of equation (A)–(C) will be asymptotically stable locally if* $(m_1+a_1+m_3+m_6)^2 = 4[(m_1+a_1)(m_3+m_6)-m_2 m_4]$.

Proof The eigenvalues of the Jacobian matrix are as follows.

$$\lambda_1 = -m_2 < 0,$$

$$\lambda_2 = \frac{-(m_1 + a_1 + m_3 + m_6) + \sqrt{(m_1 + a_1 + m_3 + m_6)^2 - 4[(m_1 + a_1)(m_3 + m_6) - m_2 m_4]}}{2}$$

$$\lambda_3 = \frac{-(m_1 + a_1 + m_3 + m_6) - \sqrt{(m_1 + a_1 + m_3 + m_6)^2 - 4[(m_1 + a_1)(m_3 + m_6) - m_2 m_4]}}{2}$$

If $(m_1 + a_1 + m_3 + m_6)^2 = 4[(m_1 + a_1)(m_3 + m_6) - m_2 m_4]$, then $\lambda_1 < 0, \lambda_2 < 0, \lambda_3 < 0$. This means that the equilibrium point is asymptotically stable locally.

2.2.2 Graphical Representation on Diabetes Model

The parameter values in a graphical representation of medicinal plants that can benefit people with diabetes mellitus are considered. Using graphs to analyze and showcase data, interpret numerical values, and improve understanding of qualitative frameworks for a diabetic patient consistently undergoing treatment with medicinal plants over a period of 30 days.

a. ***Aloe Vera (Aloe Barbadensis Miller)***

Aloe vera dietary supplements have demonstrated a positive impact on the ability of patients with type 2 diabetes and prediabetes to manage their condition (Fig. 6).

b. ***Garlic (Allium Sativum)***

Figure 8 shows the concentration of glucose levels in the blood at different intervals after garlic consumption, and the standard concentration of glucose levels in the blood. Multiple studies indicate that consuming one clove of garlic or its equivalent daily enhances insulin sensitivity and aids in managing type 2 diabetes (Fig. 7).

c. ***Garlic (Allium Sativum)***

Gurmar plants are advantageous for people with type 1 and type 2 diabetes and are effective in dealing with diabetes-related complications (Fig. 8).

d. ***Momordica Charantia (Bitter Melon)***

The above Fig. 9 illustrates that consuming bitter melon juice as a treatment greatly lowers blood glucose levels, demonstrating the usefulness of bitter melon

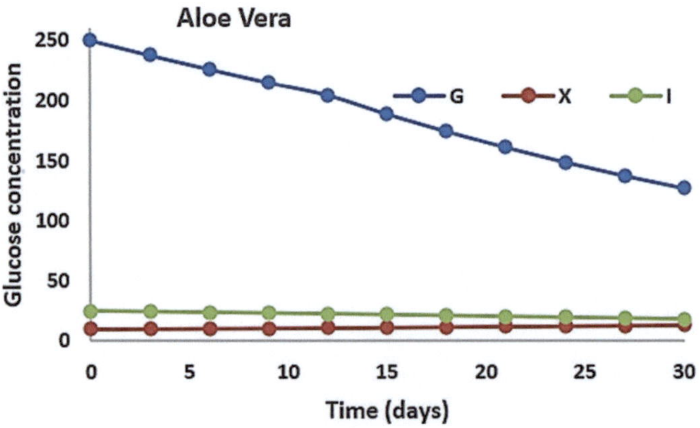

Fig. 6 Figure displaying blood glucose and insulin levels over a 30-day period for a diabetic person receiving aloe vera treatment

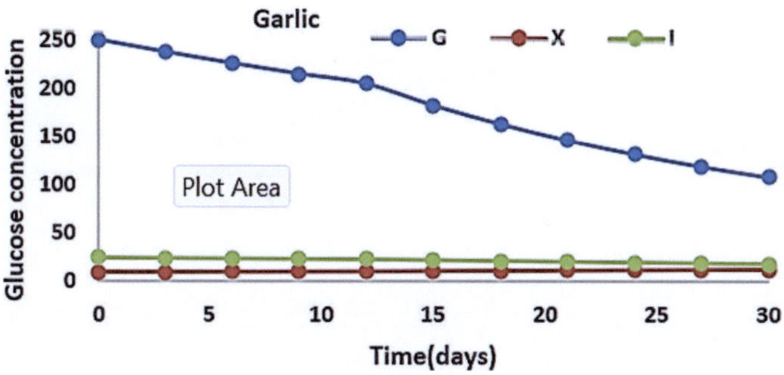

Fig. 7 Figure displaying blood glucose and insulin levels over a 30-day period for a diabetic person receiving Garlic treatment

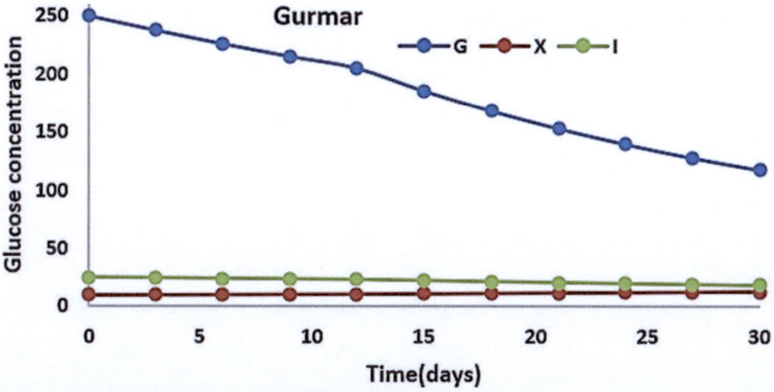

Fig. 8 Figure displaying blood glucose and insulin levels over a 30-day period for a diabetic person receiving Gurmar treatment

in managing blood glucose levels. Eating bitter melon is beneficial for both type 1 and type 2 diabetes and aids in managing diabetic complications.

e. ***Azadirachta Indica (Neem)***

Neem might lower blood sugar levels. Mixing neem with diabetes medications may result in low blood sugar levels, as shown in Fig. 10. Consuming powdered Neem leaves and seeds alongside diabetes medications might lead to a decrease in blood sugar levels.

f. ***Ocimum Sanctum (Holy Basil)***

Tulsi is a traditional Hindu herb that is effective in treating various types of diseases. Also, blood glucose levels decrease, as shown in Fig. 11.

At various points in time, the plasma insulin concentration, insulin generalization, and blood glucose level have all been assessed. MATLAB software has been used to produce the mathematical model's graphical representation. A mathematical model based on parameter values was used to create and then assess the graph (Table 1). The

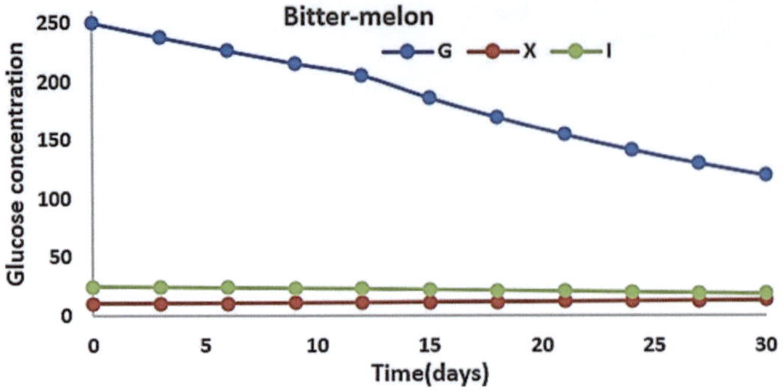

Fig. 9 Figure displaying blood glucose and insulin levels over a 30-day period for a diabetic person receiving Bitter-melon treatment

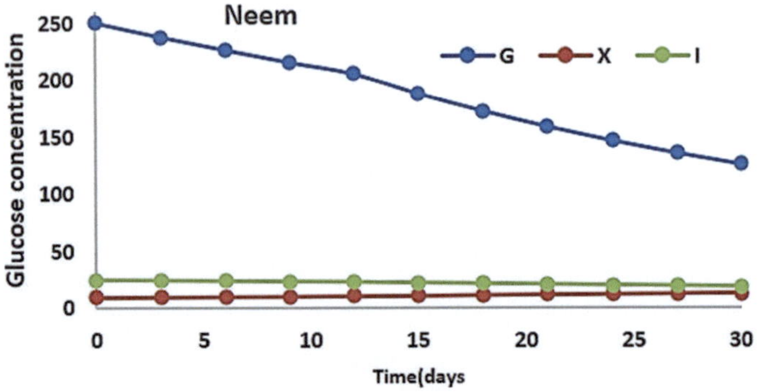

Fig. 10 Figure displaying blood glucose and insulin levels over a 30-day period for a diabetic person receiving Neem treatment

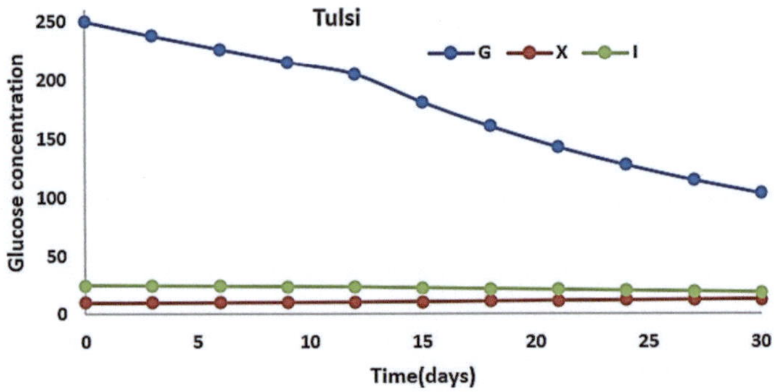

Fig. 11 Figure displaying blood glucose and insulin levels over a 30-day period for a diabetic person receiving Tulsi treatment

greatest variation in blood glucose levels is seen in all six medicinal plants (Figs. 6, 7, 8, 9, 10 and 11). Improvements were observed after utilizing the therapeutic herbs for 30 days.

Tulsi, Bitter melon, Gurmar, and Garlic had the least amount of variation in blood glucose level out of the seven therapeutic plants. After analysis, it reflects that the use of medicinal plants is very fruitful to maintain plants' Blood glucose levels in Table 2.

Table 1 Parameter estimations (Chudtong et al. 2021; Xie 2023; Lopez et al. 2020; Choudhury et al. 2018; Sugiyanto et al. 2020)

Parameter	Value
m_1	0.31700 min^{-1}
m_2	0.0123 min^{-1}
m_3	0.0000492 ($\text{min}^{-2}(\mu U/ml)$)
m_4	0.0039 (μU/ml) $\text{min}^{-2}(\text{mg/dl})^{-1}$
m_5	79.0353
m_6	0.2659
G_b	70 (mg/dl)
I_b	6 ($\mu U/ml$)
a_1	Aloe vera-0.30, Garlic-0.26, Gurmar-0.1213, Bitter-melon-0.19, Neem-0.15, Tulsi-0.30

Table 2 Evaluation of herbal remedies on blood sugar levels

Time	G (aloe vera)	G (Garlic)	G (Gurmar)	G (bitter-melon)	G (tulsi)	G (neem)
0	250	250	250	250	250	250
3	238	238	238	238	238	238
6	226	226	226	226	226	226
9	215	215	216	216	215	216
12	205	205	205	205	205	206
15	189	183	186	186	181	188
18	174	164	169	169	160	173
21	161	147	153	154	143	159
24	149	132	140	141	127	147
27	138	120	128	130	114	136
30	127	109	118	120	103	126

3 Conclusion

This study consolidates knowledge from two comprehensive investigations, highlighting the growing significance of herbal medicine in the treatment of chronic conditions like diabetes mellitus (DM) and Alzheimer's disease (AD). Leveraging insights from the existing research, it highlights how bioactive compounds in plants, Curcuma longa, Tinospora cordifolia, and Ginkgo biloba, interact with salient physiological pathways. Mathematical frameworks represent an effective lens through which these interactions can be examined to give insights into mechanisms, dosage optimization, and therapeutic efficacy. It has been shown that several pathophysiological features common to DM and AD—oxidative stress, inflammation, and insulin resistance—represent therapeutic entry points for the employment of herbal remedies. Despite optimistic perspectives, remaining challenges include a lack of standardization, imposing variability among individuals, and limitations in the regulatory framework. Those barriers require the amelioration of cross-disciplinary studies, stringent validation studies, and platform for joint policymaking insights.

In the future, the fusion of traditional herbal knowledge with cutting-edge scientific tools poses the biggest perspective-through the development of advance mathematical models that enhance the research predictions for remedy purposes and larger medical trials that would establish strong evidence concerning its efficacy and safety. Substantial alterations are needed in regulatory frameworks if innovation is to be enabled while assuring consumer safety.

Ultimately, creating space for integrative linking of traditional practices with scientific advancements has the power to revolutionize healthcare forever. When effective research supports the application of herbal medicines clinically and develops a position in modern therapy, they will undoubtedly provide a less expensive, holistic alternative for the management of complex conditions such as diabetes and Alzheimer disease.

4 Future Problems and Directions

The integration of herbal therapy within traditional healthcare faces several hurdles, first and foremost being the absence of standardization and scientific verification. One of the principal problems is that there are no uniform methodologies to assess the efficacy and safety of herbal treatments. In addition, diseases of complex nature such as diabetes mellitus (DM) and Alzheimer's disease (AD) make it harder to develop reliable models for their management. The lack of, or scant data on, pharmacokinetics and pharmacodynamics of herbal compounds also limit the capacity for accurate predictions and validation of therapeutic effects. On the other hand, it is complicated by differences in these responses as a related factor to genetic, demographic, and environmental differences among individuals taking herbal interventions. The rigorous regulatory frameworks hinder the task of integrating traditional healing arts and modern scientific approaches in a way that would not delay the adoption of herbal treatments into standard clinical practice. More sophisticated computational and mathematical models for predicting interactions between multiple targets and optimizing dosage recommendations will inform the research priorities for the foreseeable future. This potentially novel frontier combines artificial intelligence (AI) and machine learning techniques for helping to establish herbal treatments tailored to the genetic makeup and demographic factors of an individual. The validation or verification of the safety and efficacy for herbal compounds can rely upon extensive in vitro, in vivo, and clinical studies. AI can revolutionize this field by allowing for efficient high-throughput screening, increased data integration, and a better understanding of herbal-drug interactions. The collaboration of pharmacologists, computational experts, and healthcare scientists can usher a new approach and drive the development of herbal medicine. In addition, greater flexibility of regulations and increased funding are needed in order to encourage research, development, and clinical usage of herbal medicine to genuinely bridge the gap between traditional and modern medical practice.

Competing Interests The authors have no conflicts of interest to declare that are relevant to the content of this chapter.

References

Adnan M, Jeon BB, Chowdhury MHU, Oh KK, Das T, Chy MNU, Cho DH (2022) Network pharmacology study to reveal the potentiality of a methanol extract of Caesalpinia sappan L. Wood against type-2 diabetes mellitus. Life 12(2):277. https://doi.org/10.3390/life12020277

Alzheimer's Association (2022) Alzheimer's disease facts and figures. https://www.alz.org

Arora K, Khurana P (2023) Insights on early diagnosis of alzheimer disease using graph theory. Int Adv J Eng, Sci Manag 20:203–211. https://iajesm.in/admin/conference/65a3d6fa15082.pdf

Arora K, Khurana P, Kumar D (2021) Herbal medicinal based approach for Alzheimer disease-a mathematical model. Int J Curr Res Rev 13(21):53–59. https://doi.org/10.31782/IJCRR.2021.132109

Bommakanti V, Puthenparambil Ajikumar A, Sivi CM, Prakash G, Mundanat AS, Ahmad F, Haque S, Prieto MA, Rana SS (2023) An overview of herbal nutraceuticals, their extraction, formulation, therapeutic effects and potential toxicity. Separations 10(3):177. https://doi.org/10.3390/separations10030177

Cai FF, Zhou WJ, Wu R, Su SB (2018) Systems biology approaches in the study of Chinese herbal formulae. Chin Med 13:1–10. https://doi.org/10.1186/s13020-018-0221-x

Choudhury H, Pandey M, Hua CK, Mun CS, Jing JK, Kong LE, Nik A, Soohg K, Tan S, Mallikarjuna RP, Bapi G, Kesharwani P (2018) An update on natural compounds in the remedy of diabetes mellitus: a systematic review. J Tradit Complement Med 8(3):361–376. https://doi.org/10.1016/j.jtcme.2017.08.012

Chudtong M, De Gaetano A (2021) A mathematical model of food intake. Math Biosci Eng 18(2):1238–1279. https://doi.org/10.3934/mbe.2021067

Gabrielsson J, Weiner D (2001) Pharmacokinetic and pharmacodynamic data analysis: concepts and applications, 4th edn. Swedish Pharmaceutical Press

Goyal R, Singhal M, Jialal I (2023) Type 2 diabetes. StatPearls [Internet]. https://www.ncbi.nlm.nih.gov/books/NBK513253/

Hao W, Friedman A (2016) Mathematical model on Alzheimer's disease. BMC Syst Biol 10(1):108. https://doi.org/10.1186/s12918-016-0348-2
Huard B, Kirkham G (2022) Mathematical modelling of glucose dynamics. Curr Opin Endocr Metablc Res 25:100379. https://doi.org/10.1016/j.coemr.2022.100379
International Diabetes Federation (2021) IDF Diabetes Atlas. https://www.idf.org
Ismail FM, Nahar L, Sarker SD (2018) High-throughput screening of phytochemicals: application of computational methods. In: Computational phytochemistry. Elsevier, Amsterdam, The Netherlands, pp 165–192. https://doi.org/10.1016/B978-0-443-16102-5.00008-0
Kaur G, Chauhan AS, Prinja S, Teerawattananon Y, Muniyandi M, Rastogi A, Jyani G, Nagarajan K, Lakshmi PVM, Gupta A, Selvam JM (2022) Cost-effectiveness of population-based screening for diabetes and hypertension in India: an economic modelling study. Lancet Public Health 7(1):e65–e73. https://doi.org/10.1016/s2468-2667(21)00199-7
Lee S (2015) Systems biology-a pivotal research methodology for understanding the mechanisms of traditional medicine. J Pharmacopuncture 18(3):11. https://doi.org/10.3831/KPI.2015.18.020
Li W, Yuan G, Pan Y, Wang C, Chen H (2017) Network pharmacology studies on the bioactive compounds and action mechanisms of natural products for the treatment of diabetes mellitus: a review. Front Pharmacol 8:74. https://doi.org/10.3389/fphar.2017.00074
Lopez-Palau NE, Olais-Govea JM (2020) Mathematical model of blood glucose dynamics by emulating the pathophysiology of glucose metabolism in type 2 diabetes mellitus. Sci Rep 10(1):12697. https://doi.org/10.1038/s41598-020-69629-0
McGirr S, Venegas C, Swaminathan A (2020) Alzheimer's disease: a brief review. J Exp Neurol 1(3):89–98. https://doi.org/10.33696/Neurol.1.015
Muddapur UM, Manjunath S, Alqahtani YS, Shaikh IA, Khan AA, Mannasaheb BA, Yaraguppi D, More SS (2024) Exploring bioactive phytochemicals in Gymnema sylvestre: biomedical uses and computational investigations. Separations 11(2):50. https://doi.org/10.3390/separations11020050
Mukherjee PK, Kumar V, Kumar NS, Heinrich M (2008) The Ayurvedic medicine Clitoria ternatea–from traditional use to scientific assessment. J Ethnopharmacol 120:291–301. https://doi.org/10.1016/j.jep.2008.09.009
Nishanth BJ, Vijayababu P, Kurian NK (2023) Bacopa monnieri extract as a neuroprotective and cognitive enhancement agent. Int J Drug Discov Pharmacol 44–56. https://doi.org/10.53941/ijddp.2023.100015
Puri KI, Li L (2010) Mathematical modeling for the pathogenesis of Alzheimer's disease. PLoS ONE 5(12):e15176. https://doi.org/10.1371/journal.pone.0015176
Sugiyanto, Khalda AL (2020) Mathematical modeling effect habbatussauda on diabetes mellitus. Proceeding Int Conf Sci Eng 3:379–382. https://doi.org/10.14421/icse.v3.532
Tyagi K, Kumar D, Gupta R (2024) Medicinal plants approach for diabetes mellitus-a computational model. Int J Exp Res Rev 44:66–75. https://doi.org/10.52756/ijerr.2024.v44spl.006
Wang S, Ma Y, Huang Y, Hu Y, Huang Y, Wu Y (2022) Potential bioactive compounds and mechanisms of Fibraurea recisa Pierre for the treatment of Alzheimer's disease analyzed by network pharmacology and molecular docking prediction. Front Aging Neurosci 14:1052249. https://doi.org/10.3389/fnagi.2022.1052249
WHO (2020) Noncommunicable diseases. https://www.who.int/health-topics/noncommunicable-diseases#tab=tab_1
Xie X (2023) Steady solution and its stability of a mathematical model of diabetic atherosclerosis. J Biol Dyn 17(1):2257734. https://doi.org/10.1080/17513758.2023.225773
Zeng J, Jia X (2024) Systems theory-driven framework for AI integration into the holistic material basis research of traditional Chinese medicine. Engi. https://doi.org/10.1016/j.eng.2024.04.009

Enhanced Nodal Variation Approach for Fracture Analysis Using Meshless Methods

Ayush Awasthi and Mohit Pant

Abstract Current work discusses modelling and simulating of crack problems under mechanical loads with enhanced element-free Galerkin method. Approximation of values in EFGM is only dependent on the distribution of nodal data. So, a MATLAB algorithm has been developed so that region susceptible to stress singularities are given rich nodal distribution in comparison to remaining domain in a well-defined way. Proposed algorithm has been embedded to traditional EFGM algorithm for simulation of fracture problems. Effective comparison for stresses has been made using proposed algorithm and traditional EFGM for similar capacity of nodes. Results show that proposed algorithm produced more clear and accurate stress fields than traditional EFGM. Also Proposed algorithms show average relative deviation of 3.91% with the results accumulated from traditional EFGM procedure. So proposed EFG algorithm can be successfully used for enhanced fracture simulations.

Keywords Crack · Fracture · EFGM · MATLAB

1 Introduction

Role of fracture mechanics has increased with the advancement materials and their applications. Development of damage tolerant design has been the primary aim of the fracture mechanics. With the increasing demands of advancing world there has been increase in development of new materials along with increase in complexity in their applications. So proper analysis of the materials is very important to avoid any catastrophic failure. All materials have some kind of discontinuities and flaws which are susceptible to high stress singularities during their application. Proper estimation of such stresses prior to application is necessary for preventing any failure (Carlson

A. Awasthi (✉)
Mechanical Engineering, Thapar Institute of Engineering and Technology, Patiala, Punjab, India
e-mail: ayush.awasthi8080@gmail.com

M. Pant
Department of Mechanical Engineering, National Institute of Technology, Hamirpur, H.P., India

M. Ram et al. (eds.), *Advances in Mathematics for Engineering Sciences*, Mathematical Engineering, https://doi.org/10.1007/978-3-031-95693-5_12

et al. 2012). Various methods have developed over the years for modelling such fractures. Materials are subjected to various kind of mechanical and thermal loadings during its application. Stress intensity factor has been most important factor used in determining safe design in fracture mechanics (Gorbatikh et al. 2007). Over time, a range of analytical and computational techniques have been developed to model and simulate fracture phenomena. Analytical methods based on empirical formulas are limited to specific problems only, on the other hand computation methods have emerged to be more versatile tool in handling fracture problems. Finite Element Method has emerged as most common computational method used for simulation of such problems (Krimotat and Sedarat 2014; Staden et al. 2006). FEM suffers various limitation based on mesh connectivity in problems associated with large deformations, moving boundaries etc. Owing to the limitation of FEM prompted the development of advanced techniques such as extended finite element method and various meshfree techniques (Motamedi and Mohammadi 2010; Sukumar and Prévost 2003; Atluri and Zhu 2000). Meshfree method have gained importance over the years because they do not have any mesh associated with them and whole discretization and approximation process is based on nodes (Nguyen et al. 2008; Garg and Pant 2018). Various meshfree methods developed so far include diffused element method (DEM) (Nayroles et al. 1992), Reproducing kernel particle method (Chen et al. 1996), Smoothed particle hydrodynamics (Batra and Zhang 2007), Element free Galerkin method (EFGM) (Garg and Pant 2017; Belytschko et al. 1994) etc. Out of all these meshfree method Element free Galerkin method proposed by Beyltscho et al. in 1994 has emerged to be most significant in handling crack problems (Belytschko et al. 1994, 1996). Domain discretization in EFGM is based on generation of gaussian quadrature of some specific order around the nodal points. Value of field variables is approximated using MLS shape function in EFGM. Over the years EFGM has become the prominent tool used in simulating fracture problems with relatively good accuracy (Awasthi and Pant 2022; Wang 2015; Rao and Rahman 2004, 2001; Sukumar et al. 1997; Awasthi and Pant 2022). Due to integration iterations over gauss points, EFGM simulations need a significant amount of computational time (Belinha and Dinis 2006). Researchers are constantly attempting to improve the computing efficiency of this technology by introducing enhancing procedures. One of such approach is developed and implemented in the current work. Reducing the computational time with good accuracy is of primary importance. Accuracy of simulation is directly dependent on density of nodal data. High nodal density yields better results but increases computational costs. More nodal data signify more gauss points leading to tedious integration of domain.

Different domain subjected to different kinds of loads have few regions susceptible to high stresses. Cracked media have more stress singularity around the crack than rest of domain. So accurate calculation of these stress is important for a damage tolerant design. High density of nodes is necessary around the crack for better approximation results. This work encompasses for development of algorithm for modifying the nodal distribution without compromise in the accuracy. The aim of current algorithm is enhancement of computational efficiency of EFGM using limited costs. Proposed algorithm has been tested for simulation of fracture problems for effective comparison of results.

2 EFGM Formulation

2-D domain of body bounded by Γ under different loadings is shown in Fig. 1. Here, u denotes displacement vector, $\bar{t}$ represent the traction force and $\bar{n}$ is unit normal.

EFGM utilises Moving least square (MLS) approximation for evaluation of field variables. Originally MLS has been introduced by Lancaster and Salkauskas for data interpolations (Erkmen and Bradford 2010). Their concept was later utilised in EFGM for evaluation of field variable, $u^h(x)$ which is shown as:

$$u^h(x) = \sum_{j=1}^{m} p_j(x)a_j(x) = p^T(x)a(x) \tag{1}$$

Here $p^T(x)$ is full basis function & $a(x)$ represent vector of undefined coefficients:

$$p^T(x) = \left[1, x, y, xy, x^2, y^2 ......x^{k\prime}, y^{k\prime}\right] \tag{2}$$

$$a^T(x) = [a_1(x), a_2(x) ...... a_n(x)] \tag{3}$$

The shape function value is determined by evaluating $a(x)$, determined for any point x, by the minimizing weighted least square sum of difference of local approximation $(u^h(x))$ and field function parameter (u_I). Weighted least square $(L(x))$ is shown as:

$$L(x) = \sum_{I=1}^{n} w(x - x_I)\left[p^T(x)a(x) - u_I\right]^2 \tag{4}$$

Here, $w(x - x_I)$ is weight function linked to node I, with n number of nodes in the influence domain linked to point (x). After implementing minimization using

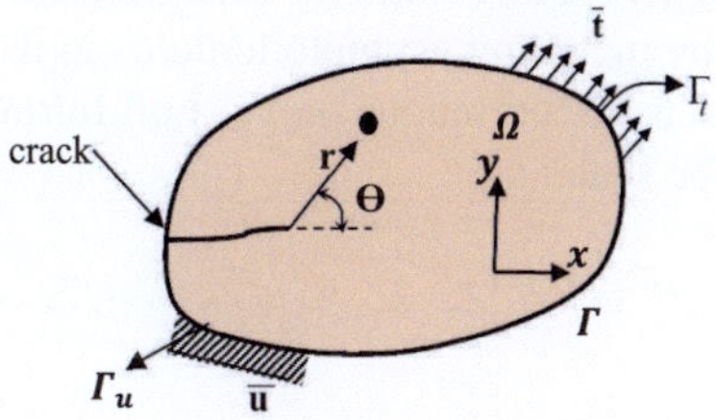

Fig. 1 Domain representation with boundary conditions

$\delta L/\delta a = 0$, equation gets modified as:

$$A(x)a(x) = B(x)u \tag{5}$$

where,

$$A(x) = \sum_{I=1}^{n} w(x - x_I)\, p(x_I) p^T(x_I) \tag{6}$$

$$B(x) = \{w(x - x_1)p(x_1),\ (x - x_2)p(x_2),\ (x - x_3)p(x_3) \ldots\ldots (x - x_n)p(x_n)\} \tag{7}$$

Substituting value of $a(x)$ from Eq. (5) into Eq. (1), the resulting approximation function is given as:

$$u^h(x) = \sum_{I=1}^{n} \varphi_I(x) u_I \tag{8}$$

Here φ_I represents mesh-free shape function, given by relation:

$$\varphi_I(x) = \sum_{j=1}^{m} p_j(x)\left(A^{-1}(x)B(x)\right)_{jI} = p^T A^{-1} B_I \tag{9}$$

Weight function $(w(x - x_I))$ preference significantly contributes to exactness of approximation. Its continuity and smoothness is in straight link with the shape function $(\varphi_I(x))$. Cubic spline function utilized for this study is represented as (Pant et al. 2011):

$$w(x - x_I) = \begin{pmatrix} 4r^3 - 4r^2 + 2/3 & r \le (1/2) \\ -4r^3/3 + 4r^2 - 4r + 4/3 & (1/2) < r \le 1 \\ 0 & r > 1 \end{pmatrix} \tag{10}$$

Here, $r = |x - x_1|/d_{mI}$, $(x - x_I)$ is nodal distance between (x_I) and evaluation point x, d_{mI} is domain of influence of node (x_I) and is taken such that the matrix is non-singular for each point in the domain.

Crack can be modelled by including some additional terms in basis function. There is discontinuity in displacement fields at crack surface which can be modelled by including asymptotic fields in the basis. This can be stated as: $u = \overline{u}$ on Γ_u and u is discontinuous on Γ_u. Full Intrinsic enrichment used for modelling of crack can be stated as:

$$P^T(x) = \left[1,\ x,\ y,\ \sqrt{r}\cos\theta/2,\ \sqrt{r}\sin\theta/2,\ \sqrt{r}\sin\theta/2\sin\theta,\ \sqrt{r}\cos\theta/2\sin\theta\right] \tag{11}$$

Interaction integral used for computing stress intensity factors in mechanical loading can be stated as (Garg and Pant 2018):

$$M^{(U)} = \int_A \left[\sigma_{ij}^{(1)} \frac{\partial u_i^{(2)}}{\partial x_1} + \sigma_{ij}^{(2)} \frac{\partial u_i^{(1)}}{\partial x_1} - W^U \delta_{ij} \right] \frac{\partial q}{\partial x_j} dA \tag{12}$$

3 Nodal Variation Algorithm

Nodal distribution plays an important role in EFGM as all approximations are based on integration over the nodal data. Gauss points are built only in accordance with these nodes. Standard EFGM utilizes unform distribution of nodes over the whole domain for approximation of results. As shown in Fig. 2a, region adjacent to crack is more susceptible for stress singularities so its correct evaluation is very important. So, idea of current algorithm is based on fact that area near to crack should be given higher significant effort for approximation than rest domain. Proposed algorithm provides higher nodal density around crack region which decreases in direction away from crack. Nodal density is variable in whole domain and is directly linked to crack location. Proposed algorithm is developed on MATLAB and focusses on mannered distribution of nodes. Aim of current logic is to increase computational efficiency of traditional EFGM by effectively management of nodal distributions. Nodal distributions are managed such that accurate solutions can be achieved using smaller number of nodes. Nodal distribution by proposed algorithm for edge crack is shown in figure Fig. 2b = Nodes are densely located along both sides of the crack, which gradually decreases towards top & bottom ends in a structured manner. The flowchart for the current algorithm is presented in Fig. 3.

The steps for the proposed algorithm are as follows:

1. Division of domain in two distinct regions along crack
   - There is overlapping of one edge of both domains near to the crack surface. Each domain overlaps with value: $(H - y) + (W/1000)$. Where, H represents total length of domain, y denotes distance of the crack and bottom edge, and W indicates the breadth of domain.
2. Generation of nodes on both the domains:
   - The domain edge near the crack has low vertical distance (v-ratio), resulting in a higher density of nodes in that region. Higher v ratio is given at edge near to crack domain for both the domains leading to generation of nodes ensuring a high density of nodes near both sides of the crack, gradually decreasing uniformly towards the top and bottom ends.
3. Generating virtual cells for the entire domain and subsequently placing Gauss points within these cells.

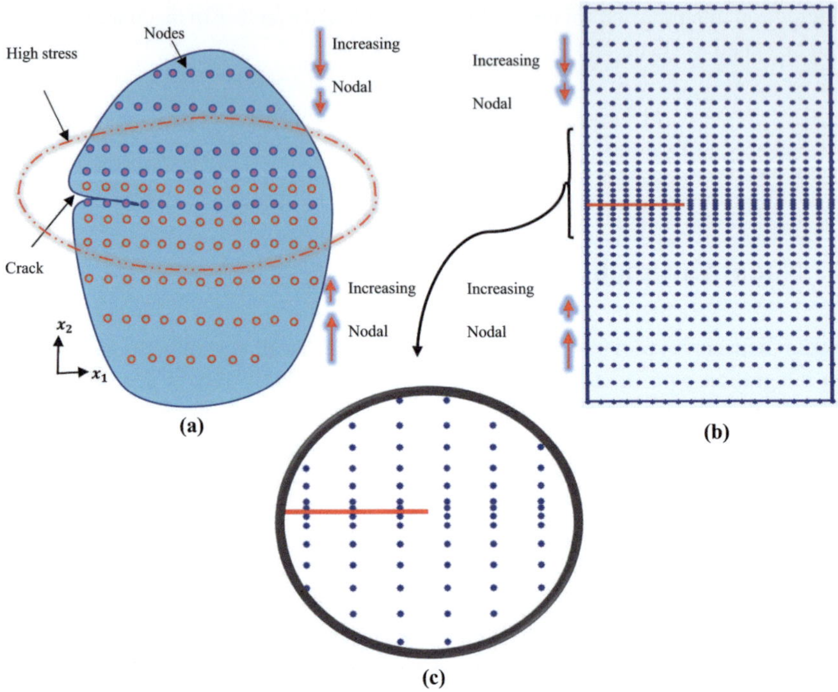

Fig. 2 **a** Domain illustration of nodal variation algorithm **b** Nodal variation algorithm in cracked domain **c** Region having high density of the nodes

4 Problem Formulation and Discussion

In the current problem rectangular plate (H*W) having edge crack under mechanical loading have been considered. There is an edge crack of size *a* located at a distance of *D*/2 from bottom of the plate, oriented parallel to the horizontal axis. Rectangular plate of height 2 units and width 1 unit is subjected to tensile loading of $\sigma_0 = 1$ unit as demonstrated in Fig. 4. Mechanical properties of plate are as follows: $E = 1 \times 10^6 units$, $\nu = 0.3$. Current problem has been simulated using proposed criterion incorporated into EFGM. Results obtained by proposed EFGM criteria have been validated by reference results (Salari-Rad et al. 2011) for (a/W = 0.4) where domain has been discretized using 20*40 (800) nodes which has been considered as ideal number of nodes in reference results. In the current problem an approach to get accurate results with lesser number of nodes by effective distribution of nodes has been done by simulating same problem with 20*25 (500) nodes. Nodal variation for this problem has been shown in Fig. 2b. These obtained results have been compared with those to traditional EFGM simulation at 800 nodes (higher number). These results have been shown in Fig. 5. From results we can see that obtained results by

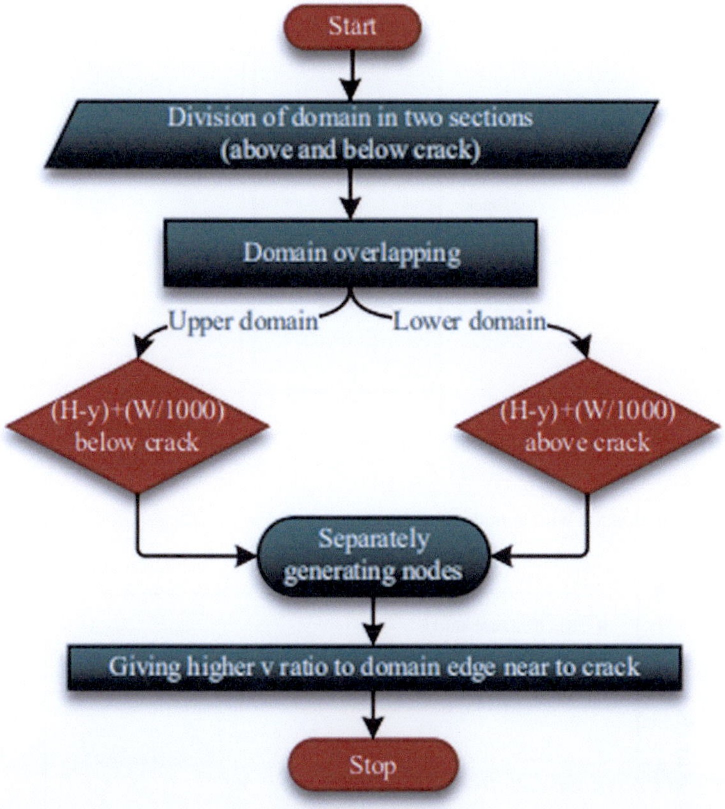

Fig. 3 Flowchart for variable nodal density criterion

current proposed criteria have good agreement with standard EFGM results. Average relative deviation of results have been calculated using:

$$relative\ deviaiton = \left|\left(\frac{K_I^{traditional\ EFGM} - K_I^{proposed\ EFGM}}{K_I^{traditional\ EFGM}}\right) \times 100\right| \quad (13)$$

Average relative deviation of 3.91% is observed using the proposed criteria. This establishes the efficiency of proposed criteria in modelling fracture simulations using prosed criteria. Also results for stress distributions σ_{xx} and σ_{yy} for one such simulation for a/W 0.4 is presented to show the efficiency of proposed criteria. The above-described problem is simulated using same number of 800 nodes by both proposed criteria and traditional EFGM. From Figs. 6 and 7 we can see that the stress fields obtained by proposed EFGM are more accurate and sharp stress fields by traditional EFGM. This establishes an effective approach for utilizing nodal distribution for better accuracy thereby enhancing the computational efficiency of traditional EFGM.

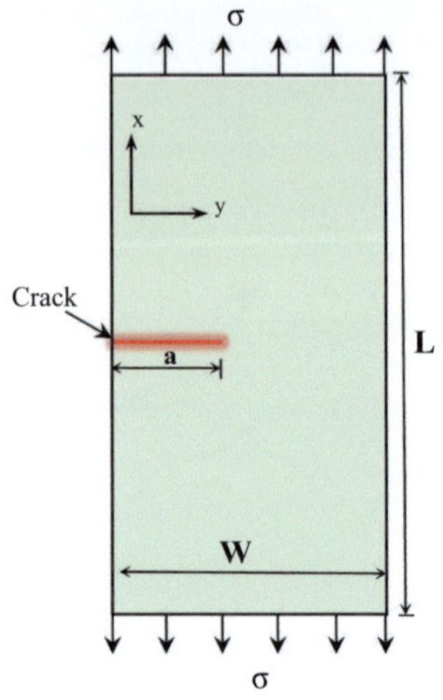

Fig. 4 Problem domain with boundary conditions

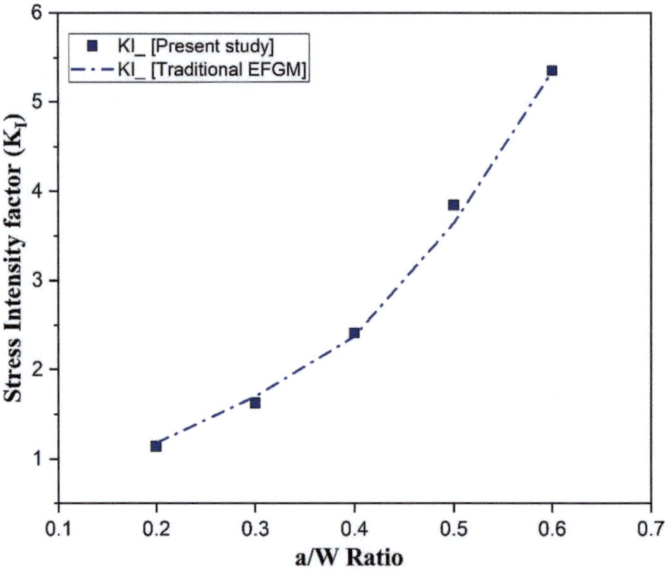

Fig. 5 Variation of SIF with a/W ratio

5 Conclusions

Current work focused on enhancement of computational efficiency of EFGM by incorporating nodal variation algorithm by developing generic MATLAB codes. Effective utilization of nodal distribution has been done by providing higher density

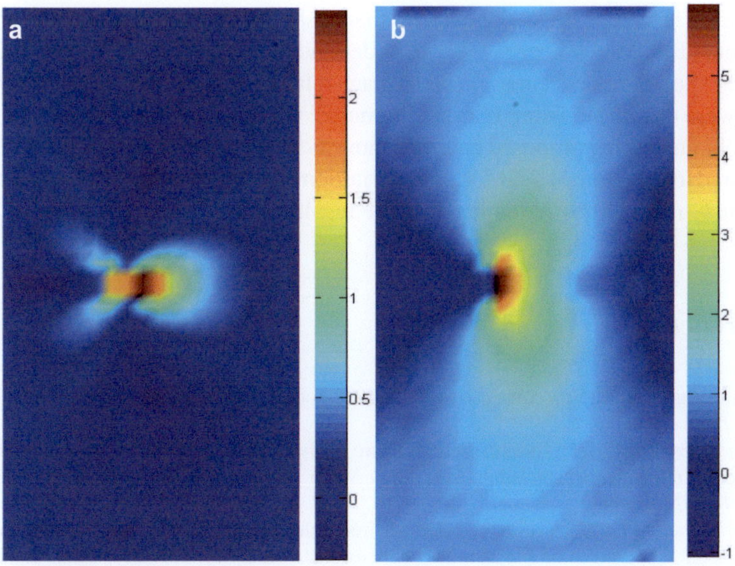

Fig. 6 a σ_{xx} without criterion **b** σ_{yy} without criterion

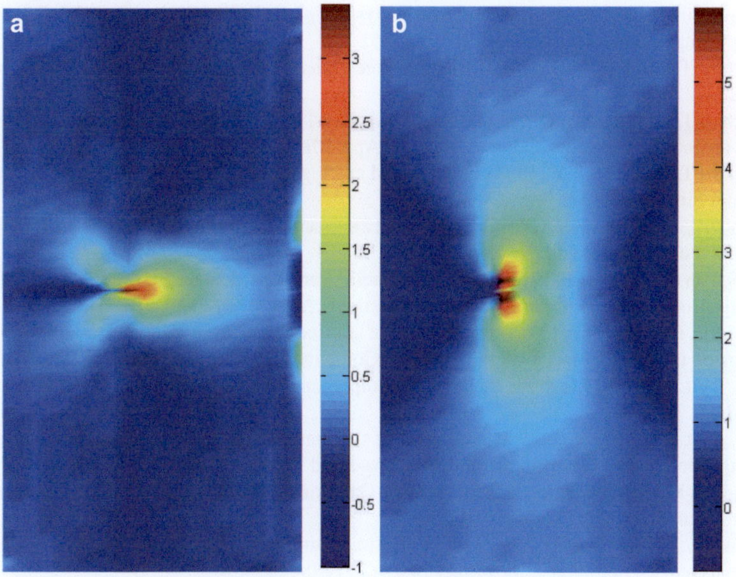

Fig. 7 **a** σ_{xx} with criterion **b** σ_{yy} with criterion

of nodes in region of stress singularities than rest of domain. The results obtained by traditional EFGM at higher number of nodes and proposed EFGM at lower numbers of nodes show similar accuracy at the average relative deviation of 3.91%. Also stress fields were calculated using both traditional EFGM and proposed approach. Stress fields obtained using proposed criterion show more accurate and sharp stress fields as compared to traditional EFGM. Current criterion effectively increases computational efficiency of EFGM by obtaining accurate results at limited number of nodes.

References

Atluri SN, Zhu T (2000) New concepts in meshless methods. Int J Numer Meth Engng 47(1−3):537–556

Awasthi A, Pant M (2022) Thermoelastic fracture analysis in orthotropic media using optimized element free Galerkin algorithm. Mech Adv Mater Struct 1−26. https://doi.org/10.1080/15376494.2022.2114039

Awasthi A, Pant M (2022) A revamped element-free Galerkin algorithm for accelerated simulation of fracture and fatigue problems in two-dimensional domains. Iran J Sci Technol-Trans Mech Eng 0123456789. https://doi.org/10.1007/s40997-021-00471-z

Batra RC, Zhang GM (2007) Search algorithm, and simulation of elastodynamic crack propagation by modified smoothed particle hydrodynamics (MSPH) method. Comput Mech 40(3):531–546

Belinha J, Dinis LMJS (2006) Elasto-plastic analysis of plates by the element free Galerkin method. Eng Comput (Swansea, Wales) 23(5):525–551

Belytschko T, Lu YY, Gu L (1994) Element free Galerkin methods. Int J Numer Methods Eng 37:229–256

Belytschko T, Krongauz Y, Organ D, Fleming M, Krysl P (1996) Meshless methods: an overview and recent developments. Comput Methods Appl Mech Eng 139(1–4):3–47

Belytschko T, Gu L, Lu YY (1994) Fracture and crack growth by element free Galerkin methods. Model Simul Mater Sci Eng 2(3A):519–34

Carlson RL, Kardomateas GA, Craig JI (2012) Fracture mechanics with an introduction to micromechanics, 3rd edn. pp 13–17

Chen JS, Pan C, Wu CT, Liu WK (1996) Reproducing kernel particle methods for large deformation analysis of non-linear structures. Comput Methods Appl Mech Eng 139(1–4):195–227

Erkmen RE, Bradford MA (2010) Elimination of slip-locking in composite beam-column analysis by using the element-free Galerkin method. Comput Mech 46(6):911–924

Garg S, Pant M (2018) Meshfree methods: a comprehensive review of applications. Int J Comput Methods 1:15(4)

Garg S, Pant M (2017) Numerical simulation of thermal fracture in functionally graded materials using element-free Galerkin method. Sadhana-Acad Proc Eng Sci. 42(3):417–431

Garg S, Pant M (2018) Accelerated element-free Galerkin method for analysis of fracture problems. J Brazilian Soc Mech Sci Eng 40(11):1–14

Gorbatikh L, Lomov S, Verpoest I (2007) On stress intensity factors of multiple cracks at small distances in 2-D problems. Int J Fract 143(4):377–384

Krimotat A, Sedarat H (2014) Structural modeling, 2nd edn. In: Bridge engineering handbook: fundamentals. pp 253–69

Motamedi D, Mohammadi S (2010) Dynamic crack propagation analysis of orthotropic media by the extended finite element method. Int J Fract 161(1):21–39

Nayroles B, Touzot G, Villon P (1992) Computational mechanics generalizing the finite element method: diffuse approximation and diffuse elements. Comput Mech 10

Nguyen VP, Rabczuk T, Bordas S, Duflot M (2008) Meshless methods: a review and computer implementation aspects. Math Comput Simul 79(3):763–813
Pant M, Singh IV, Mishra BK (2011) Evaluation of mixed mode stress intensity factors for interface cracks using EFGM. Appl Math Model 35(7):3443–3459
Rao BN, Rahman S (2001) A coupled meshless-finite element method for fracture analysis of cracks. Int J Press Vessel Pip 78(9):647–657
Rao BN, Rahman S (2004) An enriched meshless method for non-linear fracture mechanics. Int J Numer Methods Eng 59(2):197–223
Salari-Rad H, Rahimi-Dizadji M, Rahimi-Pour S, Delforouzi M (2011) Meshless EFG simulation of linear elastic fracture propagation under various loadings. Arab J Sci Eng 36(7):1381–1392
Sukumar N, Prévost JH (2003) Modeling quasi-static crack growth with the extended finite element method part I: computer implementation. Int J Solids Struct 40(26):7513–7537
Sukumar N, Moran B, Black T, Belytschko T (1997) An element-free Galerkin method for three-dimensional fracture mechanics. Comp Mecha 20:170–175
Van Staden RC, Guan H, Loo YC (2006) Application of the finite element method in dental implant research. Comput Methods Biomech Biomed Engin 9(4):257–270
Wang HS (2015) An extended element-free Galerkin method for thermo-mechanical dynamic fracture in linear and nonlinear materials. Comput Mater Sci 98:366–371

Securing and Authenticating Vedic Documents: A Blockchain-Based Approach

Nilima Patel and Mayank Aggarwal

Abstract India has a vast history of culture and thousands of books have been written. It is essential to preserve these vedic manuscripts to maintain the integrity of ancient Indian literature as they offer great knowledge about life, Ayurveda and civilizations. Tampering and deterioration with time are the significant challenges in preserving these Vedic manuscripts; it is necessary to authenticate and safeguard these manuscripts. Conventional methods like manual and physical verification processes are ineffective in capturing any alterations, forgery, and copyright protection, which can not prevent tampering. This chapter proposes a blockchain technology-based framework to verify and validate vedic texts, which can enhance the digitized vedic manuscripts in further deformation from originality. Because blockchain technology has a distinct hash value to identify its content, it offers decentralization and a digital signature which ensures that any changes are promptly communicated. Access control of the vedic manuscripts is managed by carefully crafted smart contracts, allowing only those with permission to examine and verify the documents. With the blockchain's transparency, decentralization makes traceability and unambiguous ownership attribution feasible. The suggested framework incorporates smart contracts, text encryption, hashing and digitalization.

Keywords Blockchain · Vedic document · Security · Smart contract · Hashing

1 Introduction

For several centuries, Indian Vedic knowledge has been preserved with conventional oral tradition in which one generation verbally transferred the knowledge to the next generation, and no written documents were available. With the development

N. Patel (✉) · M. Aggarwal
Computer Science and Engineering, Gurukula Kangri Deemed to be University, Haridwar, Uttarakhand, India
e-mail: 22631001@gkv.ac.in

M. Aggarwal
e-mail: mayank@gkv.ac.in

M. Ram et al. (eds.), *Advances in Mathematics for Engineering Sciences*, Mathematical Engineering, https://doi.org/10.1007/978-3-031-95693-5_13

of languages and writing skills, thousands of documents have been written, which have precious knowledge on various aspects of life. These sacred scriptures, which have been passed down over the ages, contain significant historical and cultural information. However, traditional preservation methods, such as physical storage or manual inspection, are insufficient. These techniques cannot completely prevent damage or unauthorized alterations. The government of India has taken initiatives of digitalization of these documents and texts, which are now more easily accessible because of digital technology. Still, there are drawbacks as well (Sharma et al. 2023; Jeste and Vahia 2008), some of them are integrity, authenticity, data security and central control. Blockchain technology can offer a robust and decentralized solution to these issues and ensures that Vedic texts remain authentic and protected.

Once the old historical texts are registered and placed on the cloud and powered by blockchain, data cannot be changed, and if anyone tries to alter the text, it can be easily traced and notified, and hence, the integrity of texts can be ensured. This approach begins by converting Vedic texts into a digital format. After that, cryptographic hashing is used to create a unique digital signature. Each document receives a unique hash value stored on the blockchain. This makes it easier to find the efforts if done by anyone to change the document (Arigela and Voola 2022). Also, the smart contracts written in blockchain restrict access to the texts and allow only reputable academics and organizations to view and verify them. This method keeps people from messing with historical texts without permission, stops anyone from making fake copies, and keeps a super safe and clear record of what has been happening with the documents. The probability of data loss is much less with the application of blockchain technology since it is not reliant on any one individual or organization.

Blockchain technology is even safer because the data is shared among several computers. Tracking is also done each time the document is utilized, making it obvious who owns it. Because of these characteristics, blockchain is a powerful tool for preventing the theft, unauthorized sharing, and false copying of Vedic texts. The automatic checking procedure also removes the need for human verification, which may be inaccurate (Athanere and Thakur 2022).

The proposed framework in this chapter makes Vedic texts safer and less difficult to apply through blockchain technology. Because the texts are saved in a decentralized manner, they live safely and unchanged for future generations. Smart contracts assist in controlling who can access the texts, making them useful for researchers and students. As the blockchain era keeps enhancing, future research can find better methods to save the texts and make checking the texts quicker and simpler.

2 Related Work

Preserving antique texts like Vedic manuscripts has been difficult because they can effortlessly get broken through the environment, physically managed, or human beings changing them without permission. In the past, these manuscripts were mainly stored securely in temples, libraries and colleges. They have been written on susceptible materials like palm leaves and birch bark, which can easily be harmed with moisture, bugs or getting old over the years (Adithya et al. 2023). To shield them, people have attempted to store them in special cool places and use chemical substances to stop the harm. But those steps only slow down the harm; they do not forestall the manuscripts from breaking down (Souza and Aganette 2022).

With the advancement of virtual technology, many institutions have initiated digitization projects to convert historical manuscripts into digital codecs, enhancing accessibility and lowering the hazard of physical deterioration. However, virtual archives introduce new issues, including hacking, unauthorized alterations and statistics centralization, that could lead to loss or manipulation of information (Liu et al. 2022). While traditional encryption strategies have been used to stabilize digital manuscripts, they have not offered a tamper-proof technique to verify the authenticity of texts over the years. Blockchain technology has emerged as a solution for ensuring file integrity and stopping unauthorized modifications. Studies have explored the use of blockchain for securing historical texts, leveraging cryptographic hashing and decentralized ledgers to hold an immutable document of file authenticity (Awaji et al. 2020).

Several blockchain-based frameworks for report authentication have been proposed in recent years (Chaudhari and Lakshmisudha 2024). Examined blockchain programs in educational research, demonstrating how decentralized ledgers may be used to affirm the authenticity of digital documents. Similarly, (Mohan et al. 2022) explored using smart contracts in access management, allowing only authorized people to retrieve and confirm sensitive historical documents. Additionally, (Zhong 2024) highlighted the scalability of blockchain-based preservation systems, emphasizing how decentralized storage answers can enhance long-time period record safety while maintaining accessibility. Blockchain integration in document upkeep has shown promising outcomes in numerous cultural and historical contexts. Case studies in archival control and museum facts have demonstrated that blockchain can offer a transparent, tamper-proof verification method for virtual artifacts (Satybaldy et al. 2022). These findings suggest that blockchain technology can effectively address the shortcomings of traditional and virtual preservation methods, making it a viable solution for securing and authenticating Vedic manuscripts for future generations of blockchain technology in document authentication. Table 1 provides a comparison of traditional and blockchain-based methods for vedic document preservation and highlights the variations between the two methods to ensure the safety of vedic files, and it is found that the application of blockchain offers advanced security, efficiency and transparency in document authentication and verification.

Table 1 Comparison of traditional and blockchain-based methods

Aspect	Traditional methods	Blockchain-based approach	Previous work
Authenticity verification	Relies on expert analysis; prone to human error	Uses cryptographic hashing and unique digital signatures	Raeisi-Varzaneh et al. (2024)
Tamper resistance	High risk of tampering and forgery	Immutable blockchain records prevent alterations	Thilakavathy et al. (2023)
Access control	Centralized authorities regulate access	Smart contracts automate and control user access	Satybaldy et al. (2022)
Decentralization	Stored in select libraries, temples and institutions	Distributed ledger ensures transparency and redundancy	Chavan et al. (2023)
Data integrity	Physical degradation and loss over time	Cryptographic hashing ensures document integrity	Junejo et al. (2021)
Forgery prevention	Susceptible to counterfeiting and unauthorized modifications	Hash verification detects any changes instantly	Kashyap and Subramanian (2023)
Scalability	Limited by physical storage constraints	Blockchain allows scalable digital archiving	Zhao et al. (2024)
Transparency	Limited visibility and verification of document history	Blockchain maintains an auditable, transparent record	Bris et al. (2021)
Verification process	Subjective, based on manual inspection	Objective verification through cryptographic techniques	Devanbu and Stubblebine (2000)
Interoperability	Requires physical verification across institutions	Smart contracts enable automated, cross-platform validation	Mustafa et al. (2023)
Security	Prone to theft, loss and restricted accessibility	Secure through cryptographic encryption and decentralized storage	Singh et al. (2024)
Legal recognition	Recognized but susceptible to disputes	Legally binding through digital notarization	Robert (2019)
Cost efficiency	High costs for physical storage and restoration	Reduced long-term costs through digital preservation	Wagner and Clippele (2023)

(continued)

Table 1 (continued)

Aspect	Traditional methods	Blockchain-based approach	Previous work
Time efficiency	Lengthy manual verification processes	Quick and automated authentication	Saha et al. (2023)
Use of IPFS	Not applicable	IPFS enables decentralized file storage for document integrity	Nouman et al. (2022)

3 Proposed Framework

The proposed Vedic text file management system is powered by blockchain technology to enhance the authenticity and integrity of historical texts. To observe blockchain in vedic texts upkeep, the entire text is first digitized, where bodily manuscripts are transformed into the virtual format. Then, this digitized record is processed via a cryptographic hashing mechanism to generate a completely unique hash code that serves as a virtual fingerprint of the file.

3.1 Digitization and Hashing

The first step in this system is to convert the old manuscripts into clear, high-quality digital copies. After making these digital versions, a special cryptographic hashing process is used. This process creates a unique code, called a hash value, for each document. This hash is saved safely on the blockchain. If anyone tries to change the document later, the hash will also change, making it easy to spot any unauthorized edits right away (Li et al. 2022).

To generate the hash code, the cryptographic hash function SHA-256 is used, which has a digest length of 256 bits. It is a keyless hash function where a message is processed by blocks of $512 = 16 \times 32$ bits, each block requiring 64 rounds.

$$\text{Hash function} : \text{H(x)} = \text{h}$$

In a blockchain, each block contains a reference to the hash of the previous block, forming a cryptographic chain. The hash of the current block h_i is computed based on its own data and the hash of the previous block h_{i-1}. This chaining mechanism ensures that any change in a block will alter all subsequent hashes, making tampering detectable and infeasible.

$$Block\ hash : h_i = H(Block\ Data_i + h_{i-1})$$

In Proof-of-Work (pow) consensus algorithms, mining adds a new block to the blockchain. Mining involves solving a cryptographic puzzle by finding a special number called a nonce n such that the resulting hash meets a predefined difficulty target T. The goal is to find a nonce for which the hash of the block's content is numerically less than this target.

$$Mining : \ Find\ noonce\ n\ such\ that : H(BlockData + \ h_{i-1} + n) < T$$

This process is computationally expensive and probabilistic, requiring extensive trial-and-error. However, once a valid hash is found, it is trivial for other nodes to verify. This mechanism underpins the security and consensus of decentralized blockchains like Bitcoin and Ethereum (before the switch to Proof-of-Stake). This cryptographic process ensures that blockchain data remains tamper-proof, verifiable, and secure.

3.2 Encryption and Secure Storage

To keep important information safe, the documents are locked using encryption and stored in a decentralized storage system. This means the documents are broken into secure parts and saved in many places, not just one. On the blockchain, only basic details like the document's unique code (hash value) and some extra information (metadata) are saved. This setup keeps the documents private while still proving that they are real and have not been changed (Chavan et al. 2023).

3.3 Smart Contracts for Access Control

Smart contracts are special computer programs that control who can open and read the documents. Only trusted users, like scholars or approved institutions, are allowed to unlock and see the texts. Each time someone tries to access the document, the request is safely recorded on the blockchain. This way, there is a clear and trackable record showing who accessed the documents and when, making the whole process open and trustworthy (Colin et al. 2024).

3.4 Verification and Validation

To check if a document is real and unchanged, users can compare its present hash value with the hash saved on the blockchain. If both hash values are the same, the document is original and has not been altered. However, if the hash values differ, someone might have tampered with the document. This approach ensures that digital documents live safe and proper throughout the years (Hargude 2021).

As shown in Fig. 1, once the hash is made, the document is saved in a special storage system called IPFS (Interplanetary File System). In IPFS, the document is stored securely, ensuring that it cannot be altered or damaged. It is not saved in one single place but in many places, so even if one place has a problem, the document can still be found easily. This way, the document stays safe and easy to get any time.

To make the document even more secure and clear, its unique hash value is saved on the blockchain. Along with this, the time when the document was added is also recorded. This helps prove exactly when the document was stored and stops any false claims. Smart contracts are used to control who can see or use the document. They automatically check and allow only trusted people to access it.

People can verify the document's authenticity using a simple verification system. Here, they compare the document's hash with the hash saved on the blockchain. If both matches, the document is real and unchanged. Unique cryptographic keys are used to maintain the whole lot's safety. These keys permit the most effective authorized users to deal with or look at the files securely. This method blocks all and sundry from making unwanted modifications while retaining the secure, unique, and recorded vedic texts on a gadget that no one can alter secretly. By combining digitization, sturdy cryptographic protection, blockchain's obvious nature and decentralized

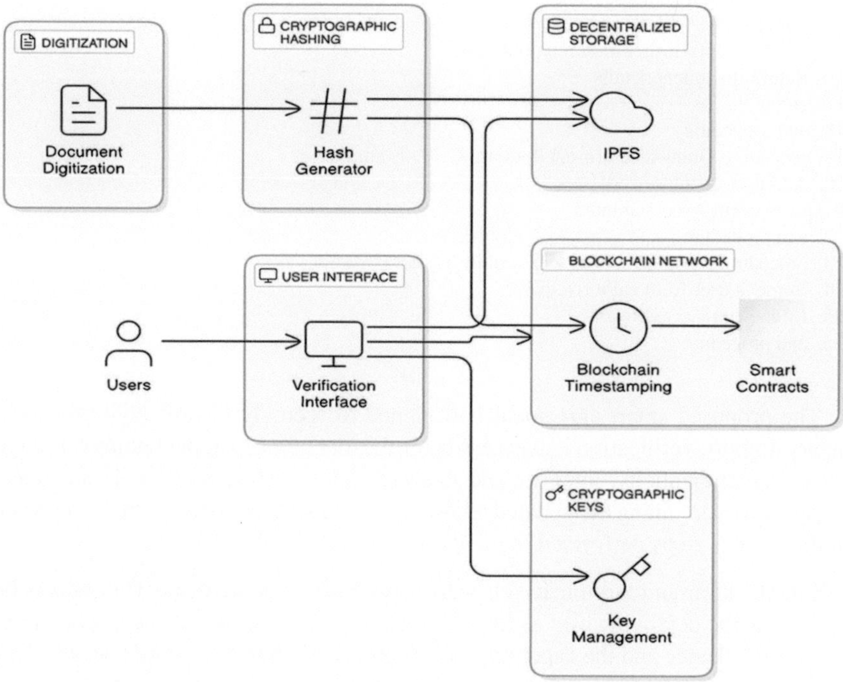

Fig. 1 Framework of vedic document digitization and storing on the blockchain

garage, this machine creates an entire and modern-day way to shield Vedic literature. It guarantees those valuable historical texts live safely and authentically, free from changes. This method facilitates holding them properly so future generations can get the right information offered by our rich Vedic literature.

4 Implementation

Algorithm: Vedic Document Preservation Using Blockchain

```
Initialize: Set admin as contract deployer
Initialize: Grant admin access to the system
Procedure UploadDocument(ipfsHash, documentHash) Require: User is authorized
if documentHash exists in storage then
Reject upload (document already exists)
else
Store (ipfsHash, documentHash, timestamp, uploader)
Emit event DocumentUploaded
end if
end procedure
procedure VerifyDocument(documentHash)
if documentHash does not exist then
Return "Document not found"
else
Retrieve (ipfsHash, timestamp, uploader)
Return document details
end if
end procedure
procedure GrantAccess(user) Require: Caller is admin
Add user to authorizedUsers
Emit event AccessGranted
end procedure
procedure RevokeAccess(user) Require: Caller is admin
Remove user from authorizedUsers
Emit event AccessRevoked
end procedure
```

The proposed smart agreement is designed to securely control digitized documents' import, verification and access control using blockchain technology. It guarantees the authenticity, integrity and managed admission to each vedic document along with maintaining the updated report of all interactions. Below is the description of the various steps performed in the implementation of the algorithm.

- Initialization and Admin Rights—At deployment, the agreement initializes by placing the deploying user as the administrator. The admin is granted full control over the device and the capability to authorize and revoke consumer access. This step establishes a trusted authority responsible for handling the report verification framework.

- Document Upload Procedure—The file upload method is initiated via the Upload Document feature. It requires the user to be authorized before intending to do so. The system then tests if the report's hash fee already exists in the garage. If the hash is determined, indicating a replica document, the add is rejected to prevent redundancy. If the file hash is particular, the machine records vital details, such as the IPFS hash, report hash, timestamp, and uploader's identity. Upon successful storage, an event is emitted to verify that the record has been uploaded.
- Document Verification Process—To verify the authenticity of any document, the Verify Document function is used. This function searches for the document's hash in the storage. If the hash is not found, the system returns a message stating "Document not found." Otherwise, the system retrieves the associated IPFS hash, timestamp and uploader details, providing verifiable proof of the document's originality and upload history.
- Access Control Management—The admin strictly controls user access through two key procedures. The Grant Access function allows the admin to add a user to the list of authorized individuals, emitting an event to indicate successful access granting. Conversely, the Revoke Access function enables the admin to remove a user's access rights, ensuring that unauthorized parties cannot interact with or manipulate the stored documents. This process is transparent and traceable, as every access change is recorded on the blockchain.

When we change Vedic manuscripts into digital format, it is very important to do it carefully. High-quality images of the original texts must be taken so that every detail is clear and correct. After making digital copies, specialists should check them well to ensure there aren't any errors. If even a small error occurs for the duration of digitization, the manuscript's meaning can be altered. So, using cautious checking and exact gear facilitates maintaining the virtual version authentic to the original. In this manner, the understanding within the manuscripts remains safe, accurate, and beneficial for future generations. To inspire research scholars working in vedic document preservation, blockchain-generated tokens may be used to present these token as special rewards. These rewards can be in the shape of academic tokens, certificates, or monetary grants. When students make a contribution with the aid of checking, studying, or verifying the texts, they are able to earn these rewards as recognition which makes them feel valued and motivated. By presenting such incentives, more excellent specialists and researchers might be inquisitive about joining the attempt, ensuring that the maintenance of historical understanding maintains in a stable and dependent way for future generations.

5 Challenges and Future Work

One principal issue in implementing blockchain in vedic documents preservation process is interoperability. Since numerous blockchain networks frequently use unique protocols, connecting them with modern structures becomes complex, hindering easy data exchange and functionality (Lai 2024). Additionally, many blockchain applications have unintuitive consumer interfaces, making them difficult for humans to apply without specialized expertise. This loss of user-friendly design can deter full-size adoption and restrict the generation's ability (Wang et al. 2018). Effective storage will become essential as the number of digitized manuscripts rises. In order to reduce on-chain record loads and enhance storage performance, researchers are looking into tactics like off-chain storage solutions, which store massive data documents outside the blockchain (Khare and Chaturvedi 2023).

Smart contracts help automate processes on the blockchain. Their efficiency can be improved by optimizing the code so they perform tasks faster and use less computational power (Odiljon and Gai 2019).

To preserve the manuscript's authenticity, future effort will involve developing sophisticated verification systems that could detect any alterations or tampering (Athanere and Thakur 2022).

6 Conclusion

Protecting Vedic manuscripts is important for maintaining authenticity and preventing unauthorized changes. Blockchain Technology provides a modern solution by providing a safe and transparent method for digitizing these ancient texts. Cryptographic hashing ensures that each document has a unique digital fingerprint so that any tampering is easily detected. Smart contracts automate verification processes so only authorized users can reach the material. Decentralized storage distributes data in many places, which reduces the risk of tampering or corruption. Using these techniques creates a strong system that protects the integrity of vedic literature for future generations. The ongoing research aims to increase the inclusion of fast Vedic literature digitization and the integration of blockchain methods to ensure that protection efforts are more efficient and accessible.

References

Adithya A et al (2023) A study on the influence of E-banking service quality on millennials: evidence from Indian banking sector. Int J Res Financ Manag 6(1). https://doi.org/10.33545/26175754.2023.v6.i1c.212

Arigela SSD, Voola P (2022) Survey on blockchain for sharing of students' credentials in education ecosystem. Int J Blockchains Cryptocurrencies 3(3). https://doi.org/10.1504/ijbc.2022.126286

Athanere S, Thakur R (2022) Blockchain based hierarchical semi-decentralized approach using IPFS for secure and efficient data sharing. J King Saud Univ-Comput Inf Sci 34(4). https://doi.org/10.1016/j.jksuci.2022.01.019

Awaji B, Solaiman E, Albshri A (2020) Blockchain-based applications in higher education: a systematic mapping study. ACM Int Conf Proceeding Ser. https://doi.org/10.1145/3411681.3411688

Bris A, Coffee JC, Lowenstein L, Roseackerman S et al (2021) Knights, raiders, and targets-the impact of the hostile takeover. J Bank Financ 37(1)

Chaudhari M, Lakshmisudha K (2024) Blockchain based document verification system. J Auton Intell 7(3). https://doi.org/10.32629/jai.v7i3.1010

Chavan DS, Dayama RN, Joshi PP, Pawar PP, Londhe PAD (2023) Document verification and validation using blockchain. Int Res J Mod Eng Technol Sci 4:1641–1645. https://doi.org/10.56726/irjmets35674

Colin LSH, Mohan PM, Pan J, Keong PLK (2024) An integrated smart contract vulnerability detection tool using multi-layer perceptron on real-time solidity smart contracts. IEEE Access 12. https://doi.org/10.1109/ACCESS.2024.3364351

Devanbu PT, Stubblebine SG (2000) Cryptographic verification of test coverage claims. IEEE Trans Softw Eng 26(2). https://doi.org/10.1109/32.841116

Hargude R, Ashutosh G, Nawale A, Adsure S, Engineer S, Department PC (2021) Verification and validation of certificate using blockchain. 9(6):2320–2882. [Online]. www.ijcrt.org

Jeste DV, Vahia IV (2008) Comparison of the conceptualization of wisdom in ancient Indian literature with modern views: focus on the Bhagavad Gita. Psychiatry 71(3). https://doi.org/10.1521/psyc.2008.71.3.197

Junejo AZ, Hashmani MA, Memon MM (2021) Empirical evaluation of privacy efficiency in blockchain networks: review and open challenges. Appl Sci 11(15). https://doi.org/10.3390/app11157013

Kashyap V, Subramanian N (2023) Secured authentication system for product verification: integrating blockchain and secure data management. In: 2023 14th international conference on computing communication and networking technologies. ICCCNT 2023. https://doi.org/10.1109/ICCCNT56998.2023.10307352

Khare V, Chaturvedi P (2023) Design, control, reliability, economic and energy management of microgrid: a review. e-Prime-Adv Electr Eng Electron Energy 5. https://doi.org/10.1016/j.prime.2023.100239

Lai Y (2024) Innovative strategies in logistics and supply chain management: navigating modern challenges. SHS Web Conf 183. https://doi.org/10.1051/shsconf/202418302020

Li L, Lu X, Wang K (2022) Hash-based signature revisited. Cybersecur 5(1). https://doi.org/10.1186/s42400-022-00117-w

Liu S, Qi S, Li Y, Wei B, Li P, Zhao J (2022) Controllable oscillated spin hall effect of Bessel beam realized by liquid crystal Pancharatnam-Berry phase elements. Light Sci Appl 11(1). https://doi.org/10.1038/s41377-022-00888-4

Mohan D, Alwin L, Neeraja P, Lawrence KD, Pathari V (2022) A private Ethereum blockchain implementation for secure data handling in internet of medical things. J Reliab Intell Environ 8(4). https://doi.org/10.1007/s40860-021-00153-2

Mustafa I, McGibney A, Rea S (2023) Smart contract life-cycle management: an engineering framework for the generation of robust and verifiable smart contracts. Front Blockchain 6. https://doi.org/10.3389/fbloc.2023.1276233

Nouman M, Ullah K, Azam M (2022) Secure digital transactions in the education sector using blockchain. EAI Endorsed Trans Scalable Inf Syst 9(35). https://doi.org/10.4108/eai.3-11-2021.171758

Odiljon A, Gai K (2019) Efficiency issues and solutions in blockchain: a survey. In: Lecture notes in computer science (including subseries lecture notes in artificial intelligence and lecture notes in bioinformatics, vol 11911. LNCS. https://doi.org/10.1007/978-3-030-34083-4_8

Raeisi-Varzaneh M, Dakkak O, Alaidaros H, Avci İ (2024) Internet of things: security, issues, threats, and assessment of different cryptographic technologies. J Commun 19(2). https://doi.org/10.12720/jcm.19.2.78-89

Robert H (2019) Legal recognition of blockchain registries and smart contracts. In: Blockchains & smart contracts legal and regulatory framework, no December

Saha S, Gupta H, Teja NR, Kothari S, Kumar DVNS (2023) An efficient blockchain and smart contracts based approach for document verification. In: 2023 international symposium on networks, computers and communications. ISNCC 2023. https://doi.org/10.1109/ISNCC58260.2023.10323829

Satybaldy A, Subedi A, Nowostawski M (2022) A framework for online document verification using self-sovereign identity technology. Sens 22(21). https://doi.org/10.3390/s22218408

Sharma S, Kadayat Y, Tyagi R (2023) Sustainable global democratic e-governance system using vedic scripture, artificial intelligence, cloud computing and augmented reality. In: Proceedings of the international conference on circuit power and computing technologies. ICCPCT 2023. https://doi.org/10.1109/ICCPCT58313.2023.10245405

Singh J, Singh G, Gupta D, Rani S, Srivastava G (2024) IoT ecosystem security via distributed ledger technology (blockchain versus IOTA): a bibliometric analysis research. J Circuits, Syst Comput 33(9). https://doi.org/10.1142/S0218126624300046

Souza LGS, Aganette EC (2022) Digital preservation: perspectives and relations with documental management and organizational culture. AtoZ 11. https://doi.org/10.5380/atoz.v11.82714

Thilakavathy P, Jayachitra S, Aeron A, Kumar N, Ali SS, Malathy M (2023) Investigating blockchain security mechanisms for tamper-proof data storage. In: 2023 international conference on communication, security and artificial intelligence. ICCSAI 2023. https://doi.org/10.1109/ICCSAI59793.2023.10421006

Wagner A, de Clippele MS (2023) Safeguarding cultural heritage in the digital era–a critical challenge. Int J Semiot Law 36(5). https://doi.org/10.1007/s11196-023-10040-z

Wang H, Zheng Z, Xie S, Dai HN, Chen X (2018) Blockchain challenges and opportunities: a survey. Int J Web Grid Serv 14(4):352. https://doi.org/10.1504/ijwgs.2018.10016848

Zhao G, He H, Di B, Guo Q (2024) BC-DERCP: blockchain-based copyright protection mechanism for digital educational resources. Educ Inf Technol. https://doi.org/10.1007/s10639-024-12612-3

Zhong XH (2024) The development and dissemination of intelligent storage management software of Chinese excellent traditional culture based on blockchain in the era of financial media. Heliyon 10(7). https://doi.org/10.1016/j.heliyon.2024.e27262

DeepFM-Driven Personalized Recommendations for Children: Tackling Data Sparsity with Negative Sampling

Ritu Rajal, Nishant Kumar, and Sanjeev Kumar

Abstract The demand for personalized recommendation systems growing rapidly during the recent years especially in entertainment and e-commerce fields. These recommendation systems face specific data consistency problems particularly when dealing with specialized groups such as young audiences. The proposed research develops an anime recommendation platform for children featuring a hybrid Deep Factorization Machines (DeepFM) and negative sampling structure which resolves data sparsity problems. The system creates a better dataset for model learning through negative sampling techniques alongside child-friendly genre dataset filtering. The implementation of DeepFM models incorporates Factorization Machines (FM) which detect low-order feature combinations with deep neural networks (DNN) for understanding high-order non-linear patterns. Anti-overfitting measures and generalization improvements are achieved through Regularization layers together with those based on Dropout. In this paper three metrics including precision, recall and Root Mean Squared Error (RMSE) are used for simulation of the model, that validate the proposed model increases recommendation accuracy while substantially decreasing user-item interactivity matrix sparsity levels. The model's ability to improve user coverage emerges through visual inspections of sparsity reduction which demonstrates its efficiency in generating specific recommendations for children. The research shows that DeepFM coupled with negative sampling represents a strong method for addressing data sparsity problems in recommendation systems because it has demonstrated potential use in additional recommendation areas.

R. Rajal (✉) · N. Kumar
Department of Computer Science and Engineering, Faculty of Engineering and Technology, Gurukula Kangri (Deemed to be University), Haridwar, India
e-mail: 20631002@gkv.ac.in

N. Kumar
e-mail: nishant@gkv.ac.in

S. Kumar
Department of Master of Computer Application, G. L. Bajaj Institute of Technology and Management, Greater Noida, India
e-mail: sanjeevnmc83@gmail.com

M. Ram et al. (eds.), *Advances in Mathematics for Engineering Sciences*, Mathematical Engineering, https://doi.org/10.1007/978-3-031-95693-5_14

Keywords DeepFM · Negative sampling · Recommendation system · Data sparsity

1 Introduction

Modern online services heavily rely on recommender systems which generate user-specific suggestions throughout e-commerce domains and streaming media services and different segments of the online world. These systems work to increase user satisfaction by managing enormous amounts of data and creating user prediction models based on previous system interactions. The main problem with recommendation systems involves data sparsity which reduces user-item engagement levels to the extent that it hinders the creation of precise suggestions. The recommendation of children's content faces special challenges because users provide minimal feedback while interacting. In data-sparse scenarios the classic recommendation approaches of collaborative filtering (CF) and content-based filtering (CBF) lose their effectiveness because they need plentiful user interaction data to discover correct patterns.

Deep learning-based recommender systems (DLRS) offer superior options as alternative solutions to handle current system limitations. Deep learning models demonstrate outstanding abilities for understanding complex user-item relations through Neural Collaborative Filtering (NCF) and Deep Factorization Machines (DeepFM) (Behera and Nain 2022). DeepFM demonstrates robustness for sparse datasets through its amalgamation of the low-order collaborative filtering in Factorization Machines (FM) with Deep Neural Networks (DNN) which learn high-order dependencies. Even so DeepFM-based models need an adequate amount of training data to function effectively and remain exposed to the risks stemming from sparse data inputs. The recommendation system relies on negative sampling as an approach to generate new training data which improves model stability when working with sparse interactions data.

The negative sampling method builds fabricated negative user-item interactions through an assumption of unmeasured relationships tending to be neutral cases. The method improves training speed while stopping overfitting on scarce positive data points allowing the model to recognize valid from invalid recommendation pairs. The application of negative sampling techniques in traditional recommendation models proved helpful to boost performance in matrix factorization alongside Word2Vec embeddings along with deep learning (DL) recommendation architectures. Research into combining DeepFM with negative sampling for children's content recommendations is lacking even though this approach could solve data sparsity issues effectively.

This chapter presented a DeepFM-based recommendation framework which uses negative sampling to boost children content recommendation quality. This approach enhances sparse learning by creating simulated negative features which DeepFM uses to discover fundamental together with complex element interconnections. Our proposed model obtains effectiveness analysis through extended testing on child-friendly anime data by measuring ranking performance through Precision@K, Recall@K and Normalized Discounted Cumulative Gain (NDCG@K).

The key contributions of our research are as follows:

1. To develop a DeepFM-based recommendation system that effectively captures both explicit and implicit user preferences in a sparse dataset.
2. Integration of negative sampling to balance positive and negative interactions, thereby improving training efficiency and model generalization.
3. Empirical evaluation of the proposed approach on a children's content dataset, demonstrating its effectiveness in tackling data sparsity.

The remaining part of this chapter is organized as follows: Sect. 2 presents a related work on recommendation system using various DL techniques. Section 3 outlines the proposed DeepFM-based recommendation framework with negative sampling in detail. Section 4 describes the experimental results and performance evaluation metrics. Finally, Sect. 5 provides the conclusion and potential future research directions.

2 Related Work

Recommender system (RS) has been studied widely in the field of online content, e-commerce and streaming platforms. To improve recommendation accuracy, various techniques such as machine learning and DL techniques are used for RS. This section reviews various research articles related to recommendation system using various DL techniques.

Niu et al. (2016) In this paper the author has presented a novel recommendation system to obtain the similarity of various user's item rating information from different online recommendation and review sites, which is based on fuses user and item information. In this model the author has used FURI which taken three parameters like user, items and user-item to forecast the item ratings. In simulation part three data set are used to predict similarity between two users the author has used the concept of Customer Relative Interest (CTRI) which outperforms than the existing models.

Gazdara and Hidri (2020), proposed a mathematical model based on the integral, linear and non- linear equation to measure the similar rating of the users. For the simulation the authors have taken 3 benchmark data set (MovieLens100K, MovielLens1M and Yahoo-Music) and the results shows that the proposed similarity measure is highly expensive as compare to the existing model, particularly in terms of accuracy.

Li and Tuzhilin (2019) In this paper the author has proposed a similarity measure recommendation system based on the multi-criteria rating. In order to map the review of different users into an existing continuous space, the variational autoencoder has used at the very first stage of the model, and then embedded vector is transform to discrete latent ratings using embedding compression method at second stage. The experimental results show that the proposed model gives better performance when applied on different multi-criteria dataset and models like KNN, SlopeOne, SVR and Aggregate etc.

Pujahari and Sisodia (2020), introduced a Probabilistic Matrix Factorization (PMF) method for CF based RS. In order to produce competent item ranking this model takes input as preference relationship, instead of user rating. The model incorporates both user and item side facts through a matrix co-factorization technique. By leveraging PMF, the proposed model collects higher-order user item connections, allowing it to effectively model the full range of desired connection. Simulation results on two 'MovieLens' dataset demonstrate that the proposed model outperform the state-of-the-art MF method in terms of recommendation quality.

Wu et al. (2021), design a deep-structure RS on high-dimensional dataset (HiDS) using deep a Deep Latent Factor Model (DLFM). Instead of using multi-layer neural network through nonlinear activation function, the author has designed a deep-structured model by sequentially connected various latent factor. As per the experimental results on four HiDS matrices, the DLFM effectively balances the computational efficiency and accuracy prediction that makes this model faster and accurate RS.

Siet et al. (2024), has presented a hybrid model for movie recommendation system using deep learning and K-means enhance the user rating. In this paper multilayer perceptron model has used to collect the user information (such as age, gender, and occupation) to predict new user and match their rating with the existing user who have similar preferences. Additionally, Additionally, in order to grouping of the similar kind of movies, the K-Means clustering is used. After that the clustering is merges with the obtained rating to improve the variety of tailored RS for the target users. The experimental results show better performance as compare to existing system in term of RMSE, MAE, precision, recall, and F1 scores.

Zhang et al. (2024) In this paper the author has proposed a novel approach based on Low-Ranking approximation method to enhance CIN network in the xDeepFM method and reducing the number of parameters. Furthermore, an attention method is accumulating after the CIN to maintain the model's accuracy. The simulation results show that the proposed model reduces the time complexity effectively while improving its accuracy.

In this paper (Haque et al. 2025), a hybrid course RS have been proposed that combine the different techniques namely Singular Value Decomposition (SVD), Non-Negative Matrix Factorization (NMF) and attention based neural network. This integration allows the model to generate highly personalized recommendations based on user connections with the application as well as intrinsic features of the course.

Although DL model has improved the accuracy in recommender system but data sparsity problem remains a challenge in some domains especially in anime recommendations. Mostly studies do not focus on the negative sampling strategies for optimizing training efficiency. Additionally, there is limited research on DeepFM for child-friendly anime recommendations, requiring a more specific approach.

3 Methodology

The proposed framework shown in Fig. 1 implements DeepFM along with negative sampling to solve the data sparsity issue affecting children's content recommendations. The model combines FM with its ability to extract low-order feature connections and DNNs for high-order feature interaction modelling to achieve effective learning in sparse datasets. The model training process gets additional enhancement through negative sampling where artificial generation of negative user-item interactions enables the model to process both positive and negative feedback.

3.1 Dataset Preprocessing

As shown in Table 1, we use the Anime dataset (Anime Recommendations Database), which consists of 12,294 anime and 73,516 users. Genre information is one of the items features available in the anime (Meng 2023) database. Every film in the database has a minimum of one genre, and every film has a combination of genres. For instance, "Cowboy Bebop" falls within the Action, Adventure, Comedy, Drama, Sci-Fi, Space genres. But the dataset used in this study is a child-friendly anime dataset, where each record represents an interaction between a user and an anime based on genres. The anime dataset contains only select titles which belong to five child-friendly genres including Kids, Adventure, Fantasy, Family and Educational. The content recommendation filter performs age-based restrictions to present suitable material for children. After filtration of child-friendly anime dataset, data preprocessing is done which aims to perform the data cleaning i.e. by removing duplicate entries, normalizing numerical attributes and encoding categorical features.

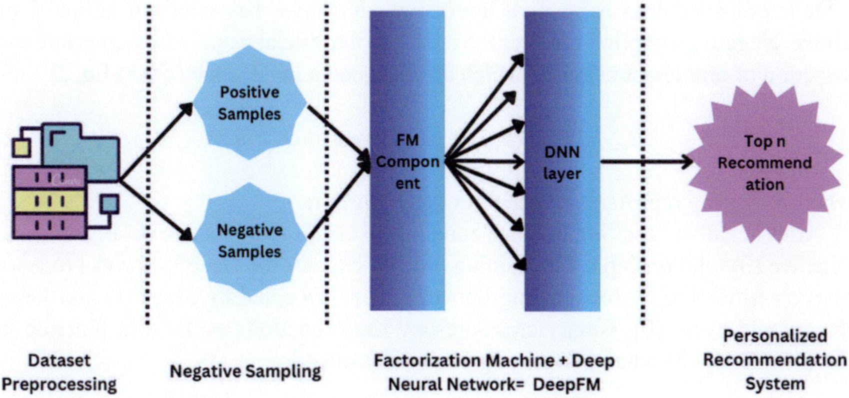

Fig. 1 Proposed framework

Table 1 Anime dataset description

Dataset	Attribute	Explanation
Anime	anime_id, name, Genre, type, episodes rating, members	12,294 anime
Rating	user_id, anime_id, rating	Ratings of 73,516 users

3.2 Proposed Framework: DPRNS

Our proposed framework is DPRNS which represents the DeepFM-Driven Personalized Recommendations with Negative Sampling. The traditional recommendation system uses explicit feedback data through ratings on items which may look like an anime's value between 1 to 10. The majority of users avoid interacting with the complete set of items while interacting with only selected items which results in a user-item interaction matrix that contains many empty spaces. Additional training on only observed interactions (positive samples) creates potential model bias because the model assumes uninteracted items contain no meaning. Negative sampling (Ma et al. 2024) represents an approach to resolve this issue. The approach assigns unobserved interactions a rating of zero through implicit feedback because it considers these interactions as negative examples. Through this technique the model becomes more proficient at differentiating relevant from irrelevant items. But how this negative sampling is applied. Let's assume that the values ranging from 1 to 10 consists of explicit user ratings in child friendly anime dataset. The positive interactions are defined in Eq. 1 as:

$$R_{u,i} = 1, if\ r_{u,i} \geq 7 \tag{1}$$

where, user u and anime i represents by the binary label $R_{u,i}$ and $r_{u,i}$ is the actual rating given by the user.

It is considered as a positive interaction, if a user has rated an anime **7 or above**. Negative sampling requires the creation of artificial negative samples through selection of uninteracted anime which are assigned a label value of 0 in Eq. 2:

$$R_{u,j} = 0, if\,(u, j) \notin\ Observed\ Interactions \tag{2}$$

where variable j represents anime not rated by user u.

After negative sampling the data contains instances that indicate positive (1) and negative (0) relationships. The training process enables the DeepFM model to learn item classification by recognizing between items accepted by users (1) and items they would avoid (0). Binary cross-entropy loss, functions as the loss function in Eq. 3 for DeepFM when performing binary classification.

$$L = -\sum_{(u,i)\in D} y_{u,i} log\hat{y}_{u,i} + (1 - y_{u,i}) log(1 - \hat{y}_{u,i}) \tag{3}$$

where the actual label $y_{u,i}$ represents either a positive class value of 1 or a negative class value of 0. The model provides a prediction through $\hat{y}_{u,i}$ represents the actual value and D is the training dataset.

Negative sampling within this system chooses random unseen user-anime combinations then assigns them zero values in the rating scale. User preferences become more effective for the DeepFM model because it distinguishes between watched (positive) and unwatched (negative) anime items. Binary cross-entropy loss function within the model ensures superior prediction optimization for watched (positive) as well as unwatched (negative) interactions which enhances recommendation quality in sparsely populated data.

The DeepFM model merges two models into one recommendation framework by binding FM that manages low-dimensional embeddings for feature interaction with DNN which learns intricate high-order interactions from raw features. The recommendation system uses DeepFM to predict user reactions to anime by analyzing both recorded user relationships and feature representations it has generated.

DeepFM has two components where first component is FM that uses embeddings for pairwise feature interactions. The FM interaction term given a feature vector x in Eq. 4 is:

$$FM_{(x)} = \sum_{i=1}^{n} w_i x_i + \sum_{i=1}^{n} \sum_{j=i+1}^{n} \left(v_i, v_j\right) x_i x_j \tag{4}$$

where weight for the i-th feature is represented by w_i, v_i is the embedding vector for the i-th feature and the dot product of embeddings that captures pairwise interactions is represented by$\left(v_i, v_j\right) = v_i^T v_j$.

DNN constitutes another system that understands complex interactions between features. The input consists of concatenated embedded features shown in Eqs. 5 and 6:

$$a^1 = f\left(W^1 x + b^1\right) \tag{5}$$

$$a^l = f\left(W^l a^{l-1} + b^l\right) for\ each\ layer\ l \tag{6}$$

$b^l andW^l$ are the biases and weights of the neural network at layer l, a^l is the output of the l-th layer and f is the activation function (e.g., ReLU).

The output from FM and DNN is combined for final prediction layer is depicted in Eq. 7:

$$\hat{y} = \sigma(FM(x) + DNN(x)) \tag{7}$$

where for ensuring the output is between 0 and 1, σ is the sigmoid activation function.

The system incorporates user and item embedding layers that are followed by a multi-layer perceptron with dropout regularization to avoid overfitting. The training period allows the model to learn various user-anime interaction patterns to achieve better prediction accuracy. The trained model functions as a component for generating recommendations to anime titles. From a provided user ID, the system executes user encoding to process entire anime databases for generating relevance scores. The model presents the anime titles with the top predicted scores to users for recommendation. The combination of DeepFM with negative sampling techniques and child-friendly filtering produces effective anime recommendations that personalize the output and resolve sparse data problems. The recommendation method provides precise results that match the preferences of youth audiences.

4 Experimental Results and Discussion

4.1 Evaluation Metrics

To evaluate the performance of recommendation system, experiments on child friendly anime dataset is performed. The different standard evaluation metrics such as Root Mean Squared Error (RMSE), log loss, binary crossentropy, precision, recall, F1-score, Hit Rate (HR), Normalized Discounted Cumulative Gain (NDCG), Mean Average Precision (MAP) and Mean Reciprocal Rank (MRR) are used shown in Table 2.

Where y_i represented the actual rating, N is the total number of ratings, predicted rating is denoted by $\hat{y}_i$, m is the number of relevant items, k referred as the number

Table 2 Performance evaluation metrics

Performance metrics	Formulas
Log loss	$-\frac{1}{N}\sum_{i-1}^{N}\left[y_i\log(\widehat{y_i}) + (1-y_i)\log\left(\hat{1}-y_i\right)\right]$
Precision	$\frac{\lvert\{relevant\_items\}\dot{\cap}\{recommended\_items@k\}\rvert}{k}$
Recall	$\frac{\lvert\{relevant\_items\}\dot{\cap}\{recommended\_items@k\}\rvert}{\lvert\{relevant\_items\}\rvert}$
F1-score	$2\times\frac{Precision@k\times Recall@k}{Precision@k+Recall@k}$
HR@k	$\frac{Number\ of\ users\ with\ at\ least\ one\ relevant\ item\ in\ top-k}{Total\ number\ of\ users}$
RMSE	$\sqrt{\frac{1}{N}\sum_{i=1}^{N}\left(y_i-\hat{y}_i\right)^2}$
NDCG@k	$NDCG@k=\frac{DCG@k}{IDCG@k}$ $DCG@k=\sum_{i=1}^{k}\frac{rel_i}{\log_2(i+1)}$
MAP@k	$\frac{1}{\lvert U\rvert}\sum_{u\epsilon U}\frac{1}{m}\sum_{i=1}^{m}Precision@i$
MRR	$\frac{1}{\lvert U\rvert}\sum_{i=1}^{\lvert U\rvert}\frac{1}{rank_i}$

of recommended items and $rank_i$ indicates the position of the first important item in the recommended list.

The evaluation metrics for proposed framework shown in Fig. 2 are used to evaluate a recommendation system's performance. Different aspects of model performance can be analyzed through these evaluation measures which examine accuracy as well as ranking quality and retrieval effectiveness. The RMSE calculates the average difference between prediction ratings and actual ratings and low numbers indicate better prediction accuracy. The prediction error falls within a moderate range because the RMSE reaches 0.6028. The classification performance of the model achieves lower uncertainty through Log Loss (1.5851) and Binary Crossentropy results (1.5623). Precision demonstrates 92.06 percent of correct recommendations from among all suggested items thus demonstrating high accuracy in model predictions. Recall (0.3608) demonstrates a low value which suggests the recommendation model is unable to retrieve most relevant recommendations. The F1-score (0.5184) shows that the model presents adequate precision despite needing to enhance recall capabilities. The ranking-based evaluation method helps identify how well recommendations perform. The recommendation model achieves a 43% success rate for relevant item recommendation through Hit Rate at 10 (HR@10). Some relevant items exist within the list according to Normalized Discounted Cumulative Gain at 10 (NDCG@10) but their placement requires improvement. The values of Mean Average Precision at 10 (MAP@10) = 0.0711 and MRR = 0.1201 show that relevant recommendations do not position consistently near the forefront of the recommendations. The model performs accurately but shows room for improvement in recall performance when generating and reordering relevant results. Improving the negative sampling methods together with attention mechanisms or optimizing hyperparameters in DeepFM and Neural Collaborative Filtering (NCF) models would create better accuracy while preserving ranking performance.

4.2 Results and Discussion

The aim of recommendation systems is to identify items which users will prefer through analysis of their past interactions as well as their personal behaviors and preferences. This chapter implements DeepFM along with Neural Collaborative Filtering (NCF) to build the recommendation system that utilizes negative sampling for better predictions in anime recommendation scenarios. A user-item matrix structure is considered the first step of this process by showing users as rows and anime items as columns. A sparse matrix structure exists at initial since recorded interactions are minimal thus resulting in challenges for delivering effective recommendation content. The recommendation model functions to predict user preferences for unknown anime which fills in missing values in the recommendation process to achieve personalized output.

DeepFM functions as an essential feature learning component because it detects low-order and high-order user-item interactions. The model contains two main

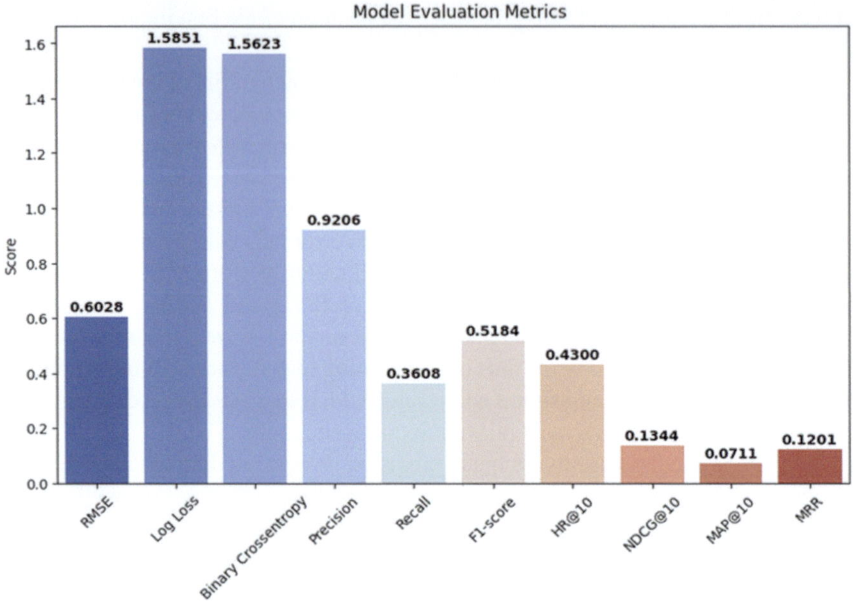

Fig. 2 Evaluation metrics for proposed framework

components: FM which analyzes pairwise feature interactions and DNN which identifies complex non-linear patterns. For predicting the final preference score represented in Eq. 8 of user-item pairs the model adds together FM representation values and DNN output:

$$\hat{y} = FM(x) + DNN(x) \tag{8}$$

Through this method the model processes both explicit patterns and extracts sophisticated user preferences directly from the provided data. The model achieves better prediction accuracy through negative sampling which selects items that users have never engaged with during training sessions. The model utilizes negative sampling to differentiate important recommendations from unimportant ones thereby preventing overfitting to the current set of user interactions. A softmax function determines recommendation selection probabilities with prediction scores acting as a factor for likelihood of recommendation. By employing this method, the model becomes more robust and addresses data sparsity problems.

For addressing the data sparsity problem, Fig. 3 represents visualization of the User-Item Interaction Matrix before and after prediction of the recommendation model through heatmap.

The left plot i.e. User-Item Matrix Before Prediction. The initial user-item interaction matrix shows its sparsity through this heatmap because no predictive model has been implemented yet. The representation shows user id in vertical rows but

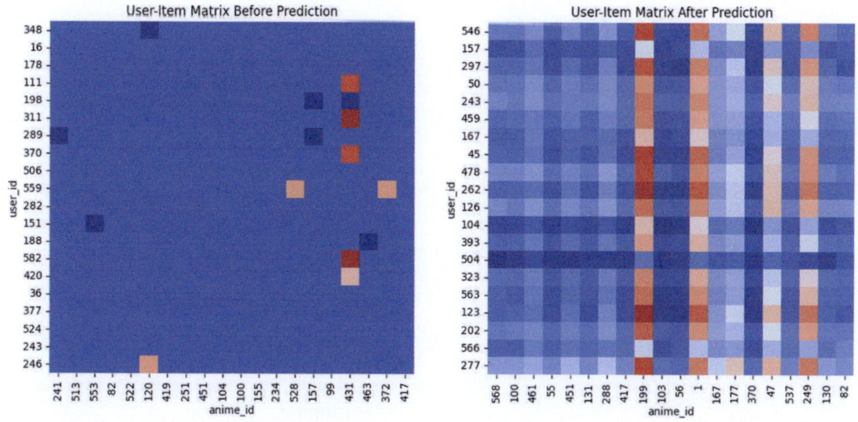

Fig. 3 Improving data sparsity problem

anime item id in horizontal columns while color intensity displays interaction levels. A majority of the deep blue cells within the matrix demonstrate the absence of user-item interactions because most users did not interact with numerous items. The heatmap mainly consists of dark blue background cells to represent missing user-item interactions until we observe the few highlighted active cells in red and light orange which indicate user feedback events on anime items. The data sparsity issue in recommender systems becomes evident through the presentation where most cell values appear deep blue for unknown user-item relationships.

The Right Plot i.e. User-Item Matrix After Prediction. This heatmap shows the predicted user-item interaction matrix that incorporates recommended data from DeepFM or NCF models. The color pattern shows uniform distribution across the matrix because the model successfully predicted previously unknown values and created a denser pattern. The model shows stronger user preferences for particular anime items throughout its red/orange colorized regions. A visual representation shows how prediction models optimize sparse recommendation datasets. The implementation of DeepFM allows predictions of missing interactions thus creating a denser and more informative user-item matrix which enhances recommendation accuracy.

After addressing the data sparsity issue, the system utilizes predicted scores to create rankings of items and present top-N recommendations to users. The evaluations of recommendation results are measured by RMSE, Precision, Recall, HR@10, MAP, MRR, F1-score and NDCG@10 to determine how effectively the system predicts user preferences. The system suggests comparable child anime items to users who previously watched child genre anime through the use of its learned embeddings. The system generates personalized recommendations for users with diverse preferences through identification of their browsing patterns. Through the incorporation of new interactions and improved ranking strategies the system learns continuously to adapt its personalized recommendations.

Table 3 Recommendations for user id 1

anime_id	name	predicted_ rating	Genre
199	Sen to Chihiro no Kamikakushi	0.087610	Adventure, drama, supernatural
121	Fullmetal Alchemist	0.081928	Action, adventure, comedy, drama, fantasy, mag…
164	Mononoke Hime	0.052496	Action, adventure, fantasy
431	Howl no Ugoku Shiro	0.044234	Adventure, drama, fantasy, romance
1	Cowboy Bebop	0.036825	Action, adventure, comedy, drama, sci-fi, space
205	Samurai Champloo	0.021197	Action, adventure, comedy, historical, samurai…
136	Hunter x Hunter	0.004063	Action, adventure, Shounen, super power
47	Akira	−0.000571	Action, adventure, horror, military, Sci-Fi, S…
223	Dragon Ball	−0.003996	Adventure, comedy, fantasy, martial arts, Shou…
392	Yuu☆Yuu☆Hakusho	−0.006381	Action, comedy, demons, fantasy, martial arts…

Table 3 represents the top-10 recommendation for user-id 1. The recommendation system performs effectively through its execution of DeepFM feature learning with NCF deep representation learning supported by negative sampling for preferring user choices. These recommendation system techniques work together to present appropriate recommendations for users who have limited tracking history while resolving sparsity issues in the data. Through the incorporation of new interactions and improved ranking strategies the system learns continuously to adapt its personalized recommendations.

5 Conclusion and Future Scope

The aim of our research is to be developed an enhanced recommendation system which utilized DeepFM, negative sampling and NCF to solve data sparsity issues while improving recommendation prediction accuracy. DeepFM's ability to discover features combined with NCF's DL approach resulted in a recommendation model which efficiently extracted low-order and high-order feature relationships to enhance its relevance capabilities. Negative sampling became essential for training efficiency and performance optimization because it enabled systems to identify true meaningful interactions from background random data. The proposed prediction method displayed high accuracy according to evaluation metrics which included

RMSE, Precision, Recall, HR@10, and NDCG@10. Predictions achieved through the proposed model demonstrated how the visualization of user item matrix predication graph represents the sparse missing values into dense data matrices thus improving recommendation relevancy for users with few historic interactions. This hybrid approach improves recommendation quality for child friendly genres which establishes it as an effective solution for personalized content recommendations within e-commerce and entertainment platforms.

The proposed model can improve the performance of the model by implementing multiple potential advancements that will strengthen RS. Better recommendation systems can be achieved through additional context such as user profiles and temporal patterns and their social network connections. Adding attention mechanisms with graph neural networks to the model will improve its ability to identify complicated relationship between users and items. Researchers focus on improving the computational efficiency of deep learning models in order to process large-scale real-world applications.

References

Behera G, Nain N (2022) DeepNNMF:deep nonlinear non-negative matrix factorization to address sparsity problem of collaborative recommender system. Int J Inf Technol (Singap) 14(7):3637–3645. https://doi.org/10.1007/s41870-022-00982-1

Gazdar A, Hidri L (2020) A new similarity measure for collaborative filtering based recommender systems ☆. Knowl-Based Syst 188:105058. https://doi.org/10.1016/j.knosys

Haque ul MM, Kotaiah B, Ahamed J (2025) Hybrid course recommendation system using SVD, NMF and attention-based neural network. Int J Inf Technol (Singapore). https://doi.org/10.1007/s41870-025-02440-0

Li P, Tuzhilin A (2019) Latent multi-criteria ratings for recommendations. In: 28th text retrieval conference, TREC 2019-proceedings. https://doi.org/10.1145/1122445.1122456

Ma H, Xie R, Meng L, Feng F, Du X, Sun X, Kang Z, Meng X (2024) Negative sampling in recommendation: a survey and future directions. Assoc Comput Mach

Meng Z (2023) Research on two popular recommendation algorithms for anime. In: Proceedings of the 3rd international conference on signal processing and machine learning, vol 6, no 1. pp 1430–1438. https://doi.org/10.54254/2755-2721/6/20230898

Niu J, Wang L, Liu X, Yu S (2016) FUIR: Fusing user and item information to deal with data sparsity by using side information in recommendation systems. J Netw Comput Appl 70:41–50. https://doi.org/10.1016/j.jnca.2016.05.006

Pujahari A, Sisodia DS (2020) Pair-wise preference relation based probabilistic matrix factorization for collaborative filtering in recommender system. Knowl-Based Syst 196. https://doi.org/10.1016/j.knosys.2020.105798

Siet S, Peng S, Ilkhomjon S, Kang M, Park DS (2024) Enhancing sequence movie recommendation system using deep learning and KMeans. Appl Sci (Switz) 14(6). https://doi.org/10.3390/app14062505

Wu D, Luo X, Shang M, He Y, Wang G, Zhou M (2021) A deep latent factor model for high-dimensional and sparse matrices in recommender systems. IEEE Trans Syst, Man, Cybern: Syst 51(7):4285–4296. https://doi.org/10.1109/TSMC.2019.2931393

Zhang Y, Wang Z, Joe I (2024) Improved xDeepFM with single value decomposition and attention mechanism. IEEE Access 12:106447–106454. https://doi.org/10.1109/ACCESS.2024.3417422

Behavioural Analytics with Digitized Sustainability

Gagan Matta, Gaurav Pant, Sanjeev Kumar Lambha, and Nishant Kumar

Abstract Digital platforms are increasingly recognized as powerful tools for promoting behavior change towards sustainability. They provide opportunities for individuals and organizations to engage in sustainable practices by facilitating the sharing of information, enabling community-building, and offering interactive features such as gamification and feedback loops. These platforms utilize data analytics and algorithms to personalize user experiences, tailoring content to specific behaviors, preferences, and sustainability goals. This level of customization enhances user engagement and commitment to sustainable actions, such as reducing waste, conserving energy, or making environmentally friendly purchasing decisions. Digital platforms can also serve as collaborative spaces for knowledge exchange, where users can learn from experts, participate in discussions, and contribute to collective problem-solving. Social media, apps, and online communities further amplify the reach and impact of sustainability initiatives by fostering peer-to-peer influence and social norms that encourage environmentally responsible behaviors. However, the effectiveness of digital platforms in driving long-term behavior change remains a subject of research, as it involves addressing challenges such as user fatigue, misinformation, and the digital divide. In conclusion, while digital platforms hold significant potential for catalyzing sustainability-focused behavior change, further exploration is needed to optimize their design, scalability, and inclusivity to ensure their meaningful and lasting impact.

Keywords Social media · Societal changes · SDGs · Environmental conservation

G. Matta (✉) · G. Pant
Hydrology and Science Communication Research Lab, Department of Zoology and Environmental Science, Gurukula Kangri (Deemed to Be University), Haridwar, India
e-mail: drgaganmatta@gkv.ac.in

S. K. Lambha · N. Kumar
Faculty of Engineering and Technology, Gurukula Kangri (Deemed to Be University), Haridwar, India

M. Ram et al. (eds.), *Advances in Mathematics for Engineering Sciences*, Mathematical Engineering, https://doi.org/10.1007/978-3-031-95693-5_15

1 Introduction

Today, the world faces various sustainability challenges that jeopardize environmental stability, social equality and economic development. They result from irresponsible consumer practices, particularly overconsumption, dependence on fossil fuels, and deforestation, leading to biodiversity loss and waste generation (Malik 2024). Climate change, resource depletion, pollution and increasing inequalities are some of the most pressing issues facing the global community today, exerting major pressures on natural ecosystems, social systems and economies. The interdependent and multifaceted character of these problems seems to require comprehensive solutions that demand a move beyond conventional sectoral monopolies and disciplines.

- **Climate Change**—We recognize that climate change is one of the most significant sustainability challenges of the twenty-first century. Increasing total emissions of greenhouse gasses (GHGs) caused mainly by the combustion of fossil fuels, agriculture and deforestation globally leads to increasing atmospheric temperatures due to human activities (Nayak et al. 2020). They are likely to breach a key 1.5 °C threshold about preindustrial levels, which the Intergovernmental Panel on Climate Change (IPCC) has warned will see entire ecosystems. Climate change has huge consequences, including droughts, floods, rising sea levels and heatwaves that significantly impact food security, water availability, health and infrastructure (Dietz et al. 2018).
- **Resource Depletion**—Overusing natural resources such as water, forests, minerals and fossil fuel is an important sustainability problem. Population growth, urbanization and economic development have increased demand for these resources to levels far above the Earth´s regenerative capacity (Matta et al. 2022a). This depletion leads to environmental degradation, such as deforestation, soil erosion, and habitat loss, and increases political instability, causing global tensions and resource wars (Matta 2020a). Water scarcity has been aggravated in many parts of the world by the depletion of freshwater resources, affecting more than two billion people globally for example.
- **Loss of Biodiversity**: Biodiversity is key to the functioning of ecosystems and the supply of ecosystem services, including pollination, water purification and climate regulation. Nonetheless, human behaviour has ramped up the speed at which species are going extinct, with it being estimated that the current extinction rate is 100 to 1,000 times higher than the natural background rate (Pereira et al. 2022). The major threats to nature are habitat destruction, pollution, overfishing and global warming/biodiversity eradication. The collapse of ecosystems and the extinction of species thus are not only environmental challenges; they also erode human well-being and prospects as biodiversity loss undermines food security, health, livelihoods (Matta and Kumar 2017a; 2017b).
- **Pollution**: In different formations, the three celebrated types are air, water, and soil, and plastic pollution poses a potential new panic for the environment and human health (Matta et al. 2018a; 2019). Air pollution, 70% of which comes

from the burning of fossil fuels, results in millions of premature deaths annually as well as millions more cases of respiratory diseases, heart conditions and cancer (Matta et al. 2015; Matta 2020c). His ill health is directly tied to water pollution; industrial waste, agricultural runoff and untreated sewage contaminate freshwater systems that cut off thousands of people from clean drinking water. Plastic pollution especially in oceans growing as it damages marine life and contaminates the food chain with potential impacts on human health (Matta et al. 2018b; 2018c).

- **Growing Disparities**: Such differences are only exacerbated by social imbalances—notably poverty, gender inequity, and uneven distribution of resources, among other challenges—compounding sustainability issues (Chancel 2020). Wealth and resources are so unequally distributed that resource depletion and climatic impact is cruelling the most vulnerable populations; citizens of developing countries. Systemic social inequalities additionally act as a barrier preventing access to education, healthcare, and sustainable livelihoods, subsequently increasing poverty and reducing the world's capacity to implement targets for global sustainability (Birkmann et al. 2022).

Change in Individual and Collective Behavior for Sustainability: Despite the pivotal role of governments, businesses, corporations and organizations in creating sustainable policies and practices to confront global sustainability challenges, change in behavior-individual as well as collective-to help transform these policy level evolutions into practice is equally important. To protect this Earth and its resources is also to protect us humans since it is ultimately our personal choices and the collective choices of communities and societies that made it so (Shrivastava et al. 2020). Sustainable behaviour—actions that support our current and future generations while not destroying or harming the Earth beyond its capacity to recover, replenish or repair itself over time—is crucial to realizing distant sustainability outcomes (Fig. 1).

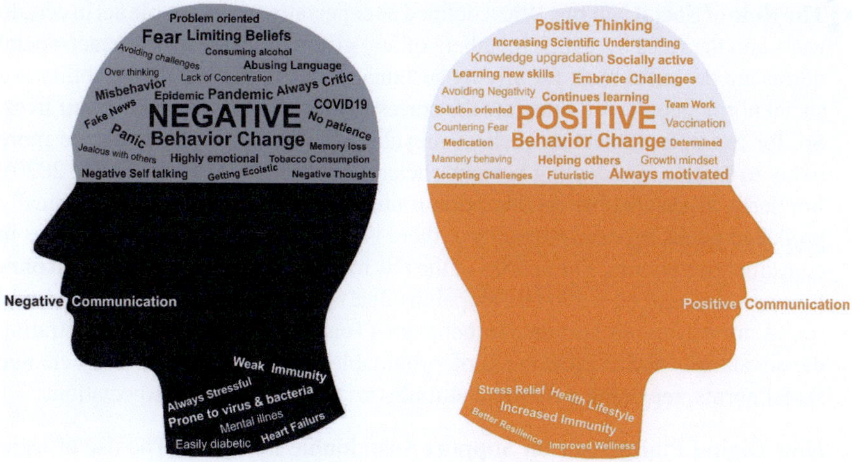

Fig. 1 Positive and negative behavior change. Copyright: Dr. Gagan Matta

- **Individual Behavior Drives**: How individual and consumptive activities (energy use, transportation modes, waste production, diet) add up with the environmental pressures. By way of illustration, energy-efficient appliances and light bulbs or renewable sources for domestic heating can go a long way to reduce GHG emissions and the household carbon footprint. In the same way, encouraging sustainability (walking, cycling or public transportation use) instead of driving a personal car can reduce air pollution and alleviate traffic jams (Rafique and Williams 2021). Behaviours that reduce waste, such as recycling, composting, and reducing single-use plastics, help to conserve resources and prevent unnecessary landfill pollution. Research in behavioural science suggests that education, awareness campaigns, social norms, and reminders can influence individual behaviour. For example, the Theory of Planned Behavior (TPB) establishes that an individual's intention to perform a behavior is a function of their behavioral beliefs, subjective norms and perceived behavioral control. These variables are manipulatable to incentivize people to do what they need to do in order to meet sustainability goals (Ajzen 2020).
- **Collective Action and Social Movements**: While we value individual behavior we see collective action and social movements as a way to create systemic change. Where individuals care enough to influence public policies, corporate practices and societal norms, the youth-led climate strikes and the resurgence of public environmental activism through groups like Extinction Rebellion and Fridays for Future show just how much attention can be drawn and pressure exerted on governments and big business to respond in ways that begin to move the mark towards sustainability (Neas et al. 2022). Community-based sustainability initiatives—like local food systems, cooperative energy projects and neighbourhood waste reduction programs—are examples of collective behavior that can come from strong community ties (McDaniel et al. 2021). All of these lead to more sustainable local practices, social capital, and resilience through stronger community ties.
- **The Role of Social Norms**: When defined as expectations that people act in certain ways and the rules by which members of any given group/society interact, social norms are powerful tools for driving sustainable behaviours. If sustainability is a social norm, people will integrate the corresponding actions more into their lives. So, for example, in societies where recycling is common, individuals are more likely to engage in recycling behavior because of social norms (Luo et al. 2020). Similarly, as sustainable consumption routines like sourcing locally or ethically produced goods become normalised: there is potential for amplifying changes in consumer behaviours. The actions of the few may also be amplified by digital platforms, media and educational campaigns that showcase creation stories to build social norms around sustainable behavior (Tomşa et al. 2021). Demonstrating the advantages and attractiveness of sustainable behavior can eventually change social norms, replacing individual attitudes with broader social expectations.

How Digital Platforms Can Support Sustainable Behavior: The rise of technology and digitization have not only transformed the systems that are powering our cities, but also informing how—we—as individuals live our lives. Social media, mobile apps, e commerce sites and online learning platforms are some of the many digital platforms that have an unprecedented power to shape individual and collective behavior at scale (Scurati et al. 2021). These tools and technologies can help these platforms achieve a broader awareness, more engagement, and better behavior change in reaching sustainability goals.

- **Reach and Influence of Digital Platforms**: Among other things, this is one of the biggest strengths of digital platforms—they now reach a global audience and can propagate information incredibly quickly. Facebook, Instagram, Twitter and TikTok are all-powerful social media platforms for promoting sustainability messages by reaching millions of active users to mobilize and increase awareness of environmental issues. These platforms can aid the voices of activists, organizations and individuals working for sustainable practices to create a ripple effect that escapes the boundaries of geography and culture (Alam 2022). The more content can be customised, the more likely individuals will engage with sustainability-related information and act upon it. For example, it could provide suggestions for sustainable products, climate-friendly lifestyle hacks or actions etc., according to a user's browsing history or interactions in a social network (Sarkar et al. 2023).
- **Promote Sustainable Consumption**: E-commerce platforms like Amazon, Etsy, and eBay heavily influence consumer behaviour. And these platforms can encourage sustainable consumption, showcasing all of the eco-friendly products available for purchase and revealing the ins-and-outs of how these items are sourced—who makes them; is the company treating its workers fairly?—and what kind of toll your purchasing habits are taking on our environment (Dartnell and Kish 2021). Others, such as Etsy, have started to take responsibility for their carbon emissions and reward sustainably-sourced sellers—urging greater awareness on the part of the consumer. Moreover digital platforms can also contribute to the circular economy, where second-hand-goods exchange, rentals and peer-to-peer share are made easier with their help. There are many such platforms, like Depop (for buying and selling), Poshmark (used clothes only for now), OLIO (sharing with your community) that lets one buy/sell or simply share pre-owned products to reduce waste and make a product go longer in its life cycle. Digital platforms can help alter consumer behavior toward sustainability by making sustainable consumption convenient and accessible (Godinho Filho et al. 2024).
- **Gamification and Behavioral Nudges**: Digital platform may integrate gamification and behavioral nudges to motivate sustainable behaviors. Gamification is the use of game design elements in a non-game context. By using gamification features in their apps and on platforms, these company have enabled their users to engage with sustainability by collaborating in tasks like energy management, carbon footprint tracking, or joining waste mitigation challenges (Kirchner-Krath 2023). EcoBuddy and JouleBug are all recent apps that leverage gamification to reward users for walking or biking to work, decreasing their household water usage or recycling. You can then share the points with your friends and accomplish some competition and social reinforcement. Behavioral nudges (reminders, prompts and feedback loops) can also help users to choose a healthier product.

For example, with energy-monitoring apps you can literally see your electricity consumption in real-time and identify ways you could suck less power (Rapp and Boldi 2023).

- **Mobilizig Communities and Social Movements**: Digital platforms can make communities move to take collective action around the issue of sustainability. Social media platforms, climate strikes, ecology groups, and agitations have been organized around them. The rise of hashtags such as #ClimateAction, #ZeroWaste or #FridaysForFuture have connected people who think alike while also adding a sense of urgency to these highly important issues (Pan et al. 2022). This can also take shape through online petitions, crowdfunding services, and digital advocacy platforms that allow citizens to support collective efforts that address sustainability issues. Platforms like Change.org connect regular folks to environmental movements, and services like Empowerment Works enable community organisers to crowd-fund in support of their local environment. They are part of a network that organizes campaigns to affect public policy and corporate practices worldwide.

How Digital Platforms Can Influence and Support Sustainable Behaviors: Multiple points of intervention on digital platforms: Information dissemination, social engagement and behavioral change techniques Digital mechanisms, with the massive reach provided by these platforms, can embed behavioral insights over time in addressing long-term changes of lasting significance in how humans and communities engage with the environment (Simeone and Scarpato (2020).

- **Raising Awareness and Educating**: The most baseline for digital platforms to underpin sustainability behaviors is to differentiate the knowledge about awareness education as well) Long-Form Content, Such as Blogs, Infographics, Videos and WebinarsCan be Promoted on Social MediaEducational content like white papers or webinars can easily be distributed on e-learning platform, social media blogs etc (Pan et al. 2022). In the online space, platforms such as Coursera and edX have hundreds of sustainability offerings that give even more people the opportunity to learn about environmental stewardship, renewable energy or sustainable business practices. Digital platforms can also show the impacts of unsustainable actions like deforestation, pollution, and loss of biodiversity by providing experiences in virtual reality (VR) or by running interactive simulations (Chew 2021). In a more immediate, emotional sense, these tools can awaken users to the pressing nature of the sustainability challenges and drive action.
- **Enabling Behavior Change**: Digital platforms can enable behavior change, by supplying users with the tools, resources, and incentives to adopt more sustainable behaviors. Using behavioral science principles, apps and platforms can be designed to nudge or positively reinforce behavior conducive to sustainability subtly. Similarly, ride-sharing apps have also included eco-friendly means of transportation through electric vehicle options or bike-sharing programs, like Lyft and Uber (Vrain et al. 2022). In addition to this, those platforms that provide live monitoring of energy use, carbon emissions or water consumption mean you can watch yourself in real-time and change your behavior. Smart Home System—The devices of IoT automate energy use at user residence and this system allows

people to monitor, control and adjust the appliance settings, lighting, or heating from a distance—which works around environmental conditions and customer needs (Hoffmann et al. 2022).

- **Promoting Social Norms**: Digital platforms also have a vital role in building social norms around sustainability. If each one of us were to see the value in sustainability everywhere we went online, these would be the norms that most individuals would adopt. But then, social media has made it possible for influencers to attract more followers by showing off how easy it is to live green and healthy by using pictures of Instagram in the gym and videos on YouTube amidst nature. Platforms may also steer social norms by including peer comparison (e.g., displaying how your use of energy compares to the rest of the community, badges and certification for sustainable achievements). Social reinforcement is the thing that can nudge and encourage behavior change, taking sustainability from being an obligation to a socially rewarded goal (Li et al. 2024).

2 Adaption to Sustainable Practices

Changing human behaviors from an individual or collective level (Behaviours of interest) in the path towards obtaining results of interest is named behavioral change. Sustainability: In sustainability, behavior change is a key driver for implementing behaviors that lower environmental impact, save resources and support socio-economic well-being (Fig. 2). As understanding grows that human activity is at the heart of current environmental challenges, averting unsustainable actions is necessary to achieve global long-term goals and standards like the Sustainable Development Goals (SDGs) (Matta et al. 2022b). In this sense, digital platforms and technologies offer a new way to generate even more effective nudges towards sustainable behaviors.

- **Theories of Behavioral Science**: This study helps us to understand the reasons behind our decisions and how we make our choices, which are very useful to market sustainability practices. This involves psychological, anthropological and contextual factors that play a role in decision-making. There are many behavioural science theories which can be particularly applicable to sustainability (Han 2021).
- **Theory of Planned Behavior (TPB)**: The theory of planned behavior, by Icek Ajzen, suggests that the intention to perform a behavior is the best predictor of actual behavior. Intention Consists of these three factors.
    - **Attitudes**: How a person evaluates the behavior personally (e.g. is recycling good?).
    - **Injunctive or Social Norm**: Predictions about what others think (e.g. What do my friends/family think I should do and them doing things in the environment-related behaviour)
    - **Perceived Behavioral Control**: This refers to the individual's beliefs about their ability to perform the behavior (*i.e.* how easy is it for me to recycle?).

Fig. 2 Applications of behavioural science to the sustainable development goals. *Source* https://www.uninnovation.network/innovation-library/united-nations-behavioural-science-report

TPB has been used extensively in sustainability to explain recycling, energy consumption, and sustainable purchasing behaviors.

- **Social Cognitive Theory (SCT)**: Introduced by Albert Bandura, SCT frames behavior as a reciprocal interaction of personal, behavioral, and environmental factors. This model emphasizes the importance of observational learning (watching others perform the behavior), self-efficacy beliefs (confidence that one can make a change) and reinforcement contingencies (rewards or punishments for engaging in the behavior) (Mehmood et al. 2024). A concrete example would be peers and co-opting factors where if you see your friends reduce their use of plastics or shift to renewable sources, you are likely to do it too.
- **Nudge Theory**: Created by Richard Thaler and Cass Sunstein, this theory rationalizes that subtle alterations in the way choices are presented to people can 'nudge' them towards better choices without compulsive requirements on their freedom of choice. Nudges are particularly used in the area of sustainability (setting default options to renewable sources for your electricity/ putting recycling bins closer by) (Rachlinski 2022).
- **Habit Formation Theory**: Habits are automatic behaviors due to repetition of the same actions in stable contexts. The Habit Formation Theory—a concept promoted by trailblazers like Wendy Wood—tells us that actions must be regularly repeated in corresponding contexts for habits to take firm hold. Digital platforms can tap into this by creating cues/triggers for engaging in sustainability (e.g., nudges to reduce water consumption).
- **Value-Belief-Norm (VBN) Theory**: The VBN theory, developed by Paul Stern, argues that values and beliefs affect motivation to perform environment-relevant behaviors in a chain of events. It suggests that those with firm pro-environmental values are prone to interpret environmental problems as threatening and experience a strong sense of moral obligation to act in an environmentally responsible fashion. This is critical for understanding how to move individuals from intrinsic motivations (e.g., wanting to help global well-being) to sustainable actions (Batool et al. 2024).

Digital Behavior Models: In an increasingly data-driven world, digital technology is almost so integral to sustainability that it has changed the behaviour models. These mechanisms leverage data-driven insights, algorithms and user interfaces to scale behavioral influence. The digital behavior models are implemented for sustainability to drive energy saving, support sustainable transportation and eco/electric purchasing.

- **Personalization and Behavioral Insights**: Digital is the best modality to deliver personalized behavioral interventions. Accordingly, platforms can infer from aggregate user data to make better recommendations and nudge us towards healthier behaviors. For example, energy-saving apps monitor household energy usage and propose tailored energy-saving ideas. In the same way, e-commerce platforms can promote eco-friendly products observant of their earlier buying decisions (Sarkar et al. 2023).

- **Gamification**—Using game-like mechanics (badges, rewards, points) has been proven to be an effective way to drive sustainable behaviors. Fitness apps with gamification, for example, which challenge users to walk or bike instead of drive, are based on digital platforms. The combination of competition and reward makes the process fun, encourages positive behavior change, and ultimately leads to long-term lifestyle changes (Kirchner-Krath 2023).
- **Social Influence**: Many digital platforms work the pull strategy by using social influence to steer behavior. Employing social proof, platforms can display to users how others are practising sustainability so that they are driven to mimic such behaviours. Energy Savings or Recycling Apps: It has been shown that apps which show how much energy a community has saved or the recycling efforts it is making can increase willingness to participate in sustainable behaviours because actions in this category are perceived as a social norm (Vrain et al. 2022).
- **Feedback Mechanisms**: Offering users real-time feedback on behaviors affects how users behave. For example, smart thermostats offer instant energy consumption feedback to consumers so that they can regulate their usage. The companies mentioned above provide example reminders, but they can be applied to a wide range of apps, such as an app that tracks carbon footprints which, throughout the user's time use, provides feedback on how they are doing in reducing daily activities to help decrease their environmental impact (Ponce et al. 2020).

Digital Affordances for Behavior Change: Digital affordances refer to the actionable properties of digital technologies that make certain behaviors easier to execute. Regarding sustainability, digital affordances offer a framework for understanding how technological platforms enable and guide behavior change.

- **Interactivity**: Interactive features like quizzes play an essential role in helping users engage with sustainability content actively on digital platforms. For example, virtual simulations can show the impacts of climate change or the environmental benefits of sustainable behaviours, leading to abstract thought processes becoming more concrete and evidence-based (Zhou 2021).
- **Accessibility**: Digital makes sustainability-related information and tools more accessible. Through apps, websites, and social platforms, users have been educated on sustainable behavior and all the easy-to-follow steps that they can incorporate into their daily activities (Pulkkinen 2024).
- **Connectivity**: Digital enables access to similar interests or advocacy groups on sustainability. The cohesiveness born from such things can cultivate peer-to-peer learning and guidance, furthering behavior change by providing group-belongingness and goal-setting.
- **Automation**: This is a strong digital affordance and essential when cognitive efforts are made to undertake sustainable behaviour. For instance, intelligent house devices can calibrate power consumption according to the user but only after recognizing the occupancy of users (Feroz et al. 2021).

User Behaviour and Technology for Sustainability: Technology greatly impacts user behaviour, specifically in the context of sustainability. Digital tools not only inform but also create the environment in which individuals operate. However, technology and behavior change are inherently linked– sometimes as solutions but often as would-be obstacles to sustainability (Khan et al. 2020).

- **Use of Facile Solution**: The availability of technology enables us to use facile solutions which were difficult or inconvenient to perform earlier. Ride-sharing apps cut down on car ownership and advance shared mobility, while online marketplaces for used goods promote recycling and reduce garbage.
- **Data-Driven Decision Making**: The progress of data analytics has led to technological implementations that give individuals and organizations the visibility needed to understand their carbon footprint. Carbon footprint calculators or monitoring energy usage with tools such as metering also want to help employees and organizations make good decisions that align with sustainability goals.
- **Potential Drawbacks–The Rebound Effect**: The rebound effect, however, is an example of why technological advancements can lead to unintended consequences. The rebound effect happens when efficiency improvements from technology result in greater consumption of other resources. By contrast, if energy-effective appliances use less power, it might seem intuitive that more will probably be bought (electricity demand), until it's a cheaper brand.

Sustainability Behavior Dimensions and Key Behaviors Targeted by Digital Platforms: Now we have technology and it is actually helping us be sustainable so we can keep the earth clean, green and alive for as long as possible. Sustainability Behaviour Dimensions and Key Behaviours Targeted via Digital Platforms: Sustainability is critical to environmental, social and economic health and thus refers to a range of sustainability-related behaviors (Tan et al. 2022). Digital platforms frequently target key sustainability dimensions:

- **Energy Conservation**: Digital platforms should engage in promoting energy conservation behaviours. While many tools exist to help users optimize their energy consumption, such as smart home technologies, energy-saving apps and usage monitors.
- **Waste Reduction**: Behavior change models use waste reduction method, e; g these platforms are recycling apps, sharing economy platforms (eg., ride-sharing and product shareing), as well as e-commerce sites that offer eco-friendly products.
- **Responsible Consumption**: There is a significant need for platforms with transparent product information, eco-labels and sustainable alternatives to promote responsibility in consumption. Online shopping sites are increasingly including filters for sustainable products, thus making it easier for people to buy more eco-friendly items.
- **Sustainable Mobility**: Apps that enable ride-sharing, bike-sharing, or public transportation allow for a shift in the way people move from one place to another and help reduce carbon emissions existing due to individual car ownership.

Sustainability Frameworks for SDGs and Digital Platforms: The Sustainable Development Goals (SDGs), established by the UN in 2015, are a holistic battery of targets focused on sustainability improvements. The 17 goals for sustainable development target various areas, such as climate action and gender equality, to realize an equitable and sustainable world by 2030 (Matta et al. 2022b).

- **Digital Platforms Aligning with SDGs**: Nowadays, digital platforms are fighting for the cause and achieving their goals. For instance, platforms supporting the adoption of renewable energy directly serve SDG 7 (Affordable and Clean Energy), while those helping to reduce waste by encouraging recycling activities either online or offline may serve SDG 12 (Responsible Consumption and Production) (Sparviero and Ragnedda (2021).
- **Measurement and Accountability**: Digital platforms enable monitoring of progress towards the SDGs. Governments, organizations and individuals track their sustainability initiatives through data analytics and reporting tools to measure their impact on global sustainability goals (Saner et al. 2020).

3 Digital Platforms for Behavior Change

Digital is shifting how we connect to different aspects of our lives, including behaviour change for sustainability. By using connectivity, personalization and automation, these platforms have opened up new ways for individuals and local communities to adapt sustainable practices, increase awareness, and drive global movements to make planet earth a better place (Al-Dmour et al., 2020). These tools play a critical role in developing strengths for several behaviors through social norms, feedback loops and incentives that promote behavioral changes over the long term:

Social Media Platforms: Awareness, Community Building, Activism And finally, social media platforms (Instagram, Facebook and Twitter) have a colossal impact when it comes to spreading awareness about sustainability issues or building communities and activism around them. Millions, or even billions of users, can access these channels anywhere in the world—making them ideal platforms for environmental campaigns and to start collective action (Matta 2018).

- **Creating Awareness**: This is perhaps the most important step of social media in sustainability; it helps get information across as fast and wide. Post facts, figures, infographics, and on-time updates about important causes like climate change, deforestation, and plastic in environmental campaigns. The use of visual formats, hashtags, and viral challenges is very powerful in spreading the environmental message through influencers, activists, and organizations. On the opposite end of the spectrum, Instagram is a visual-first platform that eco-influencers and organizations can use to post creative content such as images on sustainable living tips or green products or videos covering environment-related challenges like "Plastic-Free July. Facebook groups follow a similar premise as they help eco-friendly people come together and create communities that share ideas, experiences, or news related to sustainability (Matta 2020b). Twitter, being more fast-paced and content concise in nature, is mostly used for real-time environmental campaigns like creating awareness durin-g the course of climate summits or engaging public participation through environmental petitions.

- **Community Building**: Social media platforms can also be critical in fostering sustainability communities. For example, users can join groups or like pages on Facebook based on zero-waste living, renewable energy, or local environmental advocacy. They operate virtual communities that facilitate mutual aid, tips-sharing and grassroots action—such as tree-planting or beach clean-up. This was illustrated most vividly in campaigns such as #FridaysForFuture, where the young climate activist Greta Thunberg used Twitter, to mobilise school strikes and essentially took a journey through hashtags and user-generated content worldwide. Social media sites make people be in the middle of huge campaigns no matter where lives (Sweet et al. 2020).
- **Activism**: If you mention the most transformative effects of social media platforms, one can hardly forget how they mobilized activism. These platforms also serve as hubs for organizing protests, marches and boycotts. They facilitate fast communication and instant decision-making so much that they make coordinated actions from people to ask for sustainability policies of governments and corporations seem almost effortless. One key aspect is that the development of environmental movements like Extinction Rebellion or the People's Climate March could not have been realised without exploiting social media platforms to organize. Instagram and Twitter campaigns around these movements frequently call on people to show up in person, help people understand where protests are located, or magnify their voices via live-streaming, hashtags, and videos (Blanc 2022).

Sustainable Habit-forming Mobile Apps: Promoting Eco-Friendly Practices with Behavioural Incentives Mobile applications are an excellent medium for engaging people in sustainable practices by providing day-to-day eco-friendly task commitments and behavioural nudges. The apps are built to help consumers track their energy usage, keep tabs on their carbon footprint, and develop sustainable practices such as recycling or reducing food waste. In addition, they incent long-term behaviour change using mechanisms such as gamification, rewards and peer comparison (Ronkainen 2023).

- **Eco-Habits Tracking**: User can track their eco habits through Mobile application, which makes possible to target and reward sustainable lifestyle changes. A great example of this is the energy-saving apps JouleBug and MyEarth, which give you suggestions, both large and small, to save energy in your home, including reminders to cut back water usage and waste. It can be as simple as personalized recommendations coming from those apps, guiding you to the actions that make your impact on the planet positive and often reducing it. In simple terms, carbon footprint calculators like Oroeco and thousands of others help collect data about your daily activities (commuting, food habits and energy usage) to provide you with an estimated figure on the CO_2 burning your actions result in so that you can take steps to counterbalance the burden based on established parameters. This app aims to give users actionable advice on reducing their footprints—like switching to renewable energy or adopting a plant-based diet (Chuan et al. 2024).

- **Gamification and Rewards**: Integrating game-like features (such as points, badges, or leaderboards) in the design of reward-based interventions is one of the most effective motivators for sustainably engaging with long-term behavior change solutions. JouleBug works much in the same way with gamification, which has people doing everyday activities like biking instead of driving, reducing food waste, or using reusable bags. Gamification allows them to challenge friends or everyone else out there and win prizes, badges, and honour for their green accomplishments. Recyclebank, another app that rewards people for ecofriendly habits, takes it one step further by giving out real-life goodies. This way, users earn points by recycling or doing something eco-friendly and collect them to redeem discounts on adds to those points that can be used while purchasing eco-friendly products or getting an eco-friendly service done. Ecofy, a badge and rewards app for those more ecologically minded allows users to track their environmental contribution (Ouariachi et al. 2020).
- **Peer Comparison and Social Influence**: Most mobile applications deploying peer comparison to incentivize sustainable behavior. Oroeco is an app that allows users to track and compare their carbon footprint with friends or even just the broader community, making energy use a social activity. Within this, comparative methods tap into social cognitive theory to provide the cultural forces of group influence that make sustainability more approachable and actionable. Apps also prompt users to post their achievements on social media, spreading the word about your new consistent routines. By allowing users to share eco-friendly accomplishments or challenges on social media, some platforms like GoEco have become even more directly involved in sustainability awareness by fostering user-generated content (Boncu et al. 2022).

E-Commerce and Sustainable Consumption: Ethical Shopping and Consumer Education: Consumers now have considerably more transparency, accessibility, and opportunities for ethical shopping due to e-commerce platforms. These are websites and apps focused on sustainability, e.g. one can use these to look up if the product is environmentally friendly and second-hand purchase as well as the information on where it was sourced, made and environmental impact, helping people to make informed decisions.

- **Promoting Ethical Shopping**: Several E-commerce platforms concentrate on selling their goods through ethical and sustainable modes of production. ThredUp, Poshmark and other similar sites are rooted in second-hand clothing—a key component of the circular economy that helps to stem the demand for virgin production of fresh clothes and decreases textile waste sent to landfills. Etsy: platforms such as Etsy promote the buying and selling of handmade or eco-friendly goods from small businesses and independent sellers, helping develop local economies and benefitting from a lesser environmental footprint than typical large-scale production. Meanwhile, platforms such as DoneGood curate ethically produced products from different brands that have been vetted for factors such as fair trade or environmental and social responsibility. By providing eco-friendly versions of the most popular products, these e-commerce sites give consumers a break from bad spending habits (Kwilinski 2023).

– **Sustainability Labels and Product Transparency**: As consumers are more concerned about sustainability, labels indicating such have gained popularity in guiding consumer choice. Eco-labels are currently available on many e-commerce platforms, allowing customers to quickly identify products that fulfil environmental criteria such as organic certification, cruelty-free testing or carbon neutrality. Amazon provides a "Climate Pledge Friendly" label indicating products with sustainability certifications. Platforms like Good On You provide you with a rating that shows how sustainable fashion is, and they also show the complete details of whether the company uses sustainable materials and labor practices or how serious the brand is about reducing environmental impact. E-commerce platforms provide this information to consumers, empowering them to be more mindful of what they are buying and make purchase decisions that match their sustainability values (Duckworth et al. 2022).
– **Digital Consumer Education**: Education is one of the most important methods to encourage sustainable consumption, and many e-commerce platforms will educate consumers on their environmental costs with e-coupons. Apps like EcoCart can help shoppers estimate and neutralize the carbon output tied to their transactions by supporting projects ranging from planting trees to sustainability pushes in energy. Then there are platforms like Patagonia and Everlane that go a step further to provide detailed information about their supply chain's environmental and social standings. Ending more transparently will build knowledgeable consumers and hold brands traditionally accountable for their choices to do the right thing (Budz et al. 2023).

Learning Platforms MOOCs, YouTube and e-learning for sustainability literacy Digital learning platforms are a powerful way to foster sustainability literacy. Given the growing interest in environmental education, these websites have become hosts of online courses, webinars and multimedia resources to elevate civic awareness about sustainability topics. Be it Massive Open Online Courses (MOOCs) or YouTube tutorials, digital learning platforms have become an essential tool for disseminating knowledge and providing people with the necessary skills to take on many of the sustainability challenges across the globe.

– **Massive Open Online Courses (MOOCs)**: MOOCs, which are offered by platforms like Coursera, edX, and FutureLearn, offer free, low-cost access to university-level courses on sustainability. The subjects in these courses, from renewable energy technologies to climate change policy, are designed for all types of learners—individuals or professionals, and they help further explore the topic of sustainability. EdX, for example, provides circular economy, climate science and sustainable urban development classes from universities such as MIT, Harvard and Deft Technical University. The courses provide key capacity for addressing global sustainability challenges (Gao and Liu 2023).
– **YouTube and Educational Content Providers**: Unquestionably, one of the platforms where sustainability education is likely to thrive even more in the future is

YouTube because thousands of videos regarding zero-waste living in sustainable agriculture are available. On YouTube, channels like TED-Ed, Kurzgesagt, and The School of Life put out enjoyable content that manages to decode the seemingly incomprehensible jargon surrounding environmental problems into easily digestible nuggets. There's also the educational aspect of making sustainable choices in everyday life, such as tutorials on composting, upcycling, growing your food, and so on. That means anyone, no matter their education, can be reached with sustainability education if they have an internet connection (Zhou et al. 2020).

- **Webinars and Workshops**: Many companies now host webinars or virtual workshops about sustainability. These events bring experts in to interact live and provide real-world examples of making sustainability work at home or within your own company. Such events are often held on platforms like Zoom and Webex, so anyone from across the globe can participate.

AI/ML and IoT in digital platforms have taken the idea of having smarter ways to observe, predict, and drive sustainable behaviors of users. Powered by data insight, these platforms give personalized recommendations that increase resource utilization complemented by real-time environmental feedback.

- **Real-Time Monitoring and Feedback**: IoT-integrated data-driven platforms can monitor energy use, water consumption, and waste generation. Nest Thermostats and Philips Hue lighting systems (to name a few) use sensors to track real-time data, helping users monitor and adjust their energy consumption and reduce the environmental impact of a connected home while still keeping its inhabitants comfortable. For example, platforms like Sense deliver detailed energy use reports so homeowners can see which appliances are doing the heavy lifting and inform them whether it may be time to upgrade old appliances for energy efficiency. Not only do these tools provide data delivering actionable insights to drive sustainable behavior change (Dauvergne 2020).
- **Customized Advice Based on Personal Data**: AI-driven platforms provide personalized recommendations based on users' data. Tado, for example, uses AI algorithms to learn how households use their heating and automatically change the temperature to save energy (and ultimately money) without making the home uncomfortable. WaterSmart likewise provides homeowners data on how much water their home uses and gives them tailored ideas of how to save. Conversely, these platforms facilitate easier sustainability habits by taking the brain work of complex judgement processes and giving helpful suggestions that make sense. Data-driven platforms allow for highly individualised recommendations, ensuring sustainability efforts are as efficient and effective as possible (Pigola et al. 2021).

4 Mechanisms of Influence: How Digital Platforms Change Behavior

By now, we live in such a digital world that platforms rule the way so many of us think and act. We have good reason to believe that they can affect user behaviour—using tools like behavioural nudges, persuasion tactics, and engagement strategies. Research has shown some success in changing user behavior Toward a positive end. At the intersection of these two ideas is a new challenge for activists: to co-opt the internal logic of digital platforms and use them in service of sustainability—surreptitiously but effectively (Fig. 3). Through the integration of nudges, gamification, peer comparison and real-time feedback, these platforms drive sustainable choice end users. This migration to the digital footprint is not only about technology but also drawing on lessons from psychology, behavioral economics and social sciences to design interventions that change user behaviour sustainably (Bonina et al. 2021).

Behavioral Nudges and Persuasion: Behavior nudge is found in behavioral economics that infringement at a prompt or cue to compel a user to complete specific behavior devoid of any restriction in freedom of choice. In online environments, platforms use nudges to affect the behavior of that users in a positive way for sustainability. These include cues, defaults, and reminders to help users of digital platforms (e.g., e-commerce sites, social networks or energy management systems) make more sustainable choices (Rachlinski 2022).

- **Nudges via Defaults**: One of the most persuasive nudging tools is setting defaults. Studies suggest that people often stick to the default choices being presented to them because it would require more energy to change the defaults. The idea is also frequently used by digital platforms to make sustainable options the default. One example Hornyak gives is that an online shopping platform would automatically make the default shipping—the one that pops up when you click pay—a carbon–neutral option. Similarly, energy management software can automatically suggest lower-energy usage setups as defaults for smart home systems. In scenarios like the above, the low-key product prompts users to change their behavior into more sustainable patterns without requiring explicit persuasion or significant outright behavior changes.
- **Reminders and Cues**: Another nudge mechanism is reminders and notifications. Frequently, digital platforms are accompanied by reminders to meet sustainability milestones and nudges targeting eco-friendly action. For instance, energy monitoring apps could push notifications when users surpass the permitted energy spending limits to encourage them to reduce their usage requirements. E-commerce platforms, for example, can recommend eco-friendly alternatives at the point of checkout. These reminders are given at appropriate intervals to ensure that users' choices will be more congruent with their goals on sustainability.
- **Framing and Persuasion**: Besides nudging, digital platforms exploit framing-based persuasion. Users see it in a good and positive light when sustainable options are wrapped around the "better" or "smart" way. Labeling other products as environmentally or ethically friendly leaves the impression of moral and

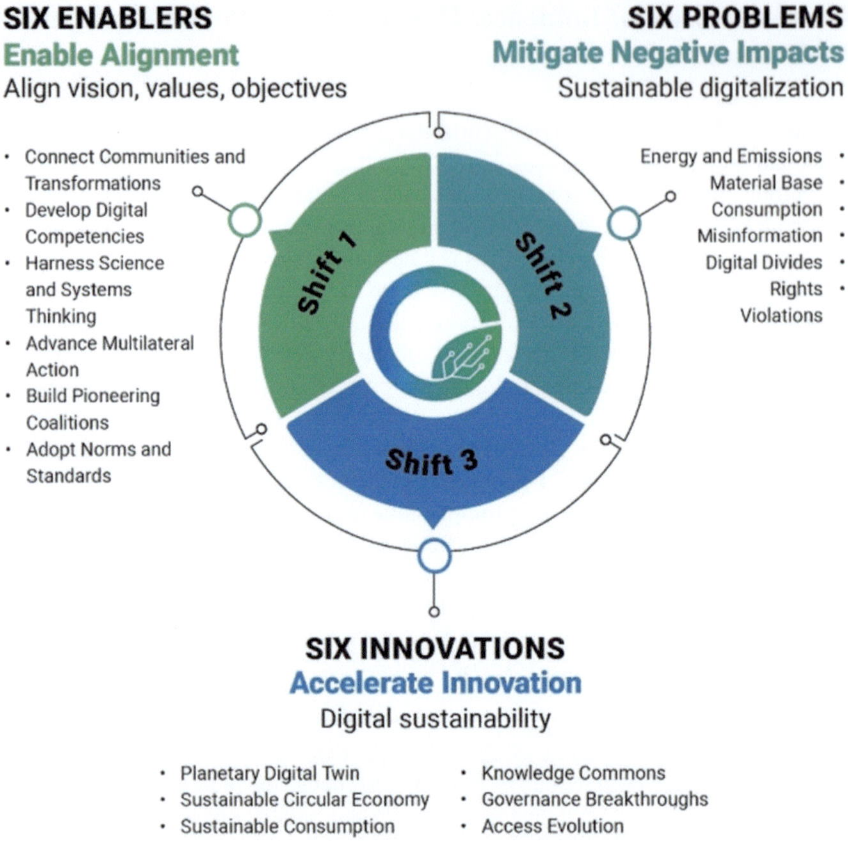

Fig. 3 Digital sustainability—problems and enablers. *Source* https://www.unep.org/technical-highlight/new-action-plan-underway-harness-digital-tools-planetary-sustainability

social responsibility, which determines consumer behaviour. Digital platforms can convince users to make greener choices, inculcating the sustainability norm while creating awareness and change around conservation efforts.

Social Influence and Peer Comparison: This behavior is part of being a social animal, but our decision-making is heavily influenced by what others are doing or saying." Social influence and peer pressure drive much human behaviour, and we can capitalise more effectively on this using digital platforms to promote sustainability. The platforms can get users to adopt more sustainable behaviours by generating a sense of 'community', reinforcing social norms or using peer pressure (Vrain et al. 2022).

– **Social Norms**: A lot of our daily behavior is governed by social norms, and very often, we don't even realize this. Showcasing others' green habits on digital fora can be used to normalize eco-friendly behavior. For example, if a user sees everyone else doing things like reducing their energy usage, recycling more, or buying sustainable products, they will be influenced to do that. With the help of social media, you can show your awareness of what is trending concerning sustainability. This will then serve a dual purpose-it would normalize behaviour (eco-friendly) and make you aware of the ongoing conversations and how important they are globally.

- **Peer Comparison and Competition**: This is another very powerful way to shape behavior. They could introduce features where users can compare themselves and their sustainability metrics to those of the community. An example of this might be an energy-saving app that demonstrates to users how they stack up against their neighbours or friends regarding energy consumption. This can be very motivating especially when paired with friendly competitiveness. By gamifying this comparison, assigning rewards or acknowledgement to those who do the best in sustainability metrics over and beyond that, you are also ensuring eco-conscious behaviors.
- **Community and Collective Action**: Digital platforms can also create a sense of working towards sustainability in tandem with joining collective action efforts. Making a point to join in groups or challenges on platforms can also be useful since these efforts can increase the overall impact of every individual action. Undertakings such as Plastic-Free July and Community-Led tree planting initiatives; if made popular across the digital platform, these challenges can generate a sort of herd participation, generating ripple effects beyond the platform (Sweet et al. 2020).

Gamification and Engagement Strategies: Gamification introduces game design elements into non-game contexts to help motivate participants. For sustainability, gamification techniques incentivize environmentally friendly actions by turning them into interactions, challenges and rewards. Digital platforms can also inject these dull systems with enjoyment and competition by adding points, leaderboards, rewards etc., to induce users into more sustainable actions (Kirchner-Krath 2023).

- **Points, Badges, and Rewards**: Most digital platforms have built-in reward systems to incentivize specific actions or outcomes; these same tools could be applied to promote sustainable behaviors. The user might get points or a badge for reducing energy, recycling, purchasing eco-friendly products etc. This serves as a feedback mechanism for users who receive tangible rewards as recognition and provides incentives to keep making eco-friendly choices. The above reward system can be readily integrated into platforms like fitness trackers, energy-saving apps or e-commerce sites.
- **Challenges and Competitions**: Another way to gamify promoting sustainability is through challenges. The app is a week of competition, like asking your friends to try and use less water. over the course of a week, you get some new product or badge. Letting users compete against each other or groups brings a new layer of

social currency, appealing to notions of social status and recognition that motivate action. Gamification techniques increase user interaction by making sustainability fun and competitive, which in turn helps grow good environmental habits (Ouariachi et al. 2020).

- **Virtual Ecosystems and Simulations**: These digital platforms are a step up from gamification as they create virtual ecosystems or simulations where users can witness the results of their actions. Immersive: Create an environment in garden or forest form that grows as the user completes various sustainable tasks (like lowering your carbon footprint or using renewable energy). These simulations give users a real and instant picture of what their actions can lead to, thereby enabling sustainability by making this concept seem more down-to-earth.

Feedback Loops and Real-Time Data: The most powerful way digital platforms drive our behavior is through feedback loops: With the conventional motivation engine—rewards, punishments, recognition—there existed one tiny problem. Users did not know how well they were performing moment-to-moment throughout their day. A Feedback loop, which could give the user immediate or near-immediate feedback to monitor their actions and adjust can be introduced as an approach. Sustainability—making available real-time data to users regarding their energy consumption, carbon emissions, or in-the-cloud offset and waste reduction (Sjödin et al. 2021).

- **Energy Usage and Carbon Savings**: Several home automation systems and energy usage monitoring apps give users a perspective of how much energy they utilize as it happens. For example, users can see how much electricity they are using in real-time to witness the impact of taking small actions—like turning off lights or adjusting their thermostat—will add up to big energy savings over time. This real-time feedback makes users aware of their consumption habits and motivates them to make changes to lessen their environmental impact (Sayed et al. 2021).
- **Increase Visual Impact**: Digital platforms can use images to help users understand the environmental impact of their actions. This could involve, for instance, an app that can display the amount of carbon saved by a user who has opted to take public transport instead of using his car (illustration) or how much waste a second individual has been able to steer away from thus far paper recycling bins at cable landing stations in Marseilles and Gizan through landfills. Feedback loops highlight the tangible impact of sustainability, which motivates users to make sustainable choices repeatedly.
- **Behavioral Adjustments**: Constant feedback also leads to behavioral adjustments over time users can follow along with chore slaves and get periodic reports, so they can become more aware of what actions lead to sustainability. This continuous feedback mechanism further cements behavioural changes as users can view the tangible impact of their actions and remain motivated to stay on track (Rapp and Boldi 2023).

Personalization and Customization: Personalizing consumers is a major method by which digital platforms manipulate behavior. This can help platforms surface sufficient content, offers, and messaging tailored to individual users to make sustainability action more relevant and effective. Personalized—Everyone is different has different preferences, value systems and lifestyles, so making sustainability data fit around their already established routines will make it much easier for them to adopt more eco-friendly behaviours (Babatunde et al. 2024).

- **Tailored Recommendations**: Many digital platforms leverage algorithms to recommend content or products that cater to the interests of their users. In the case of sustainability these suggestions help point consumers to sustainable solutions that best align with their needs. An e-commerce website could recommend fair-trade or environmentally friendly products due to a user's prior shopping history. For instance, an energy-saving app could recommend customized ways to reduce consumption through strategies tailored to a user's unique household profile.
- **Personalizing Goals and Tracking Progress**: Digital platforms permit setting personalized sustainability objectives like reducing energy consumption by x%, creating zero waste, etc., which could also be easily tracked. Because these goals are specific to every user, they also make users more likely to be engaged in sustainability efforts. How they are going is part of the personalization, too, because they could monitor their process over time and make some adjustments needed to align better with what they want to achieve.
- **Customized Messaging**: Digital platforms can change behaviors by serving personalized messages. Platforms could, for example, send out targeted push notifications or messages under users' values (and motivations). For example, if someone is really into climate change they might only get messages saying to cut down on carbon (so that neither resource overconsumption message tips them off), and vice versa for the water hawks. This will help customise messaging to make sustainability efforts by platforms more relevant and impactful.

5 Case Studies and Empirical Evidence

Technology and digital platforms are two critical tools to steer sustainable behavior in modern societies. Environmental advocates and innovators are influencing individual and collective actions toward sustainability by using the massive growth of social media, mobile applications, e-commerce platforms, and smart home integration growing. This article further explores four specific case studies as examples that apply digital tools to encourage sustainability: a social media campaign for climate change awareness, a mobile app for reducing waste, an e-commerce platform supporting ethical consumption and smart home systems promoting more energy-efficient living. These case studies provide empirical evidence of how digital interventions might influence user behavior and contribute to global initiatives addressing environmental threats.

6 Case Study 1

Climate Change Awareness Via Social Media Campaigns: The extensive reach of social media networks has provided an unimaginable stage for environmental activists that have turned awareness drives into worldwide phenomena. One good example is the "Beat Plastic Pollution" campaign initiated by United Nations Environment Program (UNEP) on World Environment Day 2018. The campaign, involving social media posts, illustrates the damage of single-use plastics on ecosystems and calls for individuals and businesses to decrease their use. Several aspects contributed to the success of "Beat Plastic Pollution" campaign. First, it enabled producing visually compelling content (e.g., infographics or short videos), which allowed the audience to share them on Twitter, Instagram or Facebook easily. It also involved the hashtag #BeatPlasticPollution which enabled the user to follow and engage in global conversation. Additionally, the participation of high-level celebrities and influencers elevated the distribution of the campaign-with millions of views and shares on their accounts. And the campaign has been such a success witness by plastic reduction behaviors that uptake across entire towns. Over the coming months, numerous countries moved to ban or reduce their use of single-use plastics; for example. For instance, the European Union enacted a directive to eliminate single-use plastics by 2021 (partly sparked by heightened public awareness around this campaign.) Millions of individual social media users took to their favorite platforms to tell the world that they are giving up plastic and shared their personal stories and photos of themselves using reusable bags, bottles, and straws. At a more scientific standard, the "Beat Plastic Pollution" marketing campaign shows how powerful social media could promote public environmental knowing and riding movements-changing behaviors. Social media platforms separate action into individual, decentralized and participatory actions taken by individuals. Yet, sustaining such behavior change post initial surge of awareness is the real challenge that entails continuous engagement and reinforcement of targeted messaging (Bagai and Henam 2021).

7 Case Study 2

Mobile App for Reduction of Waste: The mobile app designed to minimize one of the most bothering concerns of environment food wasted is one way how digital revolution takes place in this wasteful world. This is where OLIO and Too Good To Go come in, two apps that enable people to share or sell food shopping items that would end up in the bin otherwise. In this case study, we look at the OLIO app which links people with their neighbours and local businesses so that surplus food does not go to waste. Empirical evidence from user data and surveys supports that OLIO effectively reduces waste. By 2022, OLIO had shared over 56 million food portions via the app, preventing over half a million kilogrammes of food waste. This resulted in the intuitive app's success, which functions almost like a local social network, allowing people to browse (or offer) surplus food near them quickly. The community-building component of the app contributes to making sustainable choices a habit by fostering an understanding that collectively, we have a responsibility not only to our planet but perhaps even greater than this, to one another when it comes reducing food waste. Research into user behaviour indicates that OLIO has changed its users' mindset to consuming and wasting less food, which can be attributed to the app itself. According to studies, 80% users claimed they were better aware of their eating patterns and understanding what they bought post using the app. POOR IMPACT: OLIO addresses not only waste reduction but also food insecurity as they provide free food access to those who need it. But as with very digital intervention, OLIO have stagesf to become a platform for real global scale. While the team struggles with scalability in a newer app, we can see how it can come closer to achieving the mission of saving food waste internationally with increased responsibility to acquire market share. This means more awareness and collaborations with local authorities and businesses to include the app in wider waste management plans. But nonetheless, this is a powerful example of how apps are the key to changing human behavior and sustainability (de Almeida Oroski 2020).

8 Case Study 3

E-Commerce and Adhering to Ethical Consumption: E-commerce has become a strong agent of exposure by offering various products that can be sourced sustainably, pre-owned or hand made supply. But what about a global marketplace like Etsy, where people buy and sell unique handmade and vintage goods—how does this impact e-commerce on sustainability perspectives? Etsy offers a place for smaller businesses and individual creators, which allows consumers to lean toward more ethical production methods while supporting a product with a lower overall environmental impact than mass production. This official shift is backed up with empirical data supporting the notion that Etsy indeed affects how consumers shop. Last year, Etsy became the first major e-commerce shopping platform to offset 100% of its carbon emissions from shipping. Shipping is a big problem regarding the environmental impact of online shopping. That decision was highly publicized, and data from user surveys shows that consumers considered Etsy's dedication to sustainability when making their purchases. The biggest headline is that many people said they would be more likely to buy an item on Etsy than another e-commerce platform based on the environmental practices alone. Moreover, Etsy sellers often feature sustainable resources as part of their product description—like using recycled materials or implementing eco-friendly packaging. This aligns with the increasingly loud consumer desire for ethical products, especially among Gen Z and Millennials who highly value sustainability in their purchases. A financial success, the site promotes ethical consumerism, as sales of "sustainable" goods have continued to rise year-on-year. While the successful adoption of these policies is unquestionably an accomplishment, one that seems demostrative of a commitment to sustainable commerce at

Etsy and similar platforms, it also underscores another dance such companies must play between promotion sustainability on their site with growth in sugary products. As Etsy grows, it needs to protect its ethical commitment, which dominates any commercialization pressures. However, Etsy offers a window into how e-commerce platforms can influence consumer behavior and thereby promote ethical consumption (Tarakanova 2023).

9 Case Study 4

Energy Management and Automation using IoT-Enabled Smart Home Systems: The embedding of Internet of Things (IoT) technology in household systems has also directly translated to efficient energy management providing actual data on energy usage, aiding automation at home settings to act in real-time leading to reduced consumption. Connected thermostats, lights, and appliances from devices like Google Nest and Amazon Echo can help homeowners monitor their energy use through smart home systems. These claims are supported by empirical research from user data and energy audits regarding the effectiveness of smart home systems in reducing energy consumption. For example, one study by the company found that users of its Google Nest thermostats were able to lower energy use—and thus reduce their carbon dioxide emissions for heating and cooling—by up to 12%. In homes, for example, electric lighting consumes 8–15% of household electricity used and automatically switching off lights when not needed can be saved plus automatic light dimming based on daylight harvesting. Smart home systems have an upper hand even in giving live results to users about their energy consumption so that they can take some urged decisions as well. Studies in behavioral science have shown people are most motivated to change their behavior when they receive immediate feedback, which enables them to understand the direct impact of their words and actions. Also, the automation offered by smart home systems can reduce the mental load needed to enact energy-saving practices, allowing behaviors that conserve energy to be incorporated into daily regimes more easily. But, the ubiquitous implementation of smart home systems does not come without difficulty. Many smart devices are relatively expensive upfront, and security concerns around data privacy can be off-putting. Secondly, we nee d to let the public know better about where these environmental savings are coming from in a smart home system—for most people they are only convenience solutions currently-and not how we think, as outstanding sustainability ones (Hafeez et al. 2020).

10 Challenges and Barriers

The Verge: As societies worldwide are using digital platforms to enable sustainable behavior changes, many challenges and barriers come in as hurdles that can challenge progress. These hurdles span from structural and technological barriers—the digital divide, data privacy concerns—to behavioural/psychological aspects that impact meaningful interaction. Market dynamics and policy constraints also significantly influence the terrain of the sustainability initiative. Appreciating these problems would allow for insights to be drawn that are usable in the creation of strategies to increase the effectiveness and inclusivity of digital platforms designed to foster sustainable practises. The technological challenges that will be examined are diverse, such as access and literacy gaps, data privacy concerns, behavioral and psychological barriers like the intention-action gap; market and policy limitations, including greenwashing and regulatory patchwork (Costa et al. 2023).

11 Technological Challenges

- **The Digital Divide**: This is one of the main technological obstacles in creating behavior change via digital platforms, and sustainability requires users to modify their actions. This divide includes inequalities in technology access, broadband internet availability and digital literacy among different communities and geographies (Ye and Yang 2020).
- **Access Gaps**: Digital platforms are frequently more accessible to wealthy, urban societies, leading marginalized communities to be left behind. Nonexistent, poor and expensive internet infrastructure is especially common in low-income countries and rural areas, so they can also lead to barriers to sustainability interventions that depend on digital solutions. When equitable access is lacking, digital initiatives to promote sustainable practices leave these communities behind and all efforts for sustainability lack inclusivity and diversity (Helsper 2021).
- **Literacy Gaps**: On the demand side, few potential users are equipped with digital literacy skills necessary to effectively use these platforms (called the "second-level digital divide", or simply 'literacy gaps'). Even if technology is present, people might not be able to interact with these PMDs technology, make meaning out of them or take action and shift toward behaviors essential for sustainability. This literacy gap may result in obstacles to comprehending the significance of sustainability, resulting in an underutilization of resources and tools that may be available. Unfortunately, many of the people who could benefit most from sustainable behavior change initiatives are either unable to access these solutions or unaware that they exist.

- **Data Privacy**: Another major hurdle emerging in the successful implementation of digital platforms related to behavior change is data privacy. Today, users are more conscious of their data and fear its hoarding by unknown services for no good reasons.
- **Surveillance and Tracking Resistant**: Many digital platforms need to track user behavior for personalized sustainable action recommendations. Nevertheless, these surveillance features can instill ideas of resistance with users who worry about violations of their privacy. However, scepticism of digital onboarding remains as a result of concerns about misuse of information, unauthorized access and the possibility of turning it into surveillance techniques. This resistance can, in turn, discourage participation with platforms that are designed to enable the change to more beneficial habits (Kozyreva et al. 2020).

12 Behavioural/Psychological Barriers

- **The Intention-Act Gap**: in practice, digital platforms are often ineffective if targeted ads and tailored Nudges face behavioral and psychological barriers. One of these most important issues is the intention-action gap; the divide between what people say they wish to do sustainably and what they ultimately end up doing (Kilian and Mann 2021).
- **Knowledge Instead**: knowledge, but: non-existent Knowledge about sustainability is certainly increasing—however it has not only turned out that this in itself as somewhat homework on attitudes transfer restricted ends not lead to changes in behavior. Though many people want to recycle, not everyone can follow the practice. Factors like convenience, perceived efficacy, and social norms determine whether people act on their intentions. These are just some considerations that digital platforms often neglect in their attempts to translate awareness into direct action (Alshbili et al. 2021).
- **Inertia and Habitual Behaviour**: the additional psychological resistance to change. One reason it can be hard to change behaviour is that people like familiar routines and they find it difficult to incorporate new behaviours (even when they know they are ways of promoting sustainability).

    - **Habit Formation**: This has established a behavioral rut that is all too difficult to breach, one that digital platforms battle to disrupt—especially should the new offerings they introduce fail to dovetail smoothly into your existing routines. Unless these platforms deploy successful strategies to build habits on sustainable practices they might find it hard for behavioral change to last (Ronkainen 2023).

13 Market and Policy Limitations

- **Insufficient Regulation and Oversight**: The market dynamics also present formidable obstacles for digital platforms supporting sustainability. This lack of regulation and oversight can lead to what is called "greenwashing", where companies may make false claims based on the environmental attributes of a product or service without proving these claims (Kim and Ahn 2024).
- **Greenwashing Issues**: If consumers are exposed to false or exaggerated marketing on their digital platforms, they might not take any posts on sustainability seriously. The lack of clear regulatory frameworks for transparency and accountability enforcement allows companies to continue to trade off returns on capital without real sustainability implications. Companies who engage in such shady tactics are weakening the credibility of both their brands and any platforms they work with, and by doing so, these actions are proving precisely why we need to support environmentally friendly alternatives better (Free et al. 2024).
- **Inconsistent Policy Frameworks**: There is often an inconsistent policy framework across jurisdictions, making it hard for digital platforms to maneuver. This mostly depends on how effective the platforms are, which can be influenced by regulations.
- **Regional Disparities**: The definition of sustainability and the environmental issues concerns different regions might conflict; others have other regulation demands. This inconsistency could lead to user confusion and cripple scale for potentially successful behavior change initiatives Decentralized approaches to regulation also risk balkanizing the effects of regulatory measures, as digital platforms strive to adjust their strategies for different segments of a market (Scheidel et al. 2020).

14 Opportunities and Future Directions

For the first time, a confluence of emerging technologies like AI, blockchain, and immersive technologies (AR/VR) is creating unprecedented opportunities to influence sustainable behavior change across different populations. With the continuing global challenges of climate change, resource depletion, and social inequities, these technologies can be a part of innovative solutions that include collaboration between multiple stakeholders. We discuss how emerging innovations can empower sustainable behavior change, followed by a look into the emergence of collaborative platforms and citizen science initiatives. The importance of combining different disciplines for maximizing potential and strategies to scale behaviour change are also discussed.

15 Emerging Technologies in Behavior Change

- **Artificial Intelligence**: AI is the most important technology for sustainability as it can mine large amounts of data, identify trends and forecast future results to show us where we are headed. Through consumer behavior analysis, AI offers personalised suggestions to drive a change towards sustainability. For example, AI-powered platforms can recommend energy-saving products by learning about users' likes and prior actions, guiding them toward environmentally focused choices. Further, AI is also being used to optimize resource management, e.g., in energy consumption of smart grids or predicting renewable energy demand and waste management using real-time data sets. It could also use machine learning algorithms to model complex environmental systems, help inform policy decisions, and institute sustainable practices! (Chew 2021; Pigola et al. 2021)).
- **Blockchain Technology**: Blockchain technology is a transparent, decentralized yet secure method to verify tracking data that can help improve sustainability. For example, blockchain can be used to trace the origin or provenance of products and making certain these goods are sourced responsibly and sustainablyapeut-generated. Regarding consumer decisions, this level of openness provides a valuable resource for consumers to make responsible choices and promotes accountability at the production level. Blockchain can also help to encourage a decentralised energy market—one where you buy and sell renewable energy direct from your neighbour, ultimately driving consumption of more sustainable forms of energy. Through blockchain, smart contracts can automate and secure sustainability commitments among companies—encouraging them to act sustainably (Parmentola et al. 2022).
- **Augmented and Virtual Reality**: AR and VR enables users to be part of a virtual reality. Hence, it becomes an integral tool for educating people about sustainability. These tools can predict the effects of climate change. AR and VR help to illustrate what could happen if we do not live sustainably, making it systematically clearer that sustainable behaviors are necessary. At the same time, immersive experiences have the potential not just to provide eye-popping instruction about related issues but to bringing people closer, on a visceral, meaningful level of feeling, to communities that are dealing with life-destructive effects of environmental degradation. For instance, a VR experience that tells the story of people living in deforested areas can evoke emotion—helping them move from awareness to action and advocate for sustainable solutions (Abad-Segura et al. 2020).

16 Citizen Science and Collaboration Platforms

Collaborative platforms have made citizen involvement in research and sustainability initiatives a norm. Platforms like OpenStreetMap and Zooniverse allow people at all levels to contribute data, adding up to new scientific knowledge that can help inform sustainable practices.

- **Open Street Map**: Open Street is a project that creates a free and editable map of the world. Participatory methods improve data quality and accuracy, while creating community connection. Urban green spaces, waste disposal facilities, and transportation networks (among other things) can be mapped so citizens know which areas are already well-served and where good urban planning could be improved (Yeboah et al. 2021).
- **Zooniverse**: This is a platform for citizen science projects. Utilizing the combined intelligence of various participants, Zooniverse extends research capacity to address critical sustainability problems. Citizen contributions, for example, could assist us in monitoring wildlife populations, tracking climate change indicators and analysing ecological patterns. By doing so, these collaborative platforms democratize knowledge production and can also turn every citizen into an active contributor to sustainability efforts, thus breaking this sense of ownership and responsibility that each citizens must have with environment stewardship (Palmer et al. 2021).

17 Interdisciplinary Approaches

A combination of behavioral science, technology and policy making is required for behavior change to be successful. An interdisciplinary methodology in which we view technology and user behavior as inseparable can close the gap between platforms and programming, ensuring that platforms are being produced with a more holistic idea of why people use them!

- **Behavioral Insights**: A high profile example of this practice illuminating some fresh insights is how two behavioral scientists explained the psychological reasons behind decisions and behavior. Using nudges—lightweight alterations to design that shift consumer choices towards sustainability—illustrates (Boncu et al. 2022).
- **Technological Innovation**: Technologists have a big role in creating the tools and platforms to initiate behavior change. Their capabilities in AI, blockchain, and immersive technology will go a long way in designing innovative, user-friendly, and scalable solutions. Working with behavioral scientists to match technology design to human desires and needs.
- **Policymaker Engagement**: Policymakers must be involved in designing relevant policies and interventions to enable a regulatory environment supporting behavior change. Digital policies developed in collaboration with behavioral scientists and technologists can include defaults that nudge consumers to sustainable practices or use effective digital platforms more frequently. Policymakers can also help foster public–private partnerships bringing together the resources and capabilities of different stakeholders (Kim and Ahn 2024).

18 Scaling Impact Globally

The key to behaviour-change platforms making a difference on any scale is strategies to scale pilots that are successful, culturally relevant, socially acceptable and economically viable across different cultural, social and economic contexts. Key considerations include:

- **Cultural Sensitivity**: When scaling platforms internationally and adjusting to different cultural nuances that affect behavior, etc. Programes tailored according to local beliefs, customs and values are more effective. This means working with local stakeholders to see solutions as credible, feasible and relevant to the sustainability problems they address (Shin et al. 2022).
- **Social and Economic Context**: Platforms will need to account for the underlying social dynamics that determine access to technology and resources. Creating solutions which cater to different levels of digital literacy and differing resources can increase engagement. For instance, they could offer resources for offline access either by mobile technologies or a hybrid between both to mitigate the profound digital divide in options available to underserved communities.
- **Building Local Partnerships**: Developing relationships with local organizations and community leaders can help usher in effective behavior-change initiatives. These partnerships can supply credibility, expand reach, and use current networks to increase program yield. Collaborative efforts can also be used to identify and address local sustainability challenges (Temple 2023).
- **Continuous Monitoring and Learning**: Scalability to the behavior-change platforms means continuous monitoring for evaluating efficiency and impact. Data on user engagement, changes in behavior, and environmental outcomes will inform iterative improvements. Feedback integration in adaptation ensures that initiatives are tailored for a broad context and strikes the desired impact.

19 Implications for Policy and Practice

In a time when the advancement of technology is swift and almost unstoppable, we live in an age of digital platforms defining how we interact, shop and work. Nevertheless, the proliferation of these platforms has raised serious questions about sustainability; with environmental degradation, social equity and economic inequality come into play. The future of these digital environments depends on efforts from policymakers, corporations, and consumers. A couple of examples among policy recommendations are the need to develop regulatory frameworks that focus on transparency, accountability and ethics to make digital platforms a positive actor in sustainability (Martin et al. 2020).

Formulate Proper Regulatory Standards: Governments should coordinate with different authorities to bring out the regulations that should enough genuine acts on performance of data, the environment effects involved and some social duties.

Examples of such standards would be required carbon footprints and social equity metrics.

- **Promote Transparency**: Users should be able to require digital platforms disclose their algorithms, data handling practices and business models. The more transparency there is about everything going on, the easier it will be for these other platforms to determine how they wish to interact with them. Regulation would require that platforms disclose explicitly how the data is used, promoting consumer understanding of privacy and security concerns (Aloisi and De Stefano 2022).
- **Create Accountability Mechanisms**: Policymakers should introduce mechanisms to hold platforms accountable for their effects. These sanctions could take the form of fines due to involvement in unsustainable practices and deceptive marketing. The effectiveness of all these regulations will rest heavily on the setting up of independent oversight bodies to monitor and hold violators to account (Popiel 2022).
- **Incentivizing Sustainable Practices**: Regulatory frameworks should also offer clear and valuable incentives to drive platforms to enshrine sustainable business models. This would mean, for instance, tax breaks for companies that could demonstrate large reductions in their environmental impact or if they invested in social equity initiatives (Matta et al. 2022b).
- **Enabling Public–Private Partnerships**: By the nature of these types of relationships, joining forces with private companies for mutual benefit can spark innovation in sustainable practices. It means the policymakers can further this project by making pilot projects and also funding researches in the pursuit of sustainable digital solutions (Matta 2020b)

Designing for Behavior Change: Platform designers have to think through putting users first, and the intersection of fields like psychology and behavioral economics in whatever development process they are into. Best practices include:

- **User-Centric Design**: Sustainability features must also be user-friendly for every platform relying on them. This may include creating user-friendly interfaces that promote sustainable buying options—like eco products or carbon offsetting features (Ibert et al. 2022).
- **Integrating behavioral insights**: designers should be aware of why users decide. Nudges and other subtly manipulative methods used to guide users into adopting more sustainable behaviors, can also have a tremendous impact on the choices made by these users. This means that messages about sustainability can be integrated into the user journey: to motivate greener behavior, but only when relevant (Al-Dmour et al. 2020).
- **Feedback Loops**: You can offer feedback to users on their behavior which increases their engagement and in turn encourages sustainable behaviour. For example, giving platforms insights into how users consume could give them personalized advice on how to reduce their carbon footprint (Sjödin et al. 2021).
- **Community Engagement**: Including social aspects on platforms can lead to collective action toward sustainability. Sharing capabilities: features that allow

users to share their sustainability achievements or hardships, this builds a sense of communities and makes them accountable (Sweet et al. 2020).
- **Iterative Design and Testing**: Platform design should strive to improve constantly. Monitoring usability through regular user testing and feedback will allow these platforms to be iterated over time, developing to suit changing user needs and wants, thus a more effective approach in sustainability promotion.

Corporate and Consumer Roles in Sustainable Digital Solutions: The responsibilities of corporations and consumers are interlinked in the quest for sustainability within digital platforms. Corporations must take proactive steps to support sustainable digital solutions, while consumers play a vital role in adopting and promoting these practices (Herden et al. 2021).

20 Corporate Responsibilities

- **A Commitment to Sustainability**: Companies need to incorporate sustainability into the fabric of their business model—putting long-term environmental and social gains ahead of short-term profits. These might include quantifiable goals tied to sustainability and clear accountability for measuring progress.
- **Investing in Sustainable Technologies**: Companies need to invest into researching and developing technologies that increase sustainability. This involves the application of renewable energy for data centers, making supply chains greener and building digital tools that enable consumers to make sustainable choices (Wirtz et al. 2023).
- **Ethics in Business**: Ethical business practice means that workers are treated fairly and every person has equal access to digital services. Fair labour practices in the gig economy and digital inequality are vital to corporations' sustainability.
- **Engagement and Education**: Companies should educate their consumers about how we can all be effortful in taking the sustainable side. Its next steps might include citizen education about the environmental consequences of their choices and advice on them how to be more climate responsible (Herden et al. 2021).

21 Consumer Responsibilities

- **Consumer Choice**: Consumers need to be conscious of enquiring about the consequences of their digital engagements. This means being selective about the platforms they operate on, favoring those that show a real dedication to market sustainability and ethical behaviors (Lubowiecki-Vikuk et al. 2021).
- **Advocacy for Change**: Consumers also have power as they can promote sustainability and support transparency by demanding higher ethical standards from digital platforms. Taking collective action, boycotting unsustainable companies

or supporting others with a good sustainability background can make companies change their practices (Han 2021).

- **Adoption of Sustainable Practices**; One easy way individuals can influence the digital landscape is by adjusting their behaviors—purchasing eco-friendly products and using platforms committed to sustainability, or engaging in efforts likely to have a low carbon footprint (Han 2021).
- **Continuous Learning**: The second competency consumers should demonstrate to ensure they can sustain their transformation and improvement is by being information seekers paying constant attention to sustainability-related issues on Digital Platforms. In this way, they can follow sustainable values in their preferences and actions (Lubowiecki-Vikuk et al. 2021).

22 Conclusion

By investigating how digital platforms can play a role as behavior change agents in sustainability practices, we see that technology and human behavior meet to seek potential paths for sustainable development. Digital platforms, in their breadth and depth of reach, can impact individual, and subsequently collective attitudes towards more sustainable actions. By reviewing findings from many experiments and datasets, this chapter provides a comprehensive view on how those processes work in practice –confronting associated challenges and resistance but also some opportunity– if we are to reclaim our human reality out of the jaws analytics. Changing the behavior on sustainability practices is of utmost importance to curb todays pressing environmental challenges.

Nonetheless, digital platforms like social media, mobile apps and online communities offer fresh channels for driving sustainability. They help in sharing knowledge, outreach efforts and general access to resources that could give individuals as well s groups the power they need to embrace sustainable behaviors. Our findings imply that digital platforms use social norms, normative behaviours, and gamification to prompt eco-friendly behavior. For example, platforms that use peer comparisons and rewards models show that people want to reduce their energy consumption, waste and conservation behaviours. These are great innovations, but digital onboarding of sustainable practices has its own set of challenges, as it is tremendously difficult to break past the inertia. Such platforms have the potential to be useful tools, but issues such as digital inequality—the uneven distribution of access to technology—hold them back. In addition, misinformation, user dropout, and the inherent struggles of human behavior can all work together to attenuate the effectiveness of digital interventions.

It is important to recognize these barriers and how they may need to developed around. There are numerous opportunities for to improve on behavior change through digital platforms. Greater connectivity with artificial intelligence and machine learning can customize user experiences that will make sustainability practices more relatable and energetic. Secondly, using big data analytics helps identifying these behaviors and driving interventions that talk to the different segments in a language they would respond to. Additionally, partnerships among the governments, NGOs, and private sectors can also help to escalate the reach and range of digital platforms to promote sustainable behaviors. The policy and practical implications of these results are large. Policymakers must also make digital platforms an integral part of wider sustainability strategies so that they are not restricted to a specific sector only but benefit the population as a whole. To effectively transition, supporting investments in digital literacy and infrastructure will enable communities to leverage these platforms for more socio-economically equitable outcomes. Further, continued research and monitoring of digital interventions are vital to inform the development, adaptation and continuous advancement of strategies to align with user needs as they evolve.

Acknowledgements The author acknowledges NCSTC, Department of Science and Technology, Govt. of India, New Delhi, India, as a motivational force to work for science and society through science communication.

Conflict of Interest There is no conflict of Interest.

References

Abad-Segura E, Gonzalez-Zamar MD, Rosa ALDL, Morales Cevallos MB (2020) Sustainability of educational technologies: an approach to augmented reality research. Sustain 12(10):4091

Ajzen I (2020) The theory of planned behavior: frequently asked questions. Hum Behav Emerg Technol 2(4):314–324

Alam M (2022) Participatory culture and digital environmentalism. The palgrave handbook of global social change. Springer International Publishing, Cham, pp 1–13

Al-Dmour H, Salman A, Abuhashesh M, Al-Dmour R (2020) Influence of social media platforms on public health protection against the COVID-19 pandemic via the mediating effects of public health awareness and behavioral changes: integrated model. J Med Internet Res 22(8):e19996

Aloisi A, De Stefano V (2022) Your boss is an algorithm: artificial intelligence, platform work and labour. Bloomsbury Publishing

Alshbili I, Elamer AA, Moustafa MW (2021) Social and environmental reporting, sustainable development and institutional voids: evidence from a developing country. Corp Soc Responsib Environ Manag 28(2):881–895

Babatunde SO, Odejide OA, Edunjobi TE, Ogundipe DO (2024) The role of AI in marketing personalization: a theoretical exploration of consumer engagement strategies. Int J Manag Entrep Res 6(3):936–949

Bagai A, Henam SD (2021) Beating plastic pollution: UNEP's priorities and partnership in India. Environ Policy Law 51(4):265–269

Batool N, Wani MD, Shah SA, Dada ZA (2024) Theory of planned behavior and value-belief norm theory as antecedents of pro-environmental behaviour: evidence from the local community. J Hum Behav Soc Environ 34(5):693–709

Birkmann J, Liwenga E, Pandey R, Boyd E, Djalante R, Gemenne F, Wrathall D (2022) Poverty, livelihoods and sustainable development

Blanc E (2022) How digitized strategy impacts movement outcomes: Social media, mobilizing, and organizing in the 2018 teachers' strikes. Polit Soc 50(3):485–518

Boncu Ş, Candel OS, Popa NL (2022) Gameful green: a systematic review on the use of serious computer games and gamified mobile apps to foster pro-environmental information, attitudes and behaviors. Sustain 14(16):10400

Bonina C, Koskinen K, Eaton B, Gawer A (2021) Digital platforms for development: foundations and research agenda. Inf Syst J 31(6):869–902
Budz S, Bucur-Teodorescu I, Onete BC (2023) The impact of digital education towards a sustainable digital consumer in the era of circular economy. J Eng Stud Res 29(1):15–23
Chancel L (2020) Unsustainable inequalities: social justice and the environment. Harvard University Press
Chew SC (2021) Ecology, artificial intelligence, and virtual reality: life in the digital dark ages. Rowman and Littlefield
Chuan CH, Seelig M, Tsai WHS, Tao W (2024) Eco-warrior: a theory-based mobile app intervention for forming and sustaining pro-environmental habits. Mob Media Commun 20501579241271066
Climate change and its impact on biodiversity and human welfare (2022) Proc Indian Natl Sci Acad 88(2):160–171. https://doi.org/10.1007/s43538-022-00073-6
Costa F, Frecassetti S, Rossini M, Portioli-Staudacher A (2023) Industry 4.0 digital technologies enhancing sustainability: applications and barriers from the agricultural industry in an emerging economy. J Clean Prod 408:137208
Dartnell LR, Kish K (2021) Sustainable production and consumption
Dauvergne P (2020) AI in the wild: sustainability in the age of artificial intelligence. MIT Press
de Almeida Oroski F (2020) Exploring food waste reducing apps—a business model lens. Food Waste Manag: Solving Wicked Probl 367–387
Dietz S, Bowen A, Doda B, Gambhir A, Warren R (2018) The economics of 1.5 C climate change. Annu Rev Environ Resour 43(1):455–480
Duckworth JJ, Randle M, McGale LS, Jones A, Doherty B, Halford JC, Christiansen P (2022) Do front-of-pack 'green labels' increase sustainable food choice and willingness-to-pay in UK consumers? J Clean Prod 371:133466
Feroz AK, Zo H, Chiravuri A (2021) Digital transformation and environmental sustainability: a review and research agenda. Sustain 13(3):1530
Free C, Jones S, Tremblay MS (2024) Greenwashing and sustainability assurance: a review and call for future research. J Account Literature
Gao L, Liu Z (2023) Research on the sustainable development strategy of online learning: a case study of YouTube users. Sustain 15(11):8437
Godinho Filho M, Ganga GMD, Lizarelli FL, Blaz CLC, Tavares TM (2024) Circular economy via chat: Evaluation of adoption and use of whatsapp instant messaging platform for trading second-hand products. J Clean Prod 460:142510
Hafeez G, Wadud Z, Khan IU, Khan I, Shafiq Z, Usman M, Khan MUA (2020) Efficient energy management of IoT-enabled smart homes under price-based demand response program in smart grid. Sens 20(11):3155
Han H (2021) Consumer behavior and environmental sustainability in tourism and hospitality: a review of theories, concepts, and latest research. Sustain Consum Behav Environ 1–22
Helsper E (2021) The digital disconnect: the social causes and consequences of digital inequalities
Herden CJ, Alliu E, Cakici A, Cormier T, Deguelle C, Gambhir S, Edinger-Schons LM (2021) Corporate digital responsibility" new corporate responsibilities in the digital age. In: Sustainability management forum|Nachhaltigkeits management forum, vol 29, no 1. Springer Berlin Heidelberg, Berlin/Heidelberg, pp 13–29
Hoffmann S, Lasarov W, Reimers H (2022) Carbon footprint tracking apps. What drives consumers' adoption intention? Technol Soc 69:101956
Ibert O, Oechslen A, Repenning A, Schmidt S (2022) Platform ecology: a user-centric and relational conceptualization of online platforms. Global Netw 22(3):564–579
Khan MS, Saengon P, Alganad AMN, Chongcharoen D, Farrukh M (2020) Consumer green behaviour: an approach towards environmental sustainability. Sustain Dev 28(5):1168–1180
Kilian S, Mann A (2021) The phantom of the 'responsible consumer': unmasking the intention-action gap with an indirect questioning technique. Sustain 13(23):13394
Kim J, Ahn S (2024) The platform policy matrix: promotion and regulation. Policy Internet

Kirchner-Krath J (2023) Gamification for sustainable employee behavior: a design science research study

Kozyreva A, Lewandowsky S, Hertwig R (2020) Citizens versus the internet: confronting digital challenges with cognitive tools. Psychol Sci Public Interes 21(3):103–156

Kwilinski A (2023) E-commerce and sustainable development in the European union: a comprehensive analysis of SDG2, SDG12, and SDG13. In: Forum scientiae oeconomia, vol 11, no 3. Wydawnictwo Naukowe Akademii WSB, pp 87–107

Li J, Chiu DK, Ho KK, So S (2024) The use of social media in sustainable green lifestyle adoption: social media influencers and value co-creation. Sustain 16(3):1133

Lubowiecki-Vikuk A, Dąbrowska A, Machnik A (2021) Responsible consumer and lifestyle: sustainability insights. Sustain Prod Consum 25:91–101

Luo H, Zhao L, Zhang Z (2020) The impacts of social interaction-based factors on household waste-related behaviors. Waste Manage 118:270–280

Malik CDS (2024) Environmental challenges in India. In: Targets of Viksit Bharat: a roadmap of New India. pp 122–153

Martin A, Armijos MT, Coolsaet B, Dawson N, AS Edwards G, Few R, White CS (2020) Environmental justice and transformations to sustainability. Environ: Sci Policy Sustain Dev 62(6):19–30

Matta G (2020b) Science communication as a preventative tool in the COVID19 pandemic. Humanit Soc Sci Commun 7:159. https://doi.org/10.1057/s41599-020-00645-1

Matta G, Naik PK, Machell J, Kumar A, Gjyli L, Tiwari AK, Kumar A (2018a) Comparative study on seasonal variation in hydro-chemical parameters of Ganga river water using comprehensive pollution index (CPI) at Rishikesh, (Uttarakhand) India. Desalin Water Treat 118(June):87–95. https://doi.org/10.5004/dwt.2018.22487

Matta G, Kumar A, Kumar A, Naik PK, Kumar A (2018b) Assessment of heavy metal indexing on Ganga river system assessing heavy metals toxicity and ecological impact on river water quality. INAE Lett, off J Indian Natl Acad Engineering. https://doi.org/10.1007/s41403-018-0041-4

Matta G, Kumar P, Uniyal DP, Joshi DU (2022b) Communicating water, sanitation, and hygiene under sustainable development goals 3, 4, and 6 as the panacea for epidemics and pandemics referencing the succession of COVID-19 surges. Acs Esandt Water 2(5):667–689. https://doi.org/10.1021/acsestwater.1c00366

Matta G, Kumar A (2017) health risk, water hygiene, science and communication. ESSENCE Int J Env Conser Rehab 8(1):179−186

Matta G, Kumar A (2017b) Role of science and communication in health and hygiene: a case study. ESSENCE Int J Env Conser Rehab 8(2):95−101

Matta G, Srivastava S, Pandey RR, Saini KK (2015) Assessment of physicochemical characteristics of Ganga Canal water quality in Uttarakhand. Environ Dev Sustain. https://doi.org/10.1007/s10668-015-9735-x

Matta G, Kumar A, Tiwari AK, Naik PK, Berndtsson R (2018c) HPI appraisal of concentrations of heavy metals in Dynamic and static flow of Ganga river system. Environ, Dev Sustain. Springer Nature. https://doi.org/10.1007/s10668-018-01182-3

Matta G, Lu XX, Kumar A (2019) Hydrological assessment of river henwal using water quality indices with reference to planktonic composition. In: Advances in water pollution monitoring and control. Springer Publication. https://doi.org/10.1007/978-981-32-9956-6_18

Matta G, Nayak A, Kumar A, Kumar P, Kumar A, Kumar TA, Kumar NP (2020c) Evaluation of Heavy metals contamination with calculating the pollution index for Ganga river system. Taiwan Water Conserv 68(3). https://doi.org/10.6937/TWC.202009/PP_68(3).0005

Matta G, Kumar A, Nayak A, Kumar P, Kumar A, Naik PK, Singh SK (2022a) Assessing heavy metal index referencing health risk in Ganga river system. Int J River Basin Manag 1–11. https://doi.org/10.1080/15715124.2022.2098756

Matta G (2018) Environmental sustainability through scientific communication using eco-informatics. In: Kumar, Nagarthinam (ed) Exploring eco media: raising scientific temper. Eco-Visioning Forum, Department of Communication, Madurai Kamraj University, Madurai, Tamil Nadu, India, pp 193–205

Matta G (2020a) Applicability of WQI and scientific communication for conservation of river Ganga system in India. In: Naddeo V, Balakrishnan M, Choo KH (eds) Frontiers in water-energy-nexus—nature-based solutions, advanced technologies and best practices for environmental sustainability. Advances in Science, Technology and Innovation, pp 265–267. https://doi.org/10.1007/978-3-030-13068-8

McDaniel T, Soto Mas F, Sussman AL (2021) Growing connections: local food systems and community resilience. Soc Nat Resour 34(10):1375–1393

Mehmood S, Hasan Z, Ali R, Nawaz S, Amjad S (2024) Social cognitive theory in human resource management: literature review, criticism and research agenda. Bull Bus Econ (BBE) 13(2):9–13

Nayak H, Yadav SP, Yadav DK (2020) Contribution of natural and anthropogenic activities in greenhouse gases emission. Energy 4(2)

Neas S, Ward A, Bowman B (2022) Young people's climate activism: a review of the literature. Front Polit Sci 4:940876

Ouariachi T, Li CY, Elving WJ (2020) Gamification approaches for education and engagement on pro-environmental behaviors: searching for best practices. Sustain 12(11):4565

Palmer MS, Huebner SE, Willi M, Fortson L, Packer C (2021) Citizen science, computing, and conservation: how can "Crowd AI" change the way we tackle large-scale ecological challenges? Hum Comput 8(2):54–75

Pan SL, Carter L, Tim Y, Sandeep MS (2022) Digital sustainability, climate change, and information systems solutions: opportunities for future research. Int J Inf Manage 63:102444

Parmentola A, Petrillo A, Tutore I, De Felice F (2022) Is blockchain able to enhance environmental sustainability? a systematic review and research agenda from the perspective of sustainable development goals (SDGs). Bus Strateg Environ 31(1):194–217

Pereira P, Inacio M, Bogunovic I, Francos M, Barceló D, Zhao W (2022) Ecosystem services in mountain environments. Benefits Threats 177:e068

Pigola A, da Costa PR, Carvalho LC, Silva LFD, Kniess CT, Maccari EA (2021) Artificial intelligence-driven digital technologies to the implementation of the sustainable development goals: A perspective from Brazil and Portugal. Sustain 13(24):13669

Ponce P, Meier A, Méndez JI, Peffer T, Molina A, Mata O (2020) Tailored gamification and serious game framework based on fuzzy logic for saving energy in connected thermostats. J Clean Prod 262:121167

Popiel P (2022) Digital platforms as policy actors. Digital platform regulation: global perspectives on internet governance. Springer International Publishing, Cham, pp 131–150

Pulkkinen T (2024) Sustainability and professional selling: co-creation of sustainable value propositions. Master's thesis, Itä-Suomen yliopisto

Rachlinski JJ (2022) Nudges, defaults, and the problem of constructed preferences. Duke LJ 72:1731

Rafique A, Williams AP (2021) Reducing household greenhouse gas emissions from space and water heating through low-carbon technology: identifying cost-effective approaches. Energy Build 248:111162

Rapp A, Boldi A (2023) Exploring the lived experience of behavior change technologies: towards an existential model of behavior change for HCI. ACM Trans Comput-Hum Interact 30(6):1–50

Ronkainen T (2023) Principles of increasing user engagement and habit formation in social network platforms: an exploratory literature review

Saner R, Yiu L, Nguyen M (2020) Monitoring the SDGs: Digital and social technologies to ensure citizen participation, inclusiveness and transparency. Dev Policy Rev 38(4):483–500

Sarkar JG, Sarkar A, Sreejesh S (2023) Developing responsible consumption behaviours through social media platforms: sustainable brand practices as message cues. Inf Technol People 36(2):532–563

Sayed A, Alsalemi A, Himeur Y, Bensaali F, Amira A (2021) Endorsing energy efficiency through accurate appliance-level power monitoring, automation and data visualization. In: Networking, intelligent systems and security: proceedings of NISS 2021. Springer Singapore, Singapore, pp 603–617

Scheidel A, Del Bene D, Liu J, Navas G, Mingorría S, Demaria F, Martínez-Alier J (2020) Environmental conflicts and defenders: a global overview. Glob Environ Chang 63:102104

Scurati GW, Bertoni M, Graziosi S, Ferrise F (2021) Exploring the use of virtual reality to support environmentally sustainable behavior: a framework to design experiences. Sustain 13(2):943

Shin D, Chotiyaputta V, Zaid B (2022) The effects of cultural dimensions on algorithmic news: how do cultural value orientations affect how people perceive algorithms? Comput Hum Behav 126:107007

Shrivastava P, Smith MS, O'Brien K, Zsolnai L (2020) Transforming sustainability science to generate positive social and environmental change globally. One Earth 2(4):329–340

Simeone M, Scarpato D (2020) Sustainable consumption: how does social media affect food choices? J Clean Prod 277:124036

Sjödin D, Parida V, Palmié M, Wincent J (2021) How AI capabilities enable business model innovation: scaling AI through co-evolutionary processes and feedback loops. J Bus Res 134:574–587

Sparviero S, Ragnedda M (2021) Towards digital sustainability: the long journey to the sustainable development goals 2030. Digit Policy, Regul Gov 23(3):216–228

Sweet KS, LeBlanc JK, Stough LM, Sweany NW (2020) Community building and knowledge sharing by individuals with disabilities using social media. J Comput Assist Learn 36(1):1–11

Tan TM, Makkonen H, Kaur P, Salo J (2022) How do ethical consumers utilize sharing economy platforms as part of their sustainable resale behavior? the role of consumers' green consumption values. Technol Forecast Soc Chang 176:121432

Tarakanova D (2023) E-commerce entrepreneurship: a practical study of online shop development

Temple S (2023) Locally-led-development in USAID's new partnership initiative. Doctoral dissertation, University of Colorado at Denver

Tomşa MM, Romonţi-Maniu AI, Scridon MA (2021) Is sustainable consumption translated into ethical consumer behavior? Sustain 13(6):3466

Vrain E, Wilson C, Kerr L, Wilson M (2022) Social influence in the adoption of digital consumer innovations for climate change. Energy Policy 162:112800

Wirtz J, Kunz WH, Hartley N, Tarbit J (2023) Corporate digital responsibility in service firms and their ecosystems. J Serv Res 26(2):173–190

Ye L, Yang H (2020) From digital divide to social inclusion: a tale of mobile platform empowerment in rural areas. Sustain 12(6):2424

Yeboah G, Porto de Albuquerque J, Troilo R, Tregonning G, Perera S, Ahmed SAS, Yusuf R (2021) Analysis of openstreetmap data quality at different stages of a participatory mapping process: evidence from slums in Africa and Asia. ISPRS Int J Geo Inf 10(4):265

Zhou A (2021) Causal effects of affordance change on communication behavior: empirical evidence from organizational and leadership social media use. Telematics Inform 59:101549

Zhou Q, Lee CS, Sin SCJ, Lin S, Hu H, Fahmi Firdaus Bin Ismail M (2020) Understanding the use of YouTube as a learning resource: a social cognitive perspective. Aslib J Inf Manag 72(3):339–359

MIX
Papier aus verantwortungsvollen Quellen
Paper from responsible sources
FSC® C105338

If you have any concerns about our products,
you can contact us on
ProductSafety@springernature.com

In case Publisher is established outside the EU,
the EU authorized representative is:
Springer Nature Customer Service Center GmbH
Europaplatz 3, 69115 Heidelberg, Germany

Printed by Libri Plureos GmbH
in Hamburg, Germany